INTRODUCTION TO THE TAXOMETRIC METHOD

A Practical Guide

INTRODUCTION TO THE TAXOMETRIC METHOD

A Practical Guide

John Ruscio
Elizabethtown College

Nick Haslam
University of Melbourne

Ayelet Meron Ruscio
Harvard Medical School

LEA LAWRENCE ERLBAUM ASSOCIATES, PUBLISHERS
2006 Mahwah, New Jersey London

Lawrence Erlbaum Associates, Inc., Publishers
10 Industrial Avenue
Mahwah, New Jersey 07430
www.erlbaum.com

Cover design by Kathryn Houghtaling Lacey

Library of Congress Cataloging-in-Publication Data

Ruscio, John.
Introduction to the taxometric method : a practical guide / John Ruscio, Nick Haslam, Ayelet Meron Ruscio.
p. cm.
Includes bibliographical references (p.).
ISBN 0-8058-4749-9 (alk. paper)
ISBN 0-8058-5976-4 (pbk. : alk. paper)
1. Numerical taxonomy. I. Haslam, Nick, 1963– II. Ruscio, Ayelet Meron. III. Title.
QH83.R865 2006
578.01′2—dc22 2006040044
CIP

Books published by Lawrence Erlbaum Associates are printed on acid-free paper, and their bindings are chosen for strength and durability.

Printed in the United States of America
10 9 8 7 6 5 4 3 2 1

We dedicate this book to the memory of Paul Everett Meehl (1920–2003), pioneer of the taxometric method, champion of philosophical clarity and scientific rigor, and a continuing inspiration to each of us.

Contents

Preface xi

I INTRODUCTION AND BACKGROUND

1 Introduction 3

Plan of the Book *4*
Conceptual Background *5*
Distinguishing Between Taxa and Dimensions *8*
Some Possible Misunderstandings About Taxa and Dimensions *9*
Conclusions *17*

2 Why Latent Structure Matters 18

Classification *19*
Diagnosis *22*
Assessment *25*
Research *27*
Causal Explanations *30*
Lay Conceptions of Important Constructs *32*
Conclusions *34*

3 The Classification Problem 35

The Classification Problem *36*
Bimodality *37*

Finite Mixture Modeling *45*
Cluster Analysis *47*
Latent Class Analysis *49*
Dimcat *50*
Searching for Multiple Boundaries *51*
The Taxometric Method *54*
Inferential Frameworks for the Taxometric Method *58*
Conclusions *61*

II TAXOMETRIC METHOD

4 Data Requirements for Taxometrics 65

Sampling Considerations *66*
Indicator Considerations *72*
Evaluating the Data by Generating Empirical Sampling Distributions *79*
Conclusions *85*

5 Taxometric Procedures I: MAXSLOPE, MAMBAC, and L-Mode 87

MAXSLOPE *89*
MAMBAC *102*
L-Mode *116*
Conclusions *121*

6 Taxometric Procedures II: MAXCOV and MAXEIG 122

The General Covariance Mixture Theorem *123*
MAXCOV *123*
MAXEIG *136*
Evaluating the Accuracy of the Base Rate Classification Technique *146*
Blending Elements of the Traditional MAXCOV and MAXEIG Procedures *148*
Conclusions *158*

7 Consistency Tests 161

Performing Taxometric Procedures Multiple Times in Multiple Ways *162*
Examining Latent Parameters and Classified Cases *171*
Assessing Model Fit *185*
A Monte Carlo Study of MAXCOV Consistency Tests *195*
Conclusions *203*

8 Interpretational Issues 206

Graphing and Presentation of Taxometric Results *207*
The Influence of Indicator Skew *218*
The Influences of Indicator Validity and Within-Group Correlations *223*
Interpretational Safeguards *228*
Conclusions *239*

9 A Taxometric Checklist 241

Question 1: Is a Taxometric Analysis Scientifically Justified? *242*
Question 2: Are the Data Appropriate for Taxometric Analysis? *244*
Question 3: Has a Sufficient Variety of Procedures Been Implemented Properly? *252*
Question 4: Have the Results Been Presented and Interpreted Appropriately? *255*
Question 5: Are Implications of the Findings Clearly Articulated? *258*
Conclusions *259*

III APPLICATIONS AND FUTURE DIRECTIONS

10 Applications of the Taxometric Method 263

The Latent Structure of Psychopathology *264*
Normal Personality *273*
Other Latent Variables *276*
Overview of the Substantive Findings *277*
Implementation of the Taxometric Method *279*
Conclusions *286*

11 The Future of Taxometrics 287

Constructs in Traditional Domains Requiring Further Study *288*
New Domains for Taxometric Investigation *291*
New Research Questions to Be Addressed Using Taxometrics *297*
Methodological Issues for Further Study *304*
Conclusions *311*

Appendix A: Simulating Taxonic and Dimensional Comparison Data 312

Appendix B: Estimating Latent Parameters and Classifying Cases Using MAXCOV 321

Appendix C: Estimating the Taxon Base Rate Using MAXEIG 323

References 326

Author Index 337

Subject Index 343

Preface

In 1962, Paul Meehl published the first in a series of technical reports that introduced a new method for distinguishing categorical and continuous variables. These reports, printed with yellow covers and circulated among many researchers, came to be known as the *yellow monsters*. In this innovative line of work, Meehl and his collaborators at the University of Minnesota developed, evaluated, and refined a number of the data-analytic procedures that constitute Meehl's taxometric method. As it evolved over the next few decades, investigators began using taxometrics to study the latent structure of many constructs, especially in the areas of personality and psychopathology. Rather than following traditional disciplinary preferences or accepting authoritative pronouncements, researchers using the taxometric method performed empirical tests to determine whether the latent variables giving rise to observed data were categorical or continuous. In recent years, the volume of substantive and methodological taxometric research has been increasing at an accelerating pace.

This book gathers together the current state of the art in taxometric methodology, drawing from classic and contemporary sources to provide a comprehensive and accessible introduction to the method. Our intended audience includes researchers and students conducting taxometric studies, journal reviewers and editors evaluating such studies, and individuals who wish to make sense of these studies and incorporate taxometric results into their work. Interest in the taxometric method has spread to many countries and many disciplines as researchers have turned their attention to the importance of empirically evaluating latent structure and the data-analytic approaches for doing so. The taxometric method was developed by psychologists with expertise in clinical and quantitative psychology, but it is well suited to research in other social and behavioral sciences, physical sciences, education, biology, and beyond. At many universities, graduate-level courses involving psychological assessment or the classification of mental disorders have begun to incorporate instruction in the taxometric method; in some cases, entire courses are being developed to train students in taxometric methodology.

We cover a broad range of analytic techniques, describing in detail their logic and implementation as well as what is known about their performance from systematic study. We illustrate the application of taxometric analyses using a number of data sets and provide guidelines for the interpretation of results. Our overarching goals throughout the book are conceptual clarity, mathematical rigor, and accessibility to a wide audience that includes researchers new to the taxometric method as well as readers who are already familiar with some of the seminal work in this area. In a few places, technical material is placed in an appendix to facilitate an understanding of the important concepts without getting lost or sidetracked in details. We recommend that readers who initially bypass the appendixes revisit them once they firmly grasp the relevant issues.

This book is organized into three parts. The three chapters in Part I introduce background material essential to understanding the research problems that the taxometric method was designed to address. In chapter 1, we articulate the distinction between categorical and continuous data structures and discuss many potential misunderstandings of this distinction. In chapter 2, we review some of the reasons that it is important to study latent structure and explain how such studies can advance basic and applied science. In chapter 3, we discuss several methods that have been developed to distinguish categorical and continuous structure and describe key features that make the taxometric method an especially attractive tool for making this distinction.

The six chapters in Part II cover taxometric methodology. In chapter 4, we present the data requirements of taxometric analysis and introduce a technique for empirically evaluating the adequacy of data for planned analyses. In chapters 5 and 6, we focus on the nuts and bolts of the primary taxometric procedures. We discuss the logic of each procedure, review key implementation decisions, discuss the factors that can influence results, and illustrate how each procedure is performed through analyses of illustrative data sets. In chapter 7, we offer suggestions for choosing a set of taxometric procedures for a particular study and discuss strategies for obtaining additional evidence to examine the consistency of results. In chapter 8, we consider factors that can lead to interpretational ambiguity or misleading impressions and highlight methodological safeguards that can be used to prevent erroneous conclusions. Finally, in chapter 9, we work through a checklist of conceptual and methodological issues that we believe should be considered carefully and addressed explicitly in any taxometric investigation.

The two chapters in Part III of the book review applications of the taxometric method and promising directions for future taxometric research. In chapter 10, we report the conclusions of published taxometric investigations and assess the ways in which the taxometric method has been implemented. We offer general observations about the findings yielded by taxometric studies and note changes in the implementation of the method over time. In chapter 11, we explore questions central to the conduct of taxometric research in the years ahead, including which constructs and research domains are in particular need of taxometric investigation, how taxometric research might be most profitably conducted, and how the method might be evaluated, refined, and strengthened. We outline what we believe to be especially profitable avenues for future study, highlighting the primary challenges and promises that we foresee in this exciting, rapidly growing research area.

Although Meehl launched the taxometric method more than four decades ago, its popularity is a relatively recent phenomenon. In particular, the empirical evaluation of many important methodological issues is still in its infancy. Contributors to this literature vary widely in their willingness or reluctance to endorse specific approaches or to provide guidelines for taxometric research on the basis of what is often extremely limited information. We have made every effort to review the available options as comprehensively as possible and to describe the rationale for each alternative. We are explicit about the source of the recommendations that we offer, whether they stem from systematic study, preliminary testing, or our experience in performing and reviewing taxometric studies. We believe that it would be premature to devise a one-size-fits-all template for taxometric investigations. Instead, we advocate a more flexible approach that balances the available empirical evidence with reasoned judgments. Our goal is to improve a reader's ability to make informed decisions when conducting, reviewing, or reading taxometric studies.

Two additional features of this book are worthy of note. First, a unifying theme of our approach to taxometrics is the use of empirical sampling distributions. Specifically, to help determine whether data are acceptable for analysis as well as to help interpret results, we recommend that investigators generate and analyze categorical and continuous comparison data sets. By doing this in a way that reproduces important aspects of a unique set of research data, one can ask and answer the question, How would results differ if the data were categorical versus continuous? Although simulation studies can and should be performed to help address this question, the Monte Carlo literature on the taxometric method is sparse. Moreover, simulation studies often involve idealized data that differ in critical ways from research data, and virtually none of the choice points involved in implementing taxometric procedures has been studied systematically. To supplement these gaps in the literature, we recommend taking advantage of a "bootstrapping" approach that is increasingly popular for many types of data analysis. The basic idea is to tailor a small-scale simulation study to the conditions present in a particular investigation, including its unique configuration of data parameters and the particular way in which one or more analytic procedures will be performed. This approach combines rigor and feasibility in an informative and efficient manner. We explain how to use empirical sampling distributions in taxometric studies, emphasizing and illustrating the power of the approach at many points in the book.

A second feature is that we provide a suite of taxometric programs written in R, a powerful and freely available data-analytic package. Our programs were used to perform all of the analyses presented in this book, and they can be used to generate empirical sampling distributions. The current version of R, our programs, and a detailed manual are provided on the accompanying CD-ROM. Because these programs continue to evolve over time, updated versions are available on a companion Web site maintained by Lawrence Erlbaum Associates.

We are grateful to many people for their contributions to this book. We would like to thank Erlbaum senior editor Debra Reigert for her guidance in shepherding this project through the review process. Debra's keen sense for the strengths and weaknesses of the initial proposal and successive drafts proved invaluable in successfully revising and improving this work. We also thank the reviewers who committed considerable time and energy to critique drafts of some or all chapters of this book: Scott Acton, Rochester Institute of Technology; Timothy Brown, Boston University; David H. Gleaves, University of Canterbury; Eric Knowles, University of Arkansas; Todd Little, University of Kansas; and David Marcus, University of Southern Mississippi. Their detailed comments and constructive criticism led us to rethink many issues and rework many sections of the book. Finally, we are indebted to two colleagues who scrutinized a draft of this book for clarity of presentation: Michael Suvak, Department of Psychology, Boston University; and Eric Kuhn, National Center for PTSD, Palo Alto Veterans Affairs Medical Center. Michael and Eric took this charge seriously and provided us with extremely helpful feedback. Of course, any flaws that remain in the book despite the efforts of all these individuals to set us straight are our responsibility.

PART ONE

INTRODUCTION AND BACKGROUND

CHAPTER ONE

Introduction

A graduate student once sought refuge from his dissertation research by taking a vacation to India. Hoping to clear his mind of statistics and big-city pressures, he wound up in a houseboat on a remote lake in Kashmir, a peaceful spot for solitary reflection. For several days, he was the only Westerner in the vicinity, and he felt distinctly isolated. Then a young Mexican man arrived, another student, and took up residence on the neighboring boat. One evening, watching the sunset over the Himalayas, conversation turned to their work. Apologetically, our hero said he was conducting some obscure quantitative research on how to determine whether categories exist in psychological data sets. "Sounds like taxometrics," his new friend chimed in.

It may be an exaggeration to say that taxometrics has reached every corner of the globe or that it has become a common topic of conversation, but it is undeniable that the popularity of this analytic approach has increased substantially in recent years (Haslam & Kim, 2002). The volume of psychological research employing taxometric procedures is growing rapidly, and these procedures are becoming standard material in graduate-level statistics courses. The taxometric method is being brought to bear on an increasing range of research questions and problems, and the method is undergoing rapid evolution, evaluation, and refinement. But what *is* taxometrics?

Our intention in writing this book is to answer this question in a way that is conceptually clear, theoretically compelling, and—most important—practically useful. On first exposure to taxometrics, many novice researchers find it somewhat forbidding: The terminology can seem abstruse and specialized, the procedures quite different from familiar analytic approaches, the interpretation of results for complex data sets hazy, and the implications of the findings difficult to infer. This volume aims to demystify taxometric research and make it more accessible to a wider audience without sacrificing either precision or rigor. The book is not "Taxometrics for Dummies," but a clear statement of how the taxometric method can be used appropriately and fruitfully to resolve important theoretical and applied questions in the behavioral sciences. Our goal is to leave readers not only with a solid understanding of how good taxometric research may be conducted, but also with a sense for the possibilities afforded by the method and a guide for putting these possibilities into practice.

PLAN OF THE BOOK

This volume is divided into three parts. Part I, beginning with this chapter, lays out foundational issues in taxometrics, providing a rationale for the method and developing a conceptual context for later methodological material. We discuss the fundamental question that the taxometric method was designed to answer, the latent structures that are distinguished by the method, and the relevance of this structural distinction for theory, research, and practice. We then discuss the nature of the classification problem in behavioral science, review the challenges faced by classification researchers in behavioral disciplines, and introduce the taxometric method as a promising way to meet these challenges.

Part II includes an extended introduction, description, and demonstration of the taxometric method. These five chapters present an in-depth tutorial for conducting taxometric analyses, using an approach grounded in state-of-the-art empirical and simulation research. All chapters are written with an eye toward offering practical guidance on the real problems that behavioral researchers face, basing concrete suggestions on mathematical and empirical grounds when these are available and on our observations and experience when they are not. The chapters lay out the data requirements for taxometric studies, present guidelines for conducting the five most widely used taxometric procedures, and demonstrate how the findings of multiple procedures can be integrated and tested for consistency—a hallmark of the taxometric method. Special attention is given to interpreting the output of taxometric analyses, focusing on the factors that can influence the accuracy of structural inferences. The final chapter provides a comprehensive, step-by-step checklist that can be consulted to ensure that a taxometric study is properly conducted and reported. Throughout Part II, we emphasize the extent to which rigorous research provides a foundation for making informed choices when selecting or implementing taxometric procedures and consistency tests as well as when interpreting their output.

Part III concludes the book by considering what has been done, and what remains to be done, with the taxometric method. Although this volume is chiefly a guidebook for conducting new taxometric investigations, we believe that it is important for researchers to have a clear understanding of how previous studies have implemented the taxometric method and what these studies have found. Such understanding not only provides an intellectual context for future studies, but also suggests how researchers can build on existing work in more methodologically rigorous ways. To this end, we systematically review the extant taxometric literature and identify promising directions for future work.

Our review highlights the range of constructs that have received taxometric scrutiny, summarizes the investigators' conclusions about the latent structure of these constructs, and examines how taxometric practices and conventions have evolved over the past quarter century. We then focus on future priorities for taxometric research, highlighting several unresolved methodological questions, suggesting scientific applications of the method that have yet to be fully exploited, and identifying promising psychological constructs and domains that have not yet been explored in taxometric studies. We hope that this discussion gives new researchers, in particular, an inviting sense of the rich and largely untapped possibilities of taxometric investigation, motivating them to explore these possibilities for themselves and to add their contributions to the growing taxometrics literature.

CONCEPTUAL BACKGROUND

Fundamentally, taxometrics is all about the nature of variation. It begins with the simple observation that not all differences are alike. The differences between cats and dogs are not the same as the differences between hot and cold objects. The differences between gold and silver are distinct from the differences between large rocks (e.g., boulders) and small rocks (e.g., pebbles). Distinctions between branches of living organisms represent differences in quality or kind—at least when the branches represent high-level groupings such as kingdoms, phyla, classes, or orders, and sometimes less so when the branches represent low-level groupings such as genera or species—as do distinctions between chemical elements. By contrast, differences of temperature or size represent differences of quantity or degree. Some things in the world seem to fall into discrete categories. For example, an animal may be a fish or an insect, but it cannot be both a fish *and* an insect. Other things fall along a seamless dimension, differing only in their magnitude. For example, although a line can be drawn to distinguish very tall people from all other individuals, no one would view this line as anything but an arbitrary slice along an unbroken continuum of human height.

Paul Meehl (1992, 1995a), who created the taxometric method described in this book, is largely responsible for bringing this distinction between differences in kind and differences of degree to the attention of psychologists. However, even before this time, such concepts had been floating around the discipline under several other guises. Depending on the context, this distinction has been framed as one of categories versus dimensions, types versus traits, discontinuous versus continuous variation, or qualitative versus quantitative differences. Meehl instead preferred the terminology of biological classification, referring to certain sorts of categories as *taxa* (singular *taxon*) and labeling latent variables consisting of a taxon class and its remainder (the *complement* class, consisting of all individuals who do not belong to the taxon) as *taxonic*. In the broadest sense, categories that qualify as taxa are nonarbitrary, based on a distinction between category members and nonmembers that is objective (rather than subjective) and naturally occurring (rather than imposed by judgment or social convention).

Defining *taxon* more explicitly than this has proved to be rather difficult, although the concept is intuitively quite easily grasped. We later provide a mathematical definition of the taxon concept when we introduce the structural models that the taxometric method is designed to distinguish empirically. At present, however, we offer a more conceptual list of the central properties of a taxon:

1. A taxon is a latent structure. By *latent structure* we mean the fundamental nature of a construct that exists regardless of how people choose to conceptualize or measure it. *Manifest structure*, in contrast, refers to characteristics associated with observable features of the construct that depend, in part, on our theoretical assumptions and measurement decisions. An analogy with classical test theory may help to clarify this distinction. The expression $X = T + E$ represents an observed score X as the sum of a true score T plus error E. The observed score is manifest in the sense that one can directly examine this quantity. The true score is latent in the sense that it cannot be directly observed. The existence of a latent variable is inferred based on the relationships among manifest variables and invoked to explain certain patterns of relationships. For example, factor analysis is based on the premise that when a number of conceptually related items are correlated with one another, one can postulate the existence of one or more latent factors that influence scores on each manifest

item. Thus, a latent taxon is not simply a cluster of superficially similar cases. Rather, it is a grouping of cases that share an underlying commonality, a set of "deep" properties that accounts for the group's observable similarities. What makes something a cat is not its outwardly perceptible properties: A dog groomed to look like a cat and trained to purr does not thereby become a cat.

2. A taxon is a *category* with a *boundary*. It is a (latent) class that has a finite membership. In principle, the individuals belonging to a taxon could be counted. In contrast, latent dimensions do not define distinct groups of members. Instead, every individual has some degree of each dimensional property, some position along each continuum. In the animal kingdom, for example, there is not continuous variation in levels of catness. Rather, a finite set of animals are cats and the large remainder of animals are not. A taxonic *boundary* is a metaphorical notion rather than a literal line of demarcation, but it captures the fact that there is some sort of break or discontinuity between members of a taxon and its complement. Those on one side of the boundary are categorically different from those on the other side, even if in some respects they resemble one another. Some small dogs can look more like cats than like Great Danes, but they nevertheless belong with the latter, growling at the former across a deep taxonomic gulf.

3. A taxon's boundary is *nonarbitrary* or *objective*. Not all categories with boundaries are taxa. When a boundary is simply imposed on a continuum by a human classificatory decision, social convention, or naming practice, it does not constitute a true taxon. This does not mean that the boundary is necessarily unjustified or frivolous—it may, in fact, be pragmatically useful. For example, low-income individuals and premature infants probably do not belong to naturally occurring taxa, but instead fall below thresholds that have been superimposed on dimensions to facilitate social services, public health, and medical decision making. If, in contrast, a boundary represents an objective or naturally occurring discontinuity at the latent level, it is considered to demarcate a taxon. The boundary between cats and dogs, once again, is no social or linguistic artifact, but a fact about the world.

4. A taxon is (reasonably) *enduring*. This property of a taxon is perhaps more incidental than the others. However, we might hesitate to refer to a class of things (particularly a class of people) as a taxon if its membership was very unstable. Of course, the time frame by which stability should be judged varies across taxa. For example, a taxon corresponding to a personality characteristic would be expected to persist for years, rather than mere days, whereas a taxon such as influenza infection (which is qualitatively different from other viral infections that cause respiratory illnesses such as the common cold) is stable only over much briefer spans of time.

Although a taxon need not be perfectly stable, within an appropriate time frame, it should be relatively traitlike rather than fleeting.

DISTINGUISHING BETWEEN TAXA AND DIMENSIONS

The properties listed earlier help to clarify our intuitive understanding of taxa and dimensions. *Cat* is a taxon because there is a countable number of entities in the world that belong to this latent class; because there is an objective, naturally occurring category boundary between cats and noncats; and because—horror movies aside—cats do not typically transmogrify into other things. *Pet*, in contrast, is not a taxon because the boundary between pets and nonpets is based on social conventions that can vary widely across places and times so that the set of things that might be called *pets* is not a unitary latent class. Similarly, tall person is not a taxon because there is no latent class of lofty people who are categorically different from others. With no discoverable, objective boundary that cleaves the height continuum at the latent level, human height is best conceived as a dimensional construct.

With any luck, you will now be convinced that there is a meaningful conceptual distinction between taxa and dimensions. However, Meehl and his colleagues (e.g., Grove, 2004; Grove & Meehl, 1993; Meehl, 1973, 1995a; Meehl & Golden, 1982; Meehl & Yonce, 1994, 1996; Waller & Meehl, 1998) did not simply want to draw an abstract, armchair distinction between these two types of constructs. Instead, they wanted to establish that this distinction was fundamental to the study of differences between people, and that it had a host of practical and theoretical implications (to be discussed in chap. 2). Most important, they proposed that this distinction can be drawn empirically. Often, they argued, it is not obvious whether certain differences between people are taxonic. This is because differences can only be observed at the manifest level, where the opinions of theorists and laypeople and the assumptions of classification systems may lead taxonic boundaries to be drawn where none exists or may fail to recognize boundaries that do exist. Not satisfied with leaving questions of taxonicity to theoretical arguments and disciplinary preferences, Meehl and colleagues set about developing a rigorous approach that would empirically test for the existence of a taxon. They referred to their innovative methodology as the *taxometric method*. Thus, the aim of a taxometric analysis is to determine whether a latent construct is taxonic or dimensional through the rigorous use of appropriate data-analytic procedures.

These taxometric procedures, and the taxometric method as a whole, ultimately rest on a realist philosophy of science. They presume that, to a substantial extent, the latent structure of individual differences (such as personality features and psychopathological syndromes) exists independently of human efforts to classify and describe them. Taxa are not artifacts of particular discovery procedures brought into existence by the research methods. Neither are taxa mere social constructions, fabricated out of social or linguistic conventions. Because taxa are objective and discoverable, the best way to search for their existence is to employ a variety of appropriate procedures designed to detect latent boundaries. The confidence with which one can infer the existence of a taxon depends on the consistency with which multiple sources of information clearly point to its existence. When the results of different analytic procedures converge in a manner that would be extremely unlikely if no taxon existed, the independent existence of a taxon is inferred. This use of multiple procedures—dubbed "consistency testing"—is central to the taxometric method, and it receives special attention in chapter 7.

Although the taxometric method is grounded in a realist view of taxa, it is not realist in a crude or unsophisticated way. The taxometric approach does not dictate that classification systems must include all taxa and only taxa. In practice, taxonomies are ultimately cognitive and cultural products that serve our practical purposes: They do not necessarily exist in nature independent of the people who develop them. Taxonomies do not have to mirror nature; if a taxon exists, our taxonomies are not obliged to recognize it. However, taxometric researchers argue that matching our classification systems to empirically established taxa—in Plato's words, "carving nature at the joints"—will often be pragmatically as well as theoretically useful. Taxonic boundaries represent real distinctions independent of theory or fiat that are likely to have important implications for basic and applied science. If the things being classified—such as mental disorders—come in categorically different kinds, then it is possible that what is true for one kind (e.g., its causes, risk factors, developmental course, prognosis, optimal treatment) may not be true for another. Consequently, the realist view of latent structure that motivates taxometric research suggests that classification efforts should take seriously any taxa that are discovered.

SOME POSSIBLE MISUNDERSTANDINGS ABOUT TAXA AND DIMENSIONS

The previous sections have attempted to clarify the meaning of a taxon and to describe how a realist view of latent structure motivated the development of the taxometric method. It is also important at this early stage to dispel some common misconceptions about taxa and dimensions.

Misconception 1: A Taxon Should Be Readily Observable

When the taxon concept is first explained to some people, they wonder why an elaborate data-analytic apparatus would be needed to detect a taxon. Shouldn't the existence of a taxon be obvious, so that we could easily see that one group of people is different in kind from others? This misunderstanding is addressed in chapter 3, but a few remarks are relevant here. The basic point is that manifest structure is often an unreliable guide to latent structure. That is, the apparent structure of differences observed between people does not necessarily correspond to the structure that actually exists at the latent level. This can work in two directions. Sometimes, on the basis of observable features, a group of people might appear to form a tight and unified cluster, when in fact there is no discrete boundary that distinguishes them from others. However, sometimes no categories can be detected amid manifest variation, even when taxa do underlie this variation. This is especially likely where the difference between taxon and complement is subtle, imprecisely conceptualized, or poorly measured.

The latter issue is particularly relevant within the behavioral sciences. Given the intrinsic difficulty of measuring many psychological, sociological, educational, and related constructs and the relatively primitive state of assessment in many areas of these disciplines, we cannot always be confident that taxonic differences between people would be readily observable. Even where taxa do exist, current measurement tools may be unable to distinguish taxon members from complement members with high validity. For example, research indicates that effect sizes in psychology are often quite modest and studies are often underpowered (Cohen, 1962; Maxwell, 2004; Sedlmeier & Gigerenzer, 1989), suggesting that psychological taxa will rarely advertise themselves in ways that are obvious to the naked eye (Gangestad & Snyder, 1985). Therefore, it is vital to remember that manifest and latent levels of description are distinct, and that powerful quantitative tools may be needed to derive an understanding of latent structure from observable data.

Misconception 2: If a Latent Variable Is Taxonic, It Will Not Show Any Continuous Variation (and If It Does, Then It Is Not Taxonic)

This misunderstanding comes in several forms. At times it simply reflects a failure to distinguish between the manifest and latent levels of analysis. A latent variable that is taxonic can show continuous variation when it is assessed to yield manifest scores. For example, biological sex is taxonic, but manifest *indicators* or correlates of sex (e.g., masculine vs. feminine interest patterns, voice pitch, height) often evidence continuous variation. The fact that the manifest indicators of a construct can vary along a continuum does not, however, mean that the construct is underpinned by one or more latent dimensions. Similarly, the fact that a construct is taxonic does not mean that indicators must vary in a strictly categorical fashion.

A related form of this misunderstanding is that if a construct is taxonic, there should be no dimensional variation within either the taxon or its complement. In fact both meaningful and nonmeaningful variation is possible within latent classes. Nonmeaningful (manifest-level) dimensional variation may be introduced within classes in at least two ways. First, there is apt to be variation within the taxon simply as a function of random measurement error. That is, because the manifest indicators do not perfectly capture membership in the latent classes, all taxon members will not receive the same value or score, but will differ by degree in a nonsystematic fashion. Second, in addition to the unavoidable problem of measurement error, there may be systematic dimensional variation among taxon members if manifest indicators of the taxon covary for artifactual reasons. For example, indicators of a taxon might all be drawn from self-report questionnaires that share some common method variance (e.g., all may be influenced by a response bias involving the exaggeration of clinical symptoms). As a result, there might be systematic variation among taxon members due to individual differences in response bias.

It is important, however, to highlight one more possible source of dimensional variation within taxa: There may be *real* dimensional variation within a taxon or complement that is due neither to measurement error nor measurement artifacts. A certain mental disorder might be taxonic, for example, but cases within the taxon might vary systematically along one or more dimensions, such as degree of severity. Clinical features that differentiate members of the taxon from members of the complement might also reliably distinguish taxon members from one another. This sort of meaningful (latent-level) dimensional variation could occur within the taxon alone, within the complement alone, or within both latent classes (J. Ruscio & Ruscio, 2002, 2004a). Such variation may appear to be a "nuisance" from a statistical point of view (hence the widespread use of the terms *nuisance covariance* or *nuisance correlations* in the taxometric literature; we use the more familiar *within-group correlations* throughout this book), but it may nonetheless be an entirely valid component of latent structure. Indeed, because many behavioral constructs are complex and multidetermined, it is conceivable that many may have complex latent structures—a possibility that we explore further later.

In short, there may be dimensional variation within a taxon and/or complement, and its existence in no way detracts from the inherent taxonicity of the construct. A latent variable is taxonic if it contains a nonarbitrary difference in kind, regardless of whether this difference coexists with differences of degree. Although teasing apart the various potential causes of within-group indicator covariance can be an intriguing research question, one would first need to establish the existence of taxa before this question can meaningfully be raised or addressed.

Misconception 3: A Taxon Cannot Be Further Subdivided

Perhaps in part because of the previous misunderstanding—the supposition that no variation or heterogeneity is possible within taxa—people sometimes imagine that a taxon only exists at a single level. In this view, when a taxon is found, there is no point seeking lower order taxa that may be nested within it. By implication, investigations of latent structure can stop when a taxon is detected. To see why this belief may be mistaken, consider the classification of living organisms. *Mammal* is a taxon because the distinction between mammals and nonmammals meets the criteria for a categorical boundary described earlier. Within the mammal taxon, primates and rodents represent lower order taxa. The primate taxon can be divided further into the *homo* and *australopithecus* genera, with *homo* divided into species including *homo habilis* and *homo sapiens*. Thus, taxa can be nested within one another at multiple levels.

Such nesting may be less pronounced in the domains of personality or psychopathology, yet it may still exist. There is nothing to logically prevent a subtype of a taxonic mental disorder from also being taxonic. For example, schizophrenia may represent a taxon within which *paranoid type* and *catatonic type* represent nested taxa. Similarly, a personality diathesis might be taxonic, as might be the less prevalent condition for which it confers vulnerability. Conversely, if a relatively broad latent variable such as a disorder or diathesis proves to be nontaxonic, it is still possible that a narrower or less inclusive variable is taxonic. For example, a rare taxonic variant (e.g., psychotic depression) of a more common dimensional condition (e.g., major depression) might occur, although one might hesitate to call it a *subtype* if no higher order taxonic type exists. In short, detecting (or even failing to detect) a taxon may represent only the beginning of the process of mapping out the full latent structure of a construct (J. Ruscio & Ruscio, 2004a).

It is important to emphasize that this point about the potential nestedness of taxa goes double for the complement class. A taxometric analysis might reveal a qualitative boundary separating a taxon from its complement, but this does not mean that the complement is unitary or indivisible. The complement contains all cases that do not belong in the taxon, and it may well be the case that this remainder comprises a heterogeneous mix of latent classes and dimensions.

Misconception 4: Taxa Cannot Exist Because People Differ in So Many Ways That They Are Unlikely to Form Homogeneous Groups

This potential misconception builds on the previous two and incorporates an additional misunderstanding. Some researchers bristle at the notion that a taxon may exist because they misconstrue this as suggesting that one dichotomous latent variable exhaustively accounts for all individual differences between and within groups. In fact, even when a taxon exists, individuals belonging to the same latent class can differ from one another in several ways. As noted earlier, these individuals may vary along one or more latent dimensions, and a taxon or complement may be further divided into subtypes.

An additional point counters this misconception: Members of the same group can differ from one another on any number of characteristics *other* than the one whose structure is being considered. For example, people who possess XX sex chromosomes are biologically female, whereas those who possess XY chromosomes are biologically male. Setting aside the small proportion of individuals who possess other configurations of sex chromosomes, it would be silly to insist that there is no taxonic boundary in human biological sex because men and women differ greatly from one another (within and between groups) on many features other than biological sex. Individual differences among members of a latent class do not refute the existence of a genuine taxonic boundary, nor do they lessen the scientific utility of drawing a taxonic distinction between existing groups.

Misconception 5: A Taxon Is a "Natural Kind"

One common way to define a taxon has been to refer to it as a *natural kind*. This concept has been examined at great length by philosophers (e.g., Kripke, 1980). Their usual analysis holds that certain kinds of things in the world—the standard examples being chemical elements or compounds, and biological taxa—exist "in nature," independent of human classifications and naming conventions. Such natural kinds have sharp boundaries: A substance either is or is not water, and a furry animal with stripes either is or is not a tiger. Up to this point, the definition of *natural kind* would seem to match the properties of a taxon perfectly. Both concepts represent latent categories that are nonarbitrary, discovered rather than constructed, and discrete. However, the concept of natural kind typically contains two additional features that are more restrictive. First, a natural kind in the domain of living things is usually understood to be biologically based, akin to a species. Second, membership in a natural kind is usually taken to require the possession of a shared essence. What makes a substance water is having the molecular structure H_2O, and what makes a creature a tiger is having tiger DNA. Such hidden essences are necessary properties of the entities belonging to the natural kind and are causally responsible for their observable features.

These latter two properties of a natural kind—being biologically based and species-like, and having an essence—are not necessary for a category to qualify as a taxon in the taxometric sense. There is no reason that a taxon could not have an entirely environmental or learned basis, arising, for instance, through a process of social shaping or adaptation to an ecological niche. In an important article that every budding taxometric researcher should read, Meehl (1992) drew attention to such *environment-mold taxa*, using as one example the political taxon *Trotskyist*. Members of this now rare group share a tightly organized set of ideological tenets that categorically distinguish them from individuals with other beliefs, although these group differences do not arise from a biological foundation. Meehl made comparable observations about the taxonicity of certain occupational groups.

Similarly, it is not necessary to suppose that all members of a taxon must share an underlying causal essence. An essentialist view, according to which all taxon members have a single, necessary property that is causally responsible for its observed features, may be defensible in some instances (e.g., a neurological disorder that is caused by a single major gene). However, there is little reason to believe that such specific etiologies are typical of personality or psychopathology taxa. For many such taxa, the causal basis for taxon membership more plausibly corresponds to a nonessentialist view, in which taxon membership springs from multiple, interacting, and probabilistic causal factors (e.g., threshold and epigenetic effects), and members do not share any single defining characteristic. In summary, we believe that it is generally unwise to equate *taxon* with *natural kind* when the latter is understood in the normal essentialist sense.

These distinctions between the concepts of *taxon* and *natural kind* may strike some readers as conceptual hair-splitting, but we believe they lead to some important implications. In particular, the concept of taxon is broader than the concept of natural kind: All natural kinds are taxa, but not all taxa are natural kinds. In addition, when a taxon is found, one should not prejudge the issue of causation by automatically inferring that it has a biological basis or an underlying essence.

Misconception 6: If a Latent Variable Is Not Taxonic, Any Categorical Distinction Imposed on It Is Completely Arbitrary

We have repeatedly stated that a taxon possesses a nonarbitrary categorical boundary—a boundary that is not simply imposed by artifact, preference, or convention. Does this imply that categorical distinctions that are imposed on nontaxonic latent variables are arbitrary? In one sense it does, but in another important sense it does not. Such a distinction will be arbitrary in the sense that it does not correspond to an objective boundary. However, this distinction is not necessarily arbitrary in the sense of being unsystematic, careless, or unjustified. That is, categorical distinctions can sometimes be drawn on latent dimensions in ways that are well justified and pragmatically useful, even if no taxonic boundary exists (e.g., a child has to be a set height to ride a roller coaster).

It may at times make sense to derive a categorical diagnosis or dichotomous decision on the basis of scores on latent continua. This is how essential hypertension and obesity are diagnosed, for instance. Although we know of no true taxonic boundaries along the continua of blood pressure or body mass, it may nonetheless be practically important to identify individuals who exceed critical levels on these continua. If cutoffs must be drawn on dimensions for such practical purposes, there are surely better and worse ways for these cutoffs to be defined. It would be foolish, for example, to impose a cutoff on the basis of a median split, leading 50% of the population to be diagnosed with hypertension or obesity. In contrast, cutoffs could be defined in ways that are sensible, meaningful, and empirically optimized—nonarbitrary in our second sense. Actuarial data might be used to locate a point on each continuum at which (a) the health consequences of the high blood pressure or high body weight reach a threshold of clinical severity, (b) these health consequences begin to rapidly worsen (i.e., an "inflection point"; Kessler, 2002a), (c) the cost/benefit ratio of a treatment becomes satisfactory, or (d) false positive identifications (mistakenly inferring an elevated health risk where none exists) and false negative identifications (missing an elevated health risk that is present) are optimally balanced (Kraemer, Noda, & O'Hara, 2004). Although the optimal point of demarcation would likely vary according to the practical purpose at hand, the population being classified, and other relevant factors, the location of such a point would not be entirely arbitrary in any pejorative sense. Thus, there is a place for well-reasoned, empirically grounded, pragmatically useful categories in the behavioral sciences, even in the absence of taxa.

Misconception 7: A Dimensional Solution Is Not Worthy of Further Attention

One subtle but persistent practice in the taxometrics literature is the tendency to dismiss dimensional findings as less interesting, less important, or even less valid than the discovery of a taxon. This perspective is reflected in the terminology of the literature, which often refers to *nontaxonic* rather than *dimensional* findings. The apparent preference for taxonic over dimensional results may be explained by any of a number of factors, including the framing of taxometric procedures as *taxon search procedures* that achieve success only when taxa are detected; the greater ease with which taxa can be incorporated into existing categorical classification systems; or the conceptualization of dimensional structure as a kind of null hypothesis that can be rejected, but not accepted, on the basis of taxometric results.

In contrast with these views, our own perspective is that dimensional structure can be legitimately inferred from taxometric analysis, provided that the available data are shown to be capable of distinguishing taxonic from dimensional structures (see chap. 4). Under such conditions, a dimensional solution provides a meaningful and useful statement about latent structure that can serve as a springboard for important follow-up investigation. Just as a taxon may be divisible into lower order taxa, a dimensional construct may consist of multiple dimensions, some of which may subsume or be subsumed by dimensions at higher or lower orders within the broader nested construct. Although the specific analytic procedures that are used to search for hierarchically arranged types versus subtypes and higher order versus lower order factors may differ, the exploration and delineation of the full latent structural model is equally important for both structures. Moreover, investigations into the nature, causes, and correlates of a dimensional construct are arguably as valuable as those that seek to better understand a taxonic construct. Our view of the taxometric method as a tool for distinguishing taxonic from dimensional structure, rather than exclusively a taxonic search procedure, influences the manner in which we discuss taxometric results and the implications that we draw for their potential impact on theory, research, and practice. In short, we believe the proper attitude to take in taxometric research is an empirical one, rather than holding a preference for or bias toward taxonic findings.

CONCLUSIONS

In this chapter, we reviewed the distinction between taxa and dimensions, and we explored some of the conceptual implications and complexities of the taxon concept, as discussed and refined by Meehl, his colleagues, and subsequent taxometric researchers. In some ways, our elucidation of the nature of taxa and dimensions paints a more complex picture of latent structure than is usually presented in the literature. We have noted that a taxon (a) is rarely observable from manifest scores or symptoms, (b) can contain meaningful dimensional variation, (c) can be hierarchically nested within another taxon, (d) can contain individuals who differ from one another in a number of ways, and (e) need not be biologically based or involve a shared essence among individuals. We have also suggested that categorical distinctions imposed on latent dimensions may be justifiable, useful, and even necessary. Finally, we have emphasized that dimensional structure is no less important than taxonic structure, and that either outcome of a taxometric study provides a framework to be elaborated and fleshed out by explorations of more complex structures. These issues are not overly difficult and need not be confusing. They do, however, require a nuanced view of latent structure that can accommodate more complex structural models.

The distinction between taxa and dimensions highlighted in this chapter is not merely a subtle intellectual curiosity, but a truly fundamental issue in how we think about differences among individuals. Indeed the latent structure of a construct has important ramifications throughout the behavioral and social sciences. In the next chapter, we review these ramifications and consider why taxonicity matters deeply for basic and applied science, with implications for areas as diverse as personality assessment, diagnostic classification, and lay perceptions of mental disorder. Because empirical testing of latent structures has often been neglected in the behavioral disciplines, taxometrics opens up a wide array of research frontiers, allowing researchers to contribute in substantial ways to the advancement of theory and practice. It is to this diversity of issues and implications surrounding latent structure that we now turn.

CHAPTER TWO

Why Latent Structure Matters

The previous chapter provided a conceptual background for taxometrics and explained the difference between taxonic and dimensional latent structures. The next chapter begins to explore the ways in which these structures can be differentiated. But before we launch into a discussion of *how* this can be done, we must consider the fundamental question of *why* it should be done. Why does it matter whether variation along a latent variable is a matter of degree or a matter of kind?

To some readers, the answer to this question might seem self-evident. Science is all about discovering how the world is. Therefore, knowing whether mental disorders, personality characteristics, or other phenomena are better understood as categories or dimensions is intrinsically worthwhile and valuable. Determining latent structure is a basic question for behavioral science regardless of whether any practical consequences or implications follow from this determination.

Other readers will not be satisfied with this view of taxometric research as a purely curiosity-driven exercise divorced from application. They will want to know how resolving latent structure can make a difference for the sorts of practical activities and problems that behavioral scientists face. For example, if taxometric research shows a form of psychopathology to be taxonic, are there practical implications for how this condition should be classified, diagnosed, measured, investigated, or explained? The skeptical reader may well wonder whether this sort of finding makes a difference, practically speaking.

In this chapter, we hope to lay the skeptic's concerns to rest. It turns out that latent structure matters in a multitude of ways for the tasks facing both scientist and practitioner, and that taxometric research can serve many purposes, both basic and applied. Although we separately discuss the important intellectual versus practical implications of latent structure, it should be noted that many of the relevant issues overlap and interrelate across these domains.

CLASSIFICATION

Classification comes naturally to people. We are forever grouping things into categories, naming them, and making inferences about them based on the categories into which they have been placed. Classification is also, of course, a core business of science. It is hard to imagine chemistry without the periodic table of elements, physics without its classification of subatomic particles, astronomy without its nomenclature of heavenly bodies, or biology without its taxonomy of living things. This is no less true in the behavioral sciences. Although many theorists and researchers concern themselves primarily with processes and mechanisms, rather than with classification, their work rests on a bedrock of ideas about the distinctions between—and relations among—important phenomena within their area of study.

Classification is especially central to the enterprise of psychologists interested in individual differences. Clinical psychologists operate (sometimes reluctantly) within the context of psychiatric classification systems, and clinical researchers and theorists often argue that the boundaries drawn by these systems between mental disorders, or between particular disorders and normal functioning, are incorrectly located or arbitrarily imposed on latent dimensions of pathology. Personality psychologists have long been preoccupied by the structure of the trait universe, whereas the organization of intellectual abilities has been a focal task for some cognitive and educational psychologists.

Although classification is a common concern among psychologists, it is interesting to note how classifications differ across domains. Within abnormal psychology, for example, the prevailing reliance on psychiatric classification systems leads mental disorders to be represented and diagnosed as discrete, tightly bounded categories similar to classifications of disease in general medicine. These disorders are diagnosed as present or absent within any individual person, allowing estimation and monitoring of their prevalence in the general population at any particular time or over a given period. In the personality and ability domains, by contrast, classifications usually specify dimensions along which people are presumed to differ in a purely quantitative fashion. Where neuroticism or intelligence are concerned, for example, individual differences are taken to be matters of degree rather than differences in kind.

Needless to say, these differences among classification systems reflect differences in default positions about the nature of variation underlying psychological phenomena, and these positions are taxonic and dimensional, respectively. A perusal of the research literature or even an introductory textbook suggests that taxonic beliefs are held more often about psychopathology, whereas dimensional beliefs are held more often about normal personality and abilities. However, these general positions are open to empirical challenge on a construct-by-construct basis. Many structural beliefs are based on intuitive plausibility, theoretical presumptions, or barely questioned disciplinary traditions, rather than empirical tests (Meehl, 1992). Moreover, there is often disagreement about the taxonic versus dimensional nature of specific constructs. Some writers, for example, have criticized the categorical view of psychopathology embodied in standard psychiatric nosology (e.g., Widiger & Clark, 2000), whereas others have challenged the prejudice against taxa in personality psychology (Gangestad & Snyder, 1985). In short, the default positions of classification across behavioral sciences are often contentious and provide a fertile ground for empirical study.

A promising remedy for this lack of consensus is to conduct research examining the constructs in question using an approach designed to differentiate taxonic from dimensional structures. Although later chapters review the appropriateness of the taxometric method for this purpose, it is first worth considering what implications such work would have for the task of classification. In other words, how might determining the latent structure of relevant constructs affect the development and revision of classification systems? Given that the preponderance of taxometric research has been done in the field of psychology, our examples focus on the potential impact of such research on psychological classification, although similar implications would be expected for classification systems in related disciplines.

One possible outcome of taxometric investigations would be to support the taxonic view of a mental disorder and yield a taxon base rate that is consistent with the known prevalence of the disorder as it is presently diagnosed. This would bolster the status of the disorder as a discrete category and help validate its diagnostic criteria. An important additional benefit of taxometric procedures is that they can be used to identify the particular individuals who belong to a taxon. By studying the features that best distinguish members of the taxon from its complement class, the most distinctive features of the disorder can be ascertained, and the diagnostic criteria used in classification can be further improved and refined. Comparable outcomes would ensue if structural research supported the prevailing dimensional view of personality traits, with follow-up studies identifying the variables that optimally locate individuals along the uncovered latent continua.

A second possible outcome of taxometric investigations would be to challenge the prevailing structural model of a construct. For example, a mental disorder presumed to be categorical or the supposedly discrete subtypes of a disorder might prove to be dimensional. Findings such as these could call into question the current categorical representation of these disorders and suggest the need for their reevaluation or reconceptualization within the classification scheme. Follow-up work would be needed to determine the points at which distress, disability, or other relevant factors denote clinically significant disturbance (and hence require clinical attention). The dimensional nature of these disorders might be accommodated by adding severity quantifiers (e.g., *moderate* or *severe* designators, 0–8 clinician severity ratings) at the symptom or diagnosis level; by determining whether clinically meaningful forms of the disorder fall below, at, or above the current diagnostic threshold; and by making diagnostic decisions with greater emphasis on practical concerns (e.g., the level of functional impairment) or an optimized weighting of symptom severity than on the presence or absence of a designated number or combination of symptoms. A taxonic finding for a personality characteristic could have similar implications for personality taxonomies, with personality taxa requiring acknowledgment and structure-appropriate classification alongside—or in place of—standard trait dimensions.

A third possible outcome of taxometric studies that bears on psychological classification is a taxonic finding that challenges the boundaries specified by the classification system. That is, a taxon might be detected as hypothesized, but might poorly correspond to the nature, number, or combination of variables currently used to identify it. For example, the diagnostic criteria for a particular mental disorder might give equal weight to eight symptoms indicative of the condition despite possible redundancy, differential predictive power, or disparate sensitivity or specificity in detecting the disorder vis-à-vis normal functioning or neighboring disorders. Alternatively, the criteria might describe a disorder in an overly inclusive or exclusive fashion, drawing the category's boundaries too broadly or narrowly relative to the size of the underlying taxon. In such situations, taxometric findings could be used to redraw the boundaries of the diagnosed disorder to more accurately map the boundaries of its underlying taxonic structure, resulting in greater precision of classification and consequent improvement in the prediction of course, prognosis, treatment response, and other critical outcomes.

A chief impetus for taxometric research has been the desire to improve psychological classification to solve the classification problem (Meehl, 1995a) and in turn achieve greater understanding, prediction, and control over psychological phenomena. The distinction between taxonic and dimensional latent structure has many implications for psychological and related taxonomies, and a rigorous empirical method for making this distinction has a major part to play in classification research.

DIAGNOSIS

As we argued earlier, classification is an important and controversial matter in behavioral science, and one that can be clarified by taxometric research. Whether a particular condition deserves a place in taxonomies of mental disorder—and if, where, and how its boundaries should be drawn—are questions that taxometric research can help address. A thorough and systematic application of taxometric analysis to the hundreds of mental disorders now recognized in formal classification systems such as the *Diagnostic and Statistical Manual of Mental Disorders* (*DSM*; American Psychiatric Association, 1994) or *International Classification of Diseases* (*ICD*; World Health Organization, 1993) might suggest altering these systems in substantial and unexpected ways.

The aim of classification is to catalogue the kinds of entities that are presumed to exist in a particular domain, reflecting the ways in which these entities are understood to vary and interrelate. Assigning individuals to these entities is a matter of diagnosis. Just as taxometric research can clarify and refine classification systems, it can also improve diagnostic practices. How it does so depends crucially on whether the latent structure being classified is taxonic or dimensional.

Imagine that a certain mental disorder is recognized in a psychiatric classification system, which describes the characteristic clinical features of this disorder and lists its requisite diagnostic criteria. Normally a disorder is considered to be present (and a diagnosis is made) when a person is judged to have met some or all of these criteria. The precise rule that governs which or how many criteria must be met for a diagnosis to be made is usually referred to as the *diagnostic algorithm*. For example, the algorithm might specify that a particular number of symptoms (e.g., any five out of seven) must be met to pass the diagnostic threshold. The algorithm operationally defines the boundary of the diagnostic category and effectively determines its prevalence—the proportion of the population who would receive the diagnosis. If the algorithm were to be relaxed (four out of seven) or tightened (six out of seven), this proportion might change substantially.

Imagine now that the mental disorder in question represents (or is underpinned by) a taxon. This taxon has a determinate prevalence: A precise proportion of any population of interest belongs to it. The disorder's diagnostic algorithm may classify individuals accurately with respect to the underlying taxon, yielding a prevalence for the disorder that closely matches the prevalence of the taxon. However, the diagnostic algorithm may poorly identify taxon members, causing either a failure to diagnose people who have the disorder (false negatives) or erroneous diagnosis of people who do not have the disorder (false positives). Any diagnostic algorithm will make some errors, of course. However, if an algorithm is too liberal or conservative relative to the taxonic boundary, it will systematically over- or underdiagnose, potentially at high rates. The clinical implications of mislocating a diagnostic threshold in this manner may entail inefficient allocation of limited mental health resources, including unnecessary or insufficient treatment, poor distribution of disability compensation, or deficient funds for staffing and programs targeting problems whose prevalence and scope have been underestimated. There may also be important research implications: If one's control group contains many taxon members and one's disordered group includes many complement members, comparisons across these groups will be diluted, resulting in a loss of statistical power and a potential failure to detect important effects.

These adverse consequences may be minimized or avoided by empirically assessing the latent structure of the disorder in question. If a taxometric analysis suggests that the structure of the disorder is taxonic, the analysis could be used to estimate the true taxon prevalence (or taxon *base rate*) in the population sampled. The diagnosis rate (yielded by the diagnostic algorithm) could then be compared with the taxon base rate. If these rates disagree, the diagnostic algorithm could be recalibrated to improve concordance with the underlying taxon. For example, if taxon base rates consistently fell at 5%, but the five-out-of-seven algorithm yielded a rate of 10%, then the algorithm might be improved by tightening the criterial requirement to six out of seven. Follow-up analyses could determine whether the individuals diagnosed according to this new symptom threshold are the same as those who were classified into the taxon. In this way, structural investigations can help to optimize diagnostic algorithms by calibrating them to empirically derived taxonic boundaries.

Establishing that a disorder is taxonic can also lead to more radical revision of diagnostic algorithms. As indicated earlier, in addition to yielding an estimate of the taxon base rate, a taxometric analysis can allow taxon members to be identified. A sample can therefore be partitioned into individuals classified as taxon and complement members. Once this has been done, it is possible to assess how well each symptom or diagnostic criterion—or configural combination of variables—discriminates these groups. Doing so may reveal that some criteria discriminate particularly poorly or well, that some are overly stringent or overly sensitive, or that a certain combination of criteria does a better job of identifying taxon members than the existing diagnostic algorithm. One could also explore the incremental validity of each diagnostic criterion by evaluating how well it improves classification accuracy, over and above other criteria that are considered central to the definition of the disorder. Such an empirical approach could serve as a powerful tool for improving diagnostic accuracy.

The situation is very different if the latent structure of a mental disorder is found to be dimensional. In such cases, of course, there exists no objective boundary to which a diagnostic algorithm can be calibrated, no objective prevalence of the disorder independent of the diagnostic algorithm that has been devised to slice the underlying dimension(s). Consequently, the placement of the threshold separating people judged to have the disorder from those judged not to have it must be made on grounds other than a taxonic boundary. As noted in chapter 1, the absence of a taxonic boundary does not imply that a diagnostic cutoff is inappropriate or unnecessary, nor does it suggest that the placement of a diagnostic threshold must be arbitrary in the unprincipled sense. Rather, useful diagnostic cutoffs could be defined on the basis of cost/benefit calculations (Swets, Dawes, & Monahan, 2000), inflection points in the association between symptom severity and functional impairment (Kessler, 2002a), or scores maximizing the sensitivity and specificity with which individuals are identified for pragmatic purposes (Kraemer et al., 2004). Thus, methods for defining cutoffs for dimensional constructs are quite different from those appropriate for taxonic constructs; they may place greater emphasis on practical considerations and may make use of different algorithms for different purposes to balance the relative costs of false negative and false positive diagnostic decisions (Meehl & Rosen, 1955).

It should be clear by now that the process of making—and improving—diagnoses looks rather different when the relevant construct is taxonic versus dimensional. Therefore, investigations of latent structure can have major ramifications for the refinement of diagnostic procedures. To the extent that an existing diagnostic algorithm has been developed on the basis of clinical wisdom in the absence of sound structural research, taxometric analyses could play a significant role in diagnostic revision. If taxa are discovered, diagnostic criteria can be calibrated to their boundaries; if dimensions are discovered, appropriate and empirically rigorous means can be used to refine diagnostic cutoffs.

ASSESSMENT

Diagnosis involves the assignment of cases to recognized categories. As such, it represents one kind of assessment. However, the distinction between taxonic and dimensional structure also matters for other forms of assessment. For example, the manner in which psychologists assess personality—so typical that people rarely question it—is to construct a total score representing the sum (or average) of questionnaire items pertinent to the construct of interest. This score is taken to operationalize an individual's position on an underlying trait dimension, ideally doing so with high measurement precision. When we construct personality measures, we want them to differentiate people falling along the entire range of this dimension, validly distinguishing between those falling low and very low and between those falling high and very high. It should be obvious by now that this default approach to assessing personality assumes a dimensional model of personality variation.

But what if some personality characteristics are taxonic? Because it is desirable to build an assessment tool using knowledge of the latent structure of the construct that it measures (J. Ruscio & Ruscio, 2002), a measure for assessing a taxonic construct might differ in important ways from one assessing a dimensional construct. Such a measure would be used to assign people to classes rather than to quantify their levels on traits, and would therefore need to deliver class assignments as output rather than numerical scores along dimensions. Because it would be designed to discriminate optimally at the taxonic boundary, rather than make fine distinctions along the entire length of a latent trait, the resulting instrument might look quite different from a standard personality scale. It might be briefer, more focused, and scored using more complex methods than the simple summation of items. For example, Bayes' Theorem is a formula that can be used to integrate base-rate expectations with one or more pieces of pertinent evidence—such as items that validly distinguish two groups of individuals—to yield an estimate of the probability that an individual belongs to the taxon (see chap. 6 and Appendix B for details). When this probability is greater than .50, it is more likely that the individual belongs to the taxon than its complement class, hence this cutting score should maximize the accuracy with which individuals are assigned to these two classes. Using this approach with a small set of highly discriminating items can yield more valid results than any cutting score along a more traditional dimensional scale constructed by summing items (J. Ruscio & Ruscio, 2002). It should be noted, however, that questions remain about the conditions under which categorical classification is superior to dimensional assessment of taxa. An important study by Grove

(1991b) suggests that dimensional scaling can sometimes function well even for a taxonic construct.

Tailoring an assessment instrument to the latent structure of the construct being assessed can influence not only the reliability and validity of the resulting scores, but also the length of the instrument. One cannot develop an assessment tool that simultaneously maximizes the accuracy with which individuals are classified into categories *and* the precision with which individuals' scores are located along one or more dimensions (Meehl, 1992; J. Ruscio & Ruscio, 2002). This is because the maximal discriminative power of items can either be concentrated in the region of a taxonic boundary or spread across the full range of a given dimension. For example, suppose one wishes to assess individuals in a sample containing roughly equal numbers of taxon and complement members. Dichotomous items that are endorsed by almost everyone in the sample, or continuously scaled items to which nearly every individual gives a high rating, will do little to distinguish members of the two classes. Similarly, items endorsed by very few people (or given very low ratings by most people) will not discriminate between the two groups. In contrast, (valid) items endorsed by roughly half of the participants—or given more moderate ratings—may be extremely useful in classifying individuals. Thus, classification is facilitated by items that are constructed and selected to discriminate maximally in the narrow region near a taxonic boundary.

Now suppose one wishes to assess a dimensional construct in a sample of individuals who vary widely in their scores on a particular latent trait. Rather than employing items that discriminate only in a narrow range of trait levels, a useful measure would need to include a more diverse set of items capable of making finer discriminations along a broader range of values. To put it simply, for a given number of items, one cannot have it both ways. Efforts to "cover all the bases" by incorporating many items that discriminate widely across the trait levels of latent dimensions, plus additional items that discriminate in the narrow region of a potential taxon boundary, will result in inefficient measurement. If the construct is taxonic, items falling outside the immediate vicinity of the taxonic boundary provide little useful information; if the construct is dimensional, items massed at one discrimination point counter the goal of equal measurement precision at all points along each latent trait. In either case, instrument length and measurement burden needlessly increase.

Psychologists have long had a strong preference for dimensional measurement, and this preference may both reflect and propagate their common assumption that personality constructs represent latent traits (Meehl, 1992). Combined with the oft-repeated statistical claim that dimensions "retain more information" than categories, this dimensional perspective has become unquestioned common sense for many psychologists. However, if some psychological taxa exist (and taxometric evidence suggests that they do; see chap. 10) and if categorical assessment of taxa has demonstrable benefits, then this perspective must be challenged. It will become important to recognize that, in at least some instances, greater measurement efficiency may be achieved by classifying individuals into categories (using appropriate items and scoring rules) than by using dimensionally scaled instruments.

A more complex measurement situation arises when dealing with a taxonic construct that also possesses meaningful dimensional variation. Such cases might call for a combined categorical–continuous assessment approach. For example, one might first use Bayes' Theorem to assign individuals to the taxon or complement class, then use dimensionally scaled scores to represent individuals' relative positions within each class. It remains an open question whether dimensional scaling, categorical classification, or a combined approach affords greater utility for the assessment of most psychological constructs. Thus, research is needed not only to evaluate the latent structure of these constructs, but to determine the conditions under which each type of assessment yields the greatest utility for each structure.

RESEARCH

Valid assessment is a basic requirement of the research enterprise. However, latent structure is a fundamental issue for scientists in ways that go beyond assessment. Here we discuss two ways in which knowing the latent structure of a construct has important implications for empirical investigations. The first relates to how data analyses are performed; the second relates to how research questions, methods, and designs can be guided by the latent structure of the construct under study.

With respect to statistical considerations, researchers often must decide whether to dichotomize scores on a particular measure. This process involves cutting a continuous scale at some point representing either a critical value (e.g., a diagnostic threshold) or a convenient sample characteristic (e.g., a median split). Dichotomizing has several attractions. In particular, it simplifies scoring and allows variables to be submitted to popular statistical procedures such as *t* tests or analyses of variance (ANOVAs). Unfortunately, dichotomizing also has substantial disadvantages. Most notably, it can result in a severe loss of information, and hence a loss in statistical power. Indeed dichotomizing one variable may reduce the ability of a statistical test to detect an effect by as much as 40%, and dichotomizing multiple variables eliminates still more of the systematic variance to drive power lower (Cohen, 1983; MacCallum, Zhang, Preacher, & Rucker,

2002). To compensate for this reduction in power, sample size would need to be substantially increased. Clearly, dichotomization is a risky methodological choice, particularly when the sizes of effects or samples are modest.

The problems with dichotomization are directly related to latent structure. This practice often hinges on an unexamined assumption that the latent variable being assessed is (or at least might be) taxonic. For example, researchers who study depression in nonclinical settings often use a cutoff on the Beck Depression Inventory (BDI) total score to separate a group of *depressed* or *dysphoric* individuals from a group of *nondepressed* individuals. Although such practices are common, research suggesting the underlying construct to be dimensional (e.g., A. M. Ruscio & Ruscio, 2002) reveals dichotomizing to be doubly problematic: not only structurally unfounded, but also power-sapping. For this reason, Fraley and Waller (1998) argued that the practice of spurious classification—assigning cases to groups that do not correspond to latent taxa—could potentially cripple a field of research.

However, what if the latent construct is in fact taxonic? In such cases, statistical power will not necessarily be lost if a continuous indicator of the taxon is dichotomized. If this measure is cut at the correct point—at a place on the scale that closely approximates the optimal boundary between taxon and complement—then the dichotomized score may be just as powerful as the continuous scale score. In fact, if there is no meaningful dimensional variation within the taxon or complement, observed variation in scores above or below a taxonic boundary represents measurement error, and eliminating such variation through dichotomization may even enhance statistical power. Consistent with this view, some taxometric studies have found the appropriate dichotomization of taxonic constructs to yield stronger relationships with external variables than dimensional measurement of these taxa (e.g., Strube, 1989). In the taxonic situation, then, it may be preferable to dichotomize. As more becomes known about the conditions under which categorical scaling is more valid than continuous scaling of taxonic constructs, guidelines may be suggested for the measurement and statistical approaches most appropriate for different conditions. Because the advantages and disadvantages of dichotomizing are likely to depend (at least in part) on knowledge of latent structure, taxometric research has an important role to play in determining how continuous measurements should be treated in statistical analyses.

The distinction between taxonic and dimensional constructs has research implications that go well beyond issues of statistical analysis. Knowing latent structure also has profound consequences for the kinds of research questions that investigators can ask and answer. Although many examples could be mentioned, we focus here on two illustrative research questions: those pertaining to risk for psychopathology and to the detection of moderators of treatment outcome (Beauchaine, 2003).

One kind of research question that depends critically on latent structure involves vulnerability to mental disorders. Vulnerability factors, or *diatheses*, may be understood as categories or continua. If researchers want to investigate the factors that lead diatheses to be expressed or triggered, they face the problem of how to represent them. When vulnerability is understood as a continuum of risk, the preferred research design normally involves following large, unselected samples of the population longitudinally, testing for associations between continuously scored measures of the diathesis and later measures of symptoms or disorder. Because vulnerability is presumed to be continuously distributed throughout the population, there is no reason to exclude any participants from such studies.

But what if the diathesis is conceptualized as a taxon? In this case, different research questions and designs come to the fore. For example, it would become important to define the vulnerable or high-risk group and devote resources to isolating and following its members (rather than an unselected population sample) in longitudinal research. Efforts would concentrate on ascertaining the factors that contribute to symptom expression among taxon members. Attention would also turn to interactions between taxon membership and environmental influences to determine which influences trigger the emergence of disorder among taxon members, but not complement members. In summary, vulnerability research might follow very different trajectories if structural assumptions about the diathesis were dimensional or taxonic.

Treatment outcome research might also look rather different when psychopathology is understood in these two fundamentally different ways. As Beauchaine (2003) noted, a major task for clinical researchers is to identify moderator variables that influence response to treatment. Specifying these variables is critical both for identifying individuals who are unlikely to respond to a particular treatment and allocating individuals to the most appropriate treatment when alternatives are available. Ultimately, such specification is important for tailoring existing treatments to individuals who might not otherwise respond well, or for developing entirely new treatments that might suit these individuals better.

Moderator variables can be taxonic or dimensional, and this distinction may make a difference for how outcome research is conducted. If a moderator is taxonic, one can define subpopulations in outcome studies and determine whether they respond differently to intervention (e.g., a significant interaction effect for subgroup on outcome or time). However, if a moderator is dimensional, subpopulations would not normally be defined. Notably, if a moderator is actually taxonic, but is conceptualized dimensionally by researchers, its effects may be obscured and diluted in outcome research; even if it is detected, the effect may be more difficult to translate into practical guidelines for distinguishing treatment responders (taxon members) from nonresponders (complement members). In short, structural conceptualizations of treatment moderators can influence the search for these moderators, and conceptualizations based on empirical studies of structure might lead to a more powerful search for moderators in treatment outcome research than those based solely on conjecture.

Another implication of latent structure for research design concerns studies using analogue or subclinical samples, such as college students or individuals who do not meet diagnostic criteria for a particular mental disorder. The rationale for this sampling strategy is that one can study the phenomena of interest at a lower level, meaning among individuals who score lower on the relevant dimension(s). Therefore, this research strategy presumes dimensional latent structure. If a construct is taxonic, understanding it requires the study of taxon members. It is not uncommon for debates to focus on the extent to which research findings in analogue or subclinical samples generalize to the presumed higher levels of the phenomena of interest. For example, Coyne (1994) characterized studies of depressive phenomena in analogue samples as addressing distress or dysphoria, rather than depression. To quote one anonymous reviewer of this book, "we cannot learn much about men by studying only women, even those cases of 'subclinical' men (whatever that might be)." Investigators who generalize from research performed with analogue, subclinical, or other lower level samples are making an assumption regarding the dimensionality—as opposed to taxonicity—of their constructs. Empirically evaluating latent structure could help determine whether this assumption, and the generalization it supports, is warranted.

We have reviewed a number of the implications for researchers of the fundamental distinction between taxa and dimensions. Basic decisions guiding assessment and statistical analysis depend on this distinction, as do the research questions and designs that researchers may adopt. A methodological approach based on a valid differentiation between taxonic and dimensional structures may therefore lead to more focused and rapid advances in behavioral research.

CAUSAL EXPLANATIONS

Knowledge of latent structure has important implications not only for what sorts of research questions are posed and how this research is conducted, but for theories regarding the causes or origins of individual differences. Put succinctly, the causal processes that generate taxa and dimensions are often different, so the structural nature of a construct can help point to—and rule out—possible etiological models for that construct.

The causal origins of latent dimensions are relatively easy to explain. If people differ along a latent continuum, their position on this continuum is likely to represent the additive contribution of many small influences (e.g., multiple genes and/or environmental effects). The sum total of the relevant positive and negative influences that an individual has experienced (i.e., relevant genes plus the frequency or intensity of environmental factors) will determine his or her relative standing.

Differences in kind require somewhat different causal explanations. If an objective categorical boundary separates some people from others, we need to be able to account for why this discontinuity exists (Meehl, 1992). What do taxon members have—or what have they experienced—that complement members have not? There are perhaps four major kinds of explanations for such discontinuities. The first possibility is that there exists a dichotomous causal factor or specific etiology (Meehl, 1977) that all taxon members share, such as a major gene, neurochemical abnormality, or traumatic life experience. If this causal factor is present, it produces the characteristics that define the taxon; if it is absent, these characteristics do not arise. A second possible explanation that can account for taxa is a threshold model. Such models invoke some sort of vulnerability or severity dimension that has a critical point or threshold. People are continuously distributed along this dimension, but in those people who exceed the threshold value, a causal process is triggered that produces a qualitatively distinct state. For example, it could be that when variation in some psychobiological system reaches a critical limit, a cascade of pathological processes occurs, and those who exceed this limit consequently experience an emergent syndrome that is categorically distinct from normal variation at lower levels of the vulnerability dimension.

These first two causal models—the specific etiology and threshold models—each yield taxa by way of a single dichotomous cause or event (i.e., possessing or experiencing a single causal factor, exceeding a particular threshold). Additional kinds of etiological models can also be proposed that do not involve dichotomous causes. A third model that can account for taxa invokes a nonlinear interaction or other synergistic effect of two or more causal influences. For instance, a discrete kind of behavioral pattern might emerge when two particular causal factors occur in combination, their interaction giving rise to something qualitatively different from their additive effects. An example of this process in the genetic domain was described as *emergenesis* by Lykken, McGue, Tellegen, and

Bouchard (1992), where a particular trait is only observed when all of a set of necessary genes are present in a particular combination. A fourth model for explaining taxa involves a developmental bifurcation (Beauchaine, 2003). In this model, people gradually separate into discrete classes from an initially nontaxonic state of affairs through a process of divergence. Children might, for example, be channeled into categorically distinct ways of behaving by an environmental shaping process, perhaps influenced initially by dimensional differences in temperament. This causal model may be especially plausible for environment-mold taxa (Meehl, 1992).

The details of the different explanatory models described here are perhaps of secondary importance. What matters at this point is to recognize that taxonic and dimensional constructs will be explained differently. For this reason, knowledge of latent structure constrains theoretical conceptions of a construct and helps to direct the search for its causal processes. If a construct is shown to be taxonic, simple additive models or breezy statements about "a multitude of contributing factors" will not sufficiently account for the existence of a discontinuity. Similarly, if a construct is shown to be dimensional, an etiological theory involving only a single dichotomous causal factor must be mistaken. Therefore, latent structure has profound implications for the evaluation and advancement of theories, particularly causal explanations of behavioral phenomena.

LAY CONCEPTIONS OF IMPORTANT CONSTRUCTS

An often-neglected practical implication of the distinction between taxonic and dimensional structure is its potential impact on perceptions held by the lay public. Research by social and cognitive psychologists has demonstrated that laypeople make different inferences about the nature of human groups depending on how individuals are understood to differ from one another. Consequently, when professionals disseminate views about latent structure—regardless of whether these views are communicated explicitly or are empirically sound—such views are likely to influence how laypeople think.

There has been considerable concern in some quarters, for example, that the categorical view of psychopathology embodied in the *DSM* or *ICD* may medicalize the lay public's views of mental disorder (e.g., Zachar, 2000). In other words, the cultural salience of these classification systems may lead people to think of disorders as discrete entities that are biologically based and best treated through physiological means. Whether or when this is so remains to be seen, but it is clear that professional assumptions about mental disorders trickle down to laypeople in some fashion.

Recent studies shed some light on how this trickle-down process may occur. Research on psychological essentialism indicates that people regard some social groups—especially genders and ethnicities—to be akin to natural kinds (Haslam, Rothschild, & Ernst, 2000; Rothbart & Taylor, 1992). These groups, unlike those believed to be consciously chosen (e.g., political party affiliation) or socially constructed (e.g., occupation), are perceived to (a) have sharp, either–or category boundaries; (b) be natural or biologically rooted; (c) be difficult to change; and (d) have existed throughout history and across cultures. A related phenomenon has been uncovered by cognitive development theorists. They have shown not only that children hold essentialist or natural-kind beliefs about many living things and psychological traits (Gelman, 2003; Giles, 2003), but that when a social group is presented as a noun category rather than a verbal predicate (i.e., describing a common activity of group members), children infer its characteristics to be more stable over time and across situations (Gelman & Heyman, 1999). Finally, there is at least some evidence that essentialist beliefs can be activated through more direct communication of structural information (Haslam & Ernst, 2002). When participants in one study were presented with supposedly authoritative scientific evidence that disorders were more discrete than they had previously believed, they inferred that these disorders were also more biologically based and less modifiable than they had thought.

These literatures suggest that professional findings and opinions about the taxonicity of human characteristics may well affect how everyday people understand themselves and others. Viewing a social category as a label-worthy taxon may lead people to construe it as more real and enduring than if it were viewed as dimensional. In contrast, characteristics believed to be dimensional may more often be referred to by adjectives (e.g., *stupid*, *extroverted*) than by nouns, and hence may be regarded as more changeable and contextually variable. These beliefs may in turn have critical implications for help-seeking behavior, willingness to work to change problematic behavior, views toward others perceived to differ in degree or kind from oneself, and other important attitudes influencing self-perception and interpersonal action.

It should be acknowledged that these tacit appraisals and beliefs are not confined to the person on the street. There remains a lot of the layperson even in sophisticated behavioral scientists; we are all intuitive psychologists to some extent despite the transformative experiences of graduate and professional training. Regardless of their training, when behavioral scientists view a construct as taxonic, they are often more inclined to infer that it has a somatic basis and is difficult to change. For this reason as well as those previously mentioned, rigorous evaluation of latent structure is a critically important matter.

CONCLUSIONS

It should now be clear that latent structure has numerous implications for those interested in human variation, and hence that the rationale for empirically distinguishing between taxonic and dimensional structure is strong and compelling. Knowing a construct's latent structure has implications for how we develop and refine classification systems, and for how we assign people to diagnostic categories or assess their positions along latent continua. Structural knowledge can help to explain psychological variation and has implications for how the lay public thinks about important psychological and behavioral constructs. Finally, structural understanding has great practical relevance for researchers, with implications for the research questions that can and should be pursued as well as the methodological designs and statistical approaches that are used to address them. For all of these reasons, a method able to powerfully and accurately distinguish taxa from dimensions could play a fundamental role in behavioral science and practice.

We believe that the taxometric method represents a valuable technique for distinguishing taxa from dimensions. As of yet, however, we have not begun to describe the distinctive features of this method or the challenges one must confront to apply it rigorously. Testing latent structure turns out to be quite a thorny problem, and the difficulty of this classification problem is one reason that many data-analytic techniques have been developed to tackle it. In the next chapter, we lay out the classification problem, the ways in which methodologists have attempted to address it, the strengths and limitations of each approach, and the unique features of the taxometric method intended to overcome some of the limitations of alternative approaches.

CHAPTER THREE

The Classification Problem

The two previous chapters introduced the distinction between taxonic and dimensional structures, and they highlighted many reasons that behavioral scientists might wish to know the latent structure of constructs that they seek to classify, diagnose, assess, investigate, and explain. Convinced of the importance of structural knowledge for these fundamental tasks, the interested reader arrives at the next logical question: How can one determine whether a particular construct is taxonic or dimensional?

As it turns out, scientists have grappled with this classification problem for years, and debate persists about the most fruitful method to answer this complex question. The current chapter describes the conceptual and methodological challenges that researchers face when attempting to determine whether a taxonic or dimensional structural model best captures variation among individuals. We provide a broad overview of data-analytic approaches that have been used to address this problem, including examination of frequency distributions for evidence of bimodality, finite mixture modeling, cluster analysis, latent class analysis, and a more recently introduced conceptual and psychometric framework known as Dimcat. Some of the strengths and weaknesses of each approach are weighed and evaluated, and then unique features of the taxometric method are discussed in comparison with alternative methods. Throughout the chapter, we focus on the ability of each approach to distinguish taxonic from dimensional structure, rather than to address other research questions within its purview.

THE CLASSIFICATION PROBLEM

As noted in chapter 2, one of the fundamental goals of science is to understand and organize nature not only according to laws of cause and effect, but also according to latent structure. To that end, scientists working in many disciplines have developed systems to classify aspects of our universe from its largest to its smallest presently detectable features. Despite their obvious utility, these systems are continually being revised. This perpetual process of elaboration and refinement reflects the challenges involved in organizing entities into nonarbitrary categories. Valid classification is challenged by between-group similarity (e.g., organisms belonging to genetically disparate species may look and behave alike) as well as within-group variation (e.g., organisms belonging to the same species can look and behave differently from one another). It may also be undermined by limitations of available tools for reliably and validly measuring the relevant attributes of the objects under study. For example, it can be difficult to distinguish a nearby planet from a much more distant star without a telescope because their relative size and apparent brightness may be quite similar to the naked eye. Classification research in any scientific domain is likely to progress only as these, and other, obstacles can be overcome.

The classification problem poses particular challenges for psychologists and other behavioral scientists. Psychological measures are inevitably fallible, possessing some degree of measurement error. Even when taxa exist, their members are likely to overlap substantially on most observed variables or *indicators* that distinguish groups. These manifest indicators are also likely to covary within groups, making taxonic structures still more difficult to detect. The challenges of classification in psychology are apparent even in the deceptively simple task of distinguishing taxonic and dimensional structures. The issues raised next can be generalized to techniques that attempt to resolve more complex latent structures involving multiple taxa or dimensions. However, for the sake of simplicity, we restrict our discussion to the most fundamental classification problem: determining when and where a distinction can be made between two latent classes. Later, we return to the problem of evaluating more complex latent structures.

Psychologists may be surprised to find familiar procedures such as factor analysis omitted from this chapter. Although factor analysis can yield valuable information about the number and nature of dimensions underlying a construct, it begins by presuming that the construct in question is dimensional. Whereas one taxometric procedure presented in chapter 5 involves an examination of the distribution of estimated factor scores, factor analyses are ordinarily not used to make the fundamental distinction between taxonic and dimensional structure. Instead, exploratory and confirmatory factor analysis—along with other statistical techniques that are used within a taxonic or dimensional framework—are likely to play an important role at later stages of classification research, building on the foundation of initial structural investigations to more fully delineate the complete latent structure of a construct (for a more detailed treatment of this issue, see J. Ruscio & Ruscio, 2004a).

BIMODALITY

One of the earliest methods for inferring the presence or absence of a taxonic boundary involved identifying indicators of the putative taxon and examining their frequency distributions for evidence of bimodality (e.g., Harding, 1949). The rationale underlying this approach is that if two groups exist, then observed scores on measures that validly distinguish these two groups should be distributed bimodally, with each mode corresponding to the typical score of individuals in one group. Within each group, there may be true dimensional variation as well as measurement error, and the combination of these sources of variance may result in some degree of overlap in scores between members of the two groups. However, the tendency of group members to score near their mode should produce a relative scarcity of intermediate scores on the observed indicator, yielding a distribution with two distinct peaks. Following this reasoning, bimodality suggests taxonic structure, whereas the absence of bimodality suggests dimensional structure.

Although bimodality of indicator distributions can provide suggestive evidence for or against the existence of a taxonic boundary, there are several problems with drawing structural inferences on this basis (Murphy, 1964). First, even when two groups are mixed in a sample of data, they may not be sufficiently separated along the indicator to yield a bimodal distribution. If within-group indicator variation is substantial (e.g., due to measurement error), there may be no discernible dip in the distribution between the modes of the two groups, causing the distribution to appear unimodal. Murphy (1964) has shown that the observed scores for two groups must differ by at least 2.00 within-group *SD*s (e.g., a Cohen's *d* of 2.00) before bimodality will emerge—a large effect size that may be difficult to achieve with many psychological and behavioral variables.

Notably, this minimal degree of separation presumes that the two groups are of equal size (i.e., their base rates within the sample are each .50) and that scores are normally distributed within each group. Departures from this idealized scenario often necessitate even larger between-group differences for bimodality to be detected. For example, all else being equal, groups of unequal size require greater separation to yield a bimodal distribution. Likewise, if one or both of the within-group score distributions are skewed, greater between-group separation may be required, although this depends on the amount and direction of the skew. Finally, within-group kurtosis can also influence the appearance of bimodality, with platykurtic distributions requiring larger between-group separation for bimodality to be revealed.

In summary, the sensitivity of an observed score distribution to the presence of two groups depends on many factors, including the separation between groups' scores, the relative group sizes, the shapes of the score distributions within groups, and the reliability and validity of the scores. As any of these factors deviate from ideal values, the likelihood of unimodality increases—and along with it the likelihood of failing to detect a taxonic boundary if one exists.

A second problem with drawing structural inferences from indicator distributions is that there are a number of ways that one can obtain a bimodal distribution even in the absence of taxonic structure (Haslam, 1999). In other words, several factors can result in a mistaken inference of taxonicity when in fact only dimensional variation exists. First, latent dimensions may give rise to bimodal manifest distributions strictly as a result of sampling error. Indicator distributions may appear bimodal simply because of the "lumpiness" of chance, especially in small samples. To avoid identification of spurious latent classes, classification research often requires larger samples of data than are necessary to address other types of research questions.

Third, artifactual bimodality may be produced by sample bias. Selective sampling of cases from the extremes of latent continua can give rise to a bimodal distribution of observed scores (Haslam, 1999). For example, researchers sometimes apply stringent operational definitions to obtain one sample of individuals who clearly meet the diagnostic criteria for a particular disorder and a second sample of individuals who meet few or none of these criteria. Although such case-control designs can be valuable for examining certain research questions, their selective exclusion of cases with intermediate severity levels may yield an artificially bimodal distribution of scores on one or more indicators.

Fourth, bimodality may be a consequence of observer bias. Beauchaine and Waters (2003) have shown that raters' beliefs about the structure of the construct underlying the variables that they are observing can influence the way they make their ratings. For example, raters who believe that the individuals being observed fall into two classes may polarize their ratings to such an extent that the distributions of their ratings appear bimodal. Although an analogous dimensional bias was not examined by Beauchaine and Waters, it is possible that raters' assumption of dimensional structure may similarly bias their judgments, leading to unimodal distributions of ratings even when a taxonic boundary exists.

A fifth and final complicating factor in the search for bimodality was highlighted by Grayson (1987) and Golden (1991). These authors observed that a bimodal distribution of scores may result from the selection and aggregation of items with similar difficulty levels. That is, even when individuals vary dimensionally, bimodality can arise when a measure's items are maximally sensitive to a constrained range of trait levels.

Consider, for example, a standardized test such as the Graduate Record Examination (GRE). The GRE includes a pool of items written to provide good discriminating power across a wide range of ability levels. This breadth of item difficulty allows the GRE to ensure comparable measurement precision for all examinees. It also results in a unimodal distribution of GRE scores. If, however, a significant number of GRE items were written to maximally discriminate within a very narrow range of ability levels, individuals with trait levels above the typical item difficulty threshold would respond correctly to most items, whereas individuals with trait levels below the difficulty threshold would respond incorrectly to most items. In this way, items with highly similar difficulty levels can yield bimodal score distributions, regardless of the true structure of the underlying construct.

To demonstrate the limitations of drawing structural inferences based on inspections of indicator distributions, we applied this approach to eight artificial data sets that are used throughout this book to illustrate the taxometric method. These data sets are described only briefly here; more detailed descriptions of the data are provided later.

The first four data sets were taxonic; each possessed four indicators (manifest-level, measured variables that each validly assesses the latent variable) and an N of 1,000 cases. However, the data sets differed in the relative size of their taxon and complement groups, the within-group indicator distributions and correlations, and the between-group separation (separation is traditionally referred to as *indicator validity* in the taxometric literature and is expressed using the metric of Cohen's d). The first sample of taxonic data (TS#1) consisted of two equal-sized groups (i.e., a taxon base rate of $P = .50$) that differed by $d = 1.00$. To generate these data, scores on each indicator were sampled at random from a unit normal distribution, a constant (+1) was added to each indicator for one half of all cases to represent the taxon (with the remaining half of cases representing the complement), and scores were then restandardized in the full sample. Thus, the taxon (the higher scoring half of the sample) systematically differed on each indicator from the complement (the lower scoring half of the sample), and the constant used to create this between-group difference was chosen to achieve $d = 1.00$. (Of course ordinary sampling error in the creation of each data set caused the sample statistics to differ in small, unsystematic ways from these specified population parameters.)

The second sample of taxonic data (TS#2) consisted of two equal-sized groups that differed by $d = 2.00$. These data were generated using the same technique described earlier, with a larger constant (+2) added to each indicator to achieve a greater separation between groups. The third sample of taxonic data (TS#3) comprised a taxon that was one third the size of the complement (i.e., a taxon base rate of $P = .25$), with the groups once again differing by $d = 2.00$. The unequal group size was achieved by adding a constant to each indicator for only one quarter of the cases.

Because each of these samples of taxonic data involved a number of idealizations, the fourth sample (TS#4) was created to reflect some of the common complexities of actual research data. Whereas each of the first three samples contained no systematic within-group correlations and included indicators that were normally distributed (within groups) along continuous scales, our final sample of taxonic data possessed nontrivial within-group correlations, variable indicator validities, ordered categorical response scales, and skewed distributions. Variable and nontrivial within-group correlations were generated through loadings onto a latent factor within each group. The validity with which each indicator separated these two groups was varied by adding different constants to scores for taxon members on each indicator. Indicators were positively skewed by raising the initial values to a power greater than 1, with each indicator being skewed to a different extent. Finally, indicator distributions were cut into differing numbers of ordered categorical values. The taxon base rate was set at $P = .30$. Thus, our four illustrative samples of taxonic data vary in a number of important ways, including different taxon base rates, indicator validities, within-group correlations, and indicator distributions.

In addition to these four samples of taxonic data, four comparison samples of dimensional data (DS#1–DS#4) were also created. Using a variant of the technique described in Appendix A, each set of dimensional data was generated such that the distribution of scores on each indicator reproduced the score distribution of the corresponding indicator in the comparison taxonic data set. In addition, the full-sample indicator correlation matrixes of the taxonic data sets were reproduced in the dimensional data sets. Bartholomew (1987) has shown that any variance–covariance matrix can be reproduced equally well using a structural model with either m latent classes or $m - 1$ latent dimensions. Thus, the correlation matrix for any taxonic data set (with two latent classes) can be reproduced using one latent dimension. As a result, the four dimensional data sets directly paralleled the four taxonic data sets (DS#1 provided a comparison for TS#1, DS#2 for TS#2, DS#3 for TS#3, and DS#4 for TS#4), with full-sample indicator distributions and correlation matrixes reproduced. (Indeed the fact that these—and any other—indicator distributions can be held constant across taxonic *and* dimensional data sets should give one pause about drawing structural inferences from manifest score distributions.)

Table 3.1 summarizes the actual indicator validities in each taxonic data set, Table 3.2 shows the full-sample and within-group indicator correlation matrixes, and Fig. 3.1 presents the full-sample and within-group indicator distributions; a bar chart is used to show the indicator distributions for TS#4 and DS#4 because these varied along ordered categorical scales. As can be seen in Fig. 3.1, none of the indicator distributions in any of the four taxonic data sets provided compelling evidence of bimodality. For TS#1, there was no hint of bimodality despite the presence of two equal-sized groups separated by what is, in the social and behavioral sciences, conventionally considered to be a large effect size (d = 1.00). For TS#2, with two equal-sized latent classes separated by d = 2.00, the first signs of bimodality emerged for some of the indicators. However, even here it would be difficult to describe these distributions as clearly bimodal; they could just as readily be interpreted as unimodal with minor fluctuations near the peak due to sampling error. For TS#3, with its equally large d = 2.00, the unequal group sizes gave rise to unimodal distributions. For TS#4, whose indicator distributions were all positively skewed (in the full

TABLE 3.1
Indicator Validities in Each Taxonic Data Set

Data Set and Indicator		*Full Sample M (SD)*	*Taxon M (SD)*	*Complement M (SD)*	*Indicator Validity (Cohen's d)*
TS#1	1	.00 (1.00)	.44 (.92)	–.44 (.88)	.98
	2	.00 (1.00)	.41 (.90)	–.41 (.92)	.91
	3	.00 (1.00)	.48 (.87)	–.48 (.89)	1.09
	4	.00 (1.00)	.43 (.92)	–.43 (.89)	.95
TS#2	1	.00 (1.00)	.72 (.69)	–.72 (.70)	2.06
	2	.00 (1.00)	.73 (.69)	–.73 (.68)	2.12
	3	.00 (1.00)	.69 (.75)	–.69 (.71)	1.89
	4	.00 (1.00)	.71 (.71)	–.71 (.71)	1.99
TS#3	1	.00 (1.00)	1.15 (.77)	–.38 (.75)	2.04
	2	.00 (1.00)	1.15 (.75)	–.38 (.75)	2.03
	3	.00 (1.00)	1.12 (.78)	–.37 (.76)	1.96
	4	.00 (1.00)	1.07 (.78)	–.36 (.79)	1.83
TS#4	1	2.69 (.94)	3.68 (.97)	2.26 (.50)	2.10
	2	3.29 (1.42)	4.88 (1.32)	2.60 (.77)	2.36
	3	3.86 (1.90)	5.83 (1.73)	3.01 (1.23)	2.01
	4	4.45 (2.36)	7.02 (1.88)	3.35 (1.56)	2.21

Note. TS#1 was generated with N = 1,000 and P = .50; TS#2 was generated with N = 1,000 and P = .50; TS#3 was generated with N = 1,000 and P = .25. The parameters for TS#4 are more complex; see text for details.

TABLE 3.2
Indicator Correlations in Each Data Set

#1				#2				#3				#4			
DS	*1*	*2*	*3*	*DS*	*1*	*2*	*3*	*DS*	*1*	*2*	*3*	*DS*	*1*	*2*	*3*
2	.22			2	.51			2	.43			2	.56		
3	.17	.19		3	.51	.48		3	.42	.47		3	.57	.59	
4	.18	.16	.17	4	.52	.52	.49	4	.43	.44	.41	4	.54	.59	.58
TS	*1*	*2*	*3*	*TS*	*1*	*2*	*3*	*TS*	*1*	*2*	*3*	*TS*	*1*	*2*	*3*
2	.16			2	.54			2	.46			2	.56		
3	.22	.20		3	.48	.48		3	.44	.41		3	.57	.58	
4	.21	.16	.17	4	.51	.52	.47	4	.45	.43	.42	4	.54	.60	.57
Taxon	*1*	*2*	*3*	*Taxon*	*1*	*2*	*3*	*Taxon*	*1*	*2*	*3*	*Taxon*	*1*	*2*	*3*
2	–.04			2	.12			2	–.05			2	.10		
3	.05	.05		3	–.03	–.09		3	–.04	–.01		3	.25	.12	
4	–.00	–.02	–.02	4	.08	.02	–.03	4	.08	.02	.03	4	.13	.10	.15
Complement	*1*	*2*	*3*	*Complement*	*1*	*2*	*3*	*Complement*	*1*	*2*	*3*	*Complement*	*1*	*2*	*3*
2	–.01			2	–.06			2	.07			2	.10		
3	–.04	–.04		3	–.01	.01		3	.04	–.04		3	.13	.21	
4	.07	–.01	–.06	4	–.08	–.02	–.04	4	.06	.03	.02	4	.05	.21	.20

Note. TS = taxonic simulation; DS = dimensional simulation. Data sets labeled #1 were generated with $N = 1000$, $P = .50$, and $d = 1.00$; data sets labeled #2 were generated with $N = 1000$, $P = .50$, and $d = 2.00$; data sets labeled #3 were generated with $N = 1000$, $P = .25$, and $d = 2.00$; the parameters for data sets labeled #4 are more complex, see text for details.

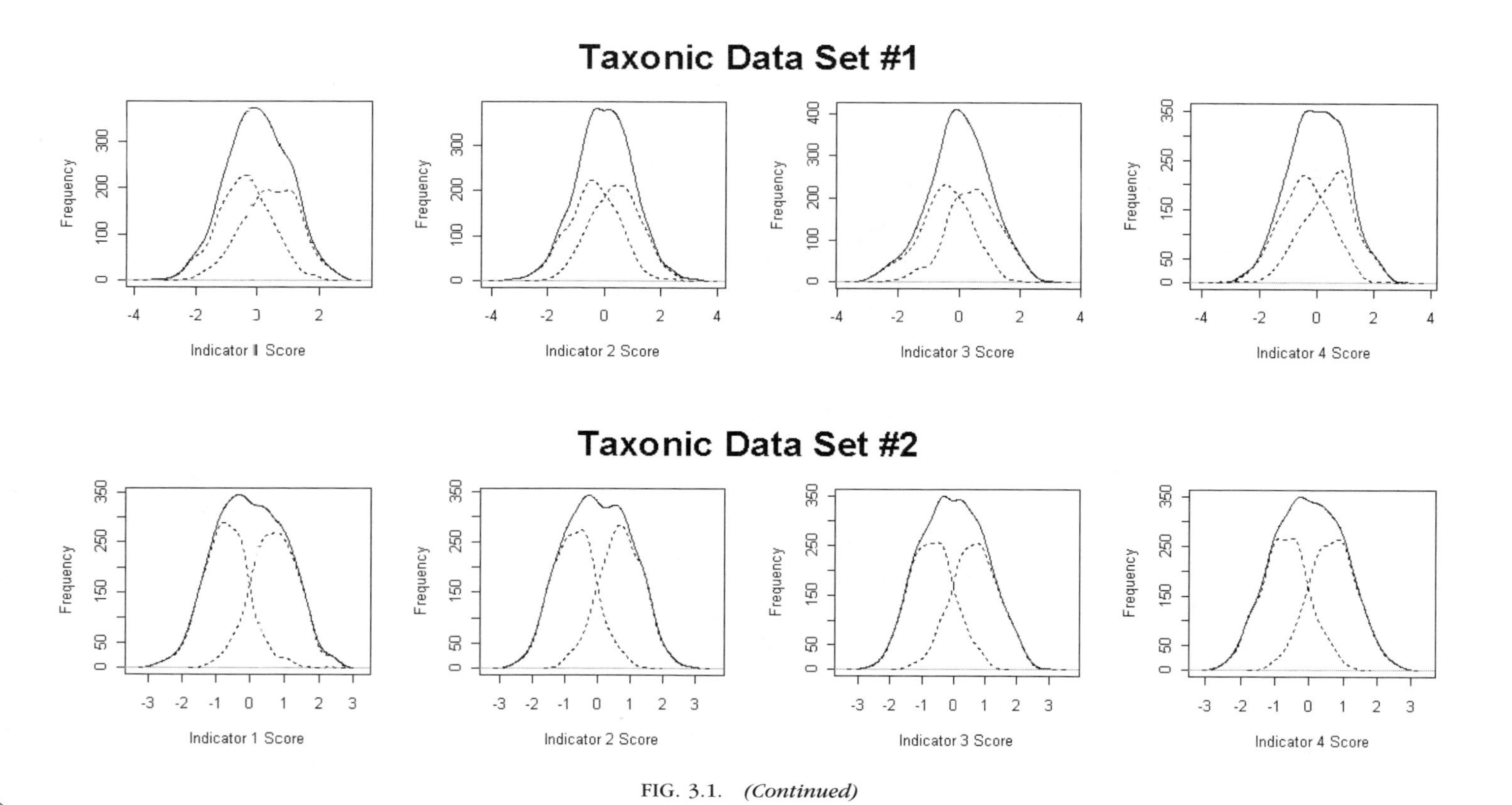

FIG. 3.1. *(Continued)*

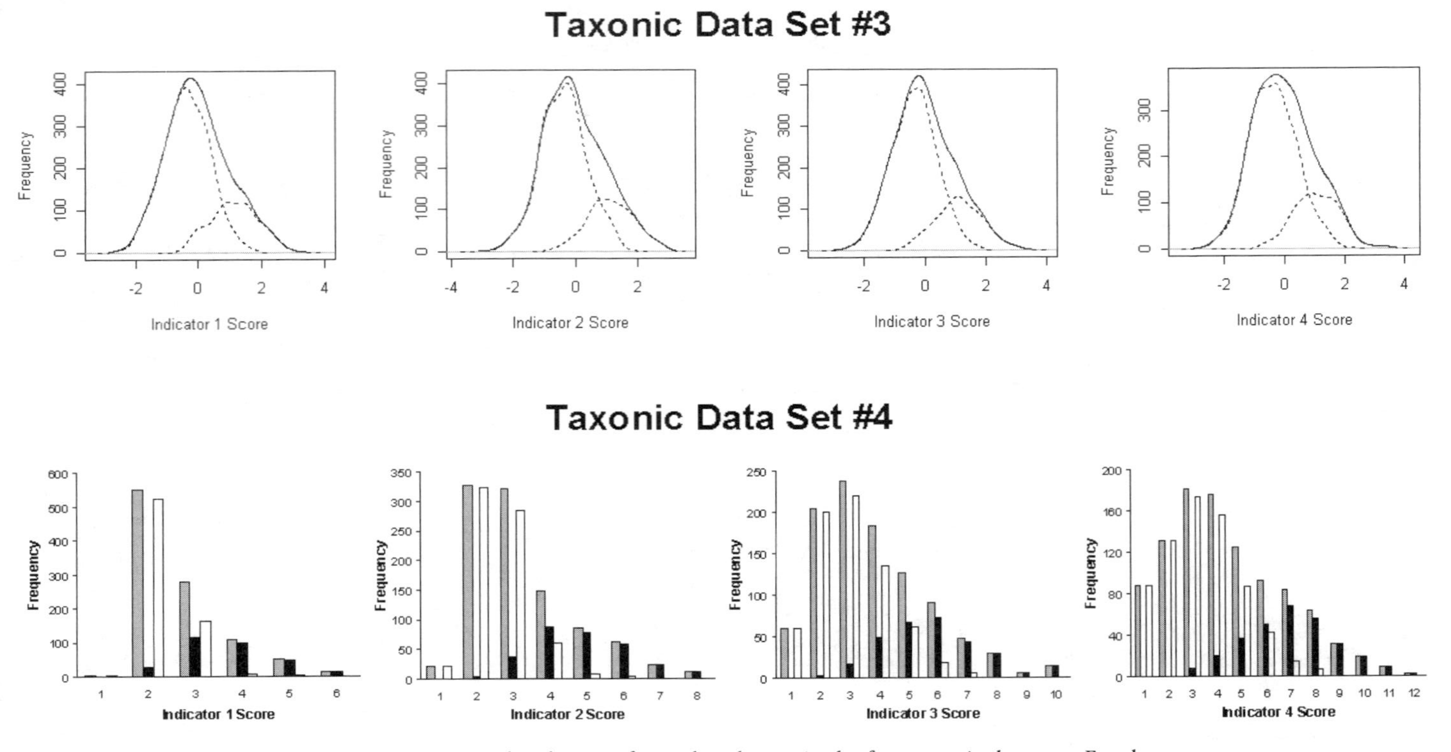

FIG. 3.1. Frequency distributions for each indicator in the four taxonic data sets. For the first three data sets, dashed lines represent the taxon and complement and solid lines represent the full sample. For the final data set, black bars represent the taxon, white bars represent the complement, and gray bars represent the full sample.

sample as well as within groups), there was no evidence of bimodality despite indicator validities that were slightly larger than those in TS#2 and TS#3. In short, simple inspection of these full-sample indicator distributions for bimodality would probably lead to the mistaken conclusion of dimensional structure for each data set, despite the relatively high validity of all indicators. However, because the full-sample indicator distributions for the four dimensional data sets were identical to those for the taxonic data sets, they would likely lead to correct inferences of dimensional structure.

In addition to its high potential for overlooking taxa, there are at least two other limitations of the search for bimodality as a test of latent structure. First, because the approach bases structural inferences on one indicator at a time, it is unclear what conclusion would be drawn if some indicators of the construct were distributed bimodally while others were distributed unimodally. Second, it may be inappropriate to evaluate the latent structure of some constructs on the basis of individual indicator distributions. For example, a constellation of symptoms, whether of similar or different latent structures, together define a mental disorder that may be either taxonic or dimensional. Discovering that anhedonia or feelings of worthlessness are distributed dimensionally in a sample does not imply that the latent structure of Major Depressive Disorder—an overarching syndrome that encompasses both symptoms—is also dimensional. Thus, for any construct that cannot be adequately represented by a single indicator, it may be unwise to reach structural conclusions on the basis of an analytic technique that examines each indicator in isolation. One could sum all available indicators and examine the composite score distribution for bimodality, but any of the alternative approaches discussed later may be more effective for multivariate analyses.

FINITE MIXTURE MODELING

Finite mixture models seek to determine the number and nature of hypothetical subgroup distributions, or *components*, that together reproduce the observed distributions of one or more indicators (e.g., Everitt & Hand, 1981; McLachlan & Basford, 1988; McLachlan & Peel, 2001). Mixture models attempt to reproduce the observed distributions by systematically varying a number of model parameters, including the number of components and possibly the base rate, mean, standard deviation, skew, or kurtosis of each. A variety of tests can be used to identify a best-fitting mixture model— one that achieves the most satisfactory reproduction of the indicator distributions. As in other statistical approaches, such as structural equation modeling, tests have been developed that take into account the number of free parameters in a mixture model so that less parsimonious models do not automatically gain an advantage (McLachlan & Peel, 2001). When finite mixture models are used to assess latent structure, the components in the best-fitting model are interpreted to represent latent classes. For example, if a two-component model better reproduces the observed score distribution than a one-component model, this suggests that the construct in question is taxonic rather than dimensional.

This statistical approach has notable advantages over a more simple inspection of distributions for evidence of bimodality. First, groups need not be separated so widely that bimodality is observed for a two-component model to yield better fit than a one-component model. Second, multivariate mixture models allow researchers to study constructs that are jointly defined by multiple indicators and therefore cannot be adequately investigated using one indicator at a time. Third, statistical tests that evaluate the fit of competing models may provide more objective structural evidence than the relatively subjective interpretation of score distributions to distinguish those that are unimodal and bimodal. Collectively, these features provide excellent justification for preferring mixture models to the more impressionistic search for bimodality in individual indicator distributions.

Despite these significant advantages, there are some limitations to the use of mixture models to distinguish taxonic from dimensional structure. The most important is that it is unclear whether such models can reliably determine the correct number of components in a mixture. McLachlan and Basford (1988) described a number of indexes for making this determination, but noted that identifying the number of components remains a "very difficult problem which has not been completely resolved" (p. 21), a conclusion echoed nearly a decade later by Bock (1996) and more recently repeated verbatim by McLachlan and Peel (2001, p. 175). Indeed using several simulated data sets, McLachlan and Peel demonstrated that many of the available indexes tend to overestimate the correct number of components. Although it would be unreasonable to demand that this problem be completely resolved—and one could argue that analogous challenges facing other approaches to evaluating latent structure have also not been *completely* resolved—the tendency toward overidentifying components underscores the need to exercise caution when using finite mixture models to determine the number of latent classes until methodologists develop and validate a more effective index.

A second limitation of mixture models is that they gauge the fit of a structural model by its ability to reproduce the manifest distribution of scores. Because manifest structure can differ from latent structure for a variety of reasons (see chap. 1), even a model that fits the manifest data well may not correctly reflect the true structure existing at the latent level. This limitation may be particularly troublesome in situations where discrepancy between manifest and latent structure is amplified, such as in small samples (i.e., the "lumpiness" of chance can substantially distort manifest distributions) or in samples selected in potentially problematic ways (e.g., selective sampling from the extremes of a distribution can yield a misleading manifest distribution).

Finally, McLachlan and Peel noted that although normal mixture models can adequately model skewed indicator distributions, the best-fitting model will often include extra (false) components to account for the skew. This may be a significant concern because non-normal distributions are ubiquitous in the behavioral sciences. For example, skew is observed even under such favorable conditions as administering rigorously developed achievement and psychometric measures to nondisturbed populations (Micceri, 1989). Many taxometric investigations have involved the study of psychopathological constructs in community or analogue populations, where the rarity of the phenomena and the nature of the measures often result in markedly skewed indicator distributions. Thus, the tendency of mixture models to uncover too many components in the presence of skewed data (and thus to favor taxonic over dimensional models) seems an important unresolved barrier to the successful application of this statistical technique.

CLUSTER ANALYSIS

Cluster analyses include a highly diverse array of data-analytic procedures that are used to determine how many relatively homogeneous groups of cases can be distinguished within a given sample. In fact, finite mixture models represent one special type of cluster analysis (McLachlan & Peel, 2001). Because a comprehensive review of cluster analysis is beyond the scope of this chapter (see Arabie, Hubert, & DeSoete, 1996; Everitt, 1993; Lorr, 1994; McLachlan & Peel, 2001), we focus on an overview of the approaches to cluster analysis that have been used most frequently in psychological research.

Several of the most popular clustering techniques involve a process analogous to using multiple indicators to create a multidimensional scatterplot, then looking to see whether cases are distributed evenly throughout this scatterplot or whether clouds of similar cases tend to clump together. When the number of cases and indicators is relatively large, the number of ways that cases can be grouped will far exceed the limits of what can feasibly be tested. Thus, many different clustering algorithms have been developed to serve as heuristic search procedures, comparing and evaluating a smaller number of distinct clustering solutions that hold the greatest promise for capturing the structure underlying the data.

Hierarchical clustering approaches, by far the most popular, involve a two-step process. First, a measure of similarity or distance is used to quantify the relations between all cases in the sample. For example, one can compute pairwise correlations among the score profiles of all cases to capture their similarity (and hence their relative closeness in the multidimensional scatterplot), or instead compute pairwise squared or unsquared Euclidean distances among all cases to capture their differences. Second, a mathematical rule is applied to parse the resulting matrix of similarity or distance values into clusters. For example, the average linkage method (Sokal & Michener, 1958) ensures that each member of a cluster is, on average, more similar to members of the same cluster than it is to members of other clusters. Another popular rule, one that tends to yield equal-sized clusters, uses Ward's (1963) criterion of minimizing within-cluster variance on the indicator variables.

When performing a hierarchical cluster analysis, one can treat each case as a cluster and then sequentially fuse similar clusters until all cases have been joined into one cluster (an agglomerative method). Alternatively, one can begin with a single cluster and then sequentially remove the least similar cases until each has been placed into its own cluster (a divisive method). Agglomerative procedures have generally been more popular in the psychological literature (Everitt, 1993; Lorr, 1994).

The advantages and disadvantages of hierarchical cluster analyses are similar to those of mixture models—which, as noted earlier, can be conceptualized as another type of cluster analysis. On the positive side, cluster analyses are multivariate, permitting the use of multiple indicators to represent and evaluate the target construct. Moreover, tests are available to aid in determining how many clusters to retain and interpret. As was the case with mixture models, the latter feature reflects both a strength and a liability. Because a hierarchical cluster analysis yields multiple solutions (e.g., a one-cluster solution, a two-cluster solution, etc. up to an N-cluster solution, in which each case is treated as a cluster), the investigator must determine the most appropriate number of clusters. When the number of clusters is the central research question, this task requires an empirical index or stopping rule to infer a construct's latent structure. Unfortunately, the available stopping rules designed to identify the appropriate number of clusters are highly fallible, and cluster analyses often overidentify clusters (e.g., Grove, 1991a; Milligan & Cooper, 1985). Stopping rules perform especially poorly in the situation that is our focal concern: distinguishing taxonic (two-cluster) from dimensional (one-cluster) latent structure (Lorr, 1994). Many authors have observed that identifying the true number of clusters remains one of the most challenging problems in cluster analysis (e.g., Everitt, 1993; Lorr, 1994; Milligan, 1996). For this and other reasons, Everitt (1993) recommended that hierarchical cluster analysis be viewed as a descriptive tool, potentially useful for summarizing data or posing a structural hypothesis. Similar recommendations may also apply to mixture modeling. Thus, the most defensible approach may be to submit the clusters or components yielded by each of these methods to another analytic procedure that can more accurately differentiate taxonic from dimensional latent structure.

LATENT CLASS ANALYSIS

Like finite mixture modeling, latent class analysis (e.g., Green, 1951; Lazarsfeld & Henry, 1968; Muthén, 2001; Uebersax, 1999) models responses to a set of categorical indicators using a categorical latent variable. Competing models containing one or more latent classes are compared to determine which can best reproduce the responses to all indicators. A number of quantitative indexes are available to evaluate the fit of competing models, and the number of latent classes in the best-fitting model is interpreted as representing the latent structure of the target construct. A variant of latent class analysis called *latent profile analysis* extends the approach to allow the use of continuous, rather than categorical, indicators (Muthén, 2001), but it has been less intensively studied and less frequently applied than latent class analysis.

The advantages of this approach are again similar to those of mixture modeling and cluster analysis, in that the technique is multivariate and incorporates statistical tests by which to compare the relative fit of competing models. However, the interpretation of results in latent class analyses can be complicated by several factors (Uebersax, 1999). As with the two previously discussed approaches, latent class analysis has a tendency to overidentify the number of latent classes. Whereas latent class analysis can be useful for assigning cases to classes once a categorical latent structure has been determined using other procedures, it is not clear whether the technique is sufficiently valid to initially determine the number of classes. An additional complication involves the method's assumption of *local independence*, which requires all indicators to be statistically independent—both pairwise and in higher order relationships—within each latent class. Such a strict assumption may be unrealistic and can affect the fit of many latent class models; one model may achieve superior fit to the data not necessarily because it better reflects the true latent structure of the target construct, but because the data better satisfy the assumption of local independence when testing this model than when testing others. Fortunately, this assumption is not required in all applications of latent class analysis. For example, models can be specified and tested in which within-group indicator correlations are estimated, rather than fixed at zero. How well this improves the determination of the number of latent classes remains to be tested. Finally, when calculating model fit using the type of maximum likelihood algorithm ordinarily employed with latent class analysis, the final solution can be sensitive to starting parameter values that the investigator must provide. If replications using different starting values converge on the same structural solution, one can have greater confidence in structural inferences drawn from the best-fitting model. However, if replications do not converge on the same structural solution, it can be difficult to determine which to trust, and hence to draw structural inference from the varied results.

DIMCAT

Recently, a conceptual and psychometric framework for distinguishing the latent structure underlying manifest categories called *Dimcat* was introduced (De Boeck, Wilson, & Acton, 2005). The name of this approach reflects its goal of differentiating dimensional and categorical latent structures. In recognition of the fact that these structures are not necessarily qualitatively distinct, but can differ as a matter of degree in several different ways, De Boeck et al. described Dimcat as a means to determine whether the best-fitting model is more consistent with a dimension-like or category-like latent structure. Category-like structures are characterized by features such as latent-level homogeneity within manifest-level categories (i.e., members of a given group differ little from one another on each indicator's underlying true score distribution) and heterogeneous factor structures across manifest categories (i.e., the nature of the latent dimensions differs across groups). Dimension-like structures are characterized by latent-level heterogeneity within manifest categories and homogeneous factor structures across these categories.

To apply the Dimcat technique, one generates and tests multiple structural models whose free parameters vary systematically. For example, one model might constrain all cases' scores on latent dimensions to a single value within each manifest category, whereas another might allow scores on the latent dimensions to vary across cases. The former model corresponds to within-category homogeneity, the latter to within-category heterogeneity. Comparing the relative fit of these models affords an inference as to how category-like or dimension-like latent structure is with regard to this characteristic. Similarly, comparing the fit of models in which within-category factor structures are constrained to equality versus free to vary in specified ways assesses latent structure according to another characteristic of category-like versus dimension-like features.

This intriguing technique addresses a research question similar to that involved in taxometric studies: distinguishing categorical from continuous models of latent structure that underlie manifest variables. One important difference, however, is that all taxometric procedures were developed for use with continuous indicators. Dimcat, in contrast, requires categorical indicators. Thus, it is unlikely that these two approaches would be equally appropriate for a particular set of research data. (As discussed in chap. 10, some taxometric procedures were adapted for use with dichotomous items in early studies, but this technique has fallen out of favor.) A second difference is that the taxometric method has been used and evaluated in a rapidly growing number of substantive and methodological investigations. De Boeck et al. (2005) presented a series of illustrative analyses to empirically demonstrate the potential utility of Dimcat. However, because it was introduced so recently, no published studies have examined the validity with which Dimcat uncovers the category-like versus dimension-like aspects of data of known latent structures. This intriguing approach certainly warrants research attention.

SEARCHING FOR MULTIPLE BOUNDARIES

Our discussion has thus far focused on the problem of differentiating two-group (taxonic) from one-group (dimensional) structural models. One feature of mixture modeling, cluster analysis, and latent class analysis is that they enable researchers to test a variety of structural models, including those with three or more groups. The search for bimodality in manifest indicator distributions can likewise be broadened to a search for multimodality. Given this apparent benefit of the aforementioned methods, our emphasis on their ability to distinguish taxonic from dimensional latent structure might be perceived as unfairly limited. More pointedly, one could suggest that our arguments stack the deck in favor of the taxometric method, which was expressly designed to differentiate only these two structural models, and that we unfairly pit this method against more versatile approaches that were designed to test a broader and more complex range of structural possibilities.

To this we would reply that the fact that a procedure *can* be used to reach complex structural inferences does not necessarily provide assurance that these inferences will be trustworthy. As was noted earlier, finite mixture modeling, cluster analysis, and latent class analysis all have a tendency to identify spurious latent classes. The risk of detecting excess classes seems particularly high when some or all of the indicators are skewed—a common concern when rare phenomena are studied.

Another, almost opposite, danger is that true latent classes can be missed when a single set of variables is used to simultaneously search for multiple latent boundaries. This is because the available indicators may not all be relevant to each of the classes being tested. At first blush, such an analysis may appear to offer a great deal of information in an analytically simple and efficient package by searching for multiple latent classes in a single sweep. However, this analytic economy may also pose a liability because a single set of variables is unlikely to be valid for detecting multiple, heterogeneous boundaries between the different classes. Boundaries represented by a subset of the variables may be missed if variables irrelevant to a particular boundary are nonetheless included in all calculations. Likewise, boundaries may be missed when variables unique to their identification are excluded from the analysis. Although analytic efficiency is a worthy goal, it may be unrealistic to expect any technique to simultaneously uncover multiple types and/or subtypes in a single analysis.

Consider a concrete, hypothetical example. Suppose that a biologist naive to all valid taxonomies studied a sample of animals including dogs, cats, hawks, and doves. These four groups can obviously be represented in a hierarchical manner: There are two types of mammals (dog and cat) and two types of birds (hawk and dove). Some of the indicators that the biologist could record may be useful for distinguishing mammals from birds. However, the same indicators may not differentiate dogs from cats or hawks from doves. Additional indicators may be useful for distinguishing dogs versus cats (but not hawks vs. doves or mammals vs. birds), and still other indicators may be useful for differentiating hawks versus doves (but not dogs vs. cats or mammals vs. birds). Thus, submitting all available indicators to a single multivariate analysis may make it difficult to identify all existing boundaries—and hence to resolve the correct latent structure.

Such an exploratory approach may be bested by an approach informed even minimally by theory about which indicators are useful for making particular structural distinctions. For example, if the biologist had a hunch that higher order groups corresponding to mammals and birds existed, he or she could perform an initial analysis using only those indicators believed to be capable of making this distinction, increasing the odds of correctly identifying the mammal and bird taxa and assigning cases to these two groups. Subsequent analyses within each of these groups could then be performed using indicators believed to differentiate hypothesized subtypes, which may increase the odds of correctly identifying the dog–cat and hawk–dove distinctions. Thus, a series of three analyses, each performed with a theoretically based subset of the available indicators, may be the best way to identify the hierarchical latent structure encompassing these four animal types.

The four-animal example illustrates an important choice that researchers inevitably face when studying constructs whose latent structures are relatively subtle or complex. Rather than submitting a broad array of available indicators to a single analysis to tease apart an unspecified number of latent classes, we suggest that it may often be desirable to focus on one putative taxonic boundary at a time using variables that sensitively and specifically demarcate that boundary (J. Ruscio & Ruscio, 2004a). This multistage approach is not limited to any particular analytic procedure, but represents an alternative way to implement any technique that evaluates latent structure, including those described earlier in this chapter.

For example, J. Ruscio and Ruscio (2004c) compared the accuracy with which cluster analyses correctly identified a simple hierarchical latent structure when performed in the traditional manner (i.e., simultaneous entry of all indicators into the analysis) versus the multistage approach described earlier. The known latent structure included two types, one of which contained two subtypes; it was therefore analogous to a sample of animals containing mammals, hawks, and doves. Some of the available indicators validly distinguished the two overarching types (akin to mammals vs. birds), but did not validly distinguish the two subtypes (akin to hawks vs. doves); other indicators validly distinguished the two subtypes, but not the two overarching types. Ruscio and Ruscio first submitted all indicators to a single hierarchical cluster analysis. This was performed using the average linkage method and Ward's method in turn, with the two-, three-, and four-cluster solutions examined for each method. None of these solutions uncovered a structure closely resembling the correct three-group configuration or achieved a classification accuracy in excess of 81%.

Next, a two-stage approach was evaluated. In the first stage, the indicators that validly distinguished the two overarching types were submitted to a cluster analysis; because a single boundary was hypothesized, the two-cluster solution was retained. In the second stage, indicators validly distinguishing the two subtypes were analyzed within the subsample of cases assigned to the overarching type in the first stage (this was analogous to searching for hawk and dove subtypes within the subset of cases identified as birds). Once again, a two-cluster solution was retained. Using this multistage approach, both clustering methods (average linkage and Ward's method) assigned cases to the three groups with just over 94% accuracy. Thus, deliberately focusing on one taxonic boundary at a time—rather than all boundaries at once—enabled these analyses to classify individuals into latent classes with a far higher degree of accuracy on the basis of the same set of indicators.

As this example demonstrates, the search for multiple latent classes can be undertaken in different ways even when a particular data-analytic technique is used. Moreover, restricting each analysis to test a single hypothesized distinction between two latent classes can, at least under some circumstances, yield results superior to those of a single-pass approach in which multiple latent classes are tested simultaneously. This suggests that the potential misidentification of complex latent structures may be less a limitation of the statistical techniques discussed thus far than a problem that can arise when they are implemented in a purely exploratory rather than theory-driven manner. It also implies that if investigators perform structural analyses in a multistage manner guided by relevant theory, with indicators carefully selected at each stage rather than concurrently submitted to a single analysis, many of the available statistical tools may perform more satisfactorily. Notably, this tactic of testing one latent boundary at a time, ideally in a theory-driven way, is the approach *required* to implement the taxometric method. It is to this analytic approach that we now turn.

THE TAXOMETRIC METHOD

The taxometric method is the final technique that we discuss in this chapter and the approach on which we focus for the remainder of this book. Although the term *taxometrics* can be used to broadly refer to the entire domain of empirical classification, it has come to be associated with the *coherent cut kinetics* taxometric method pioneered by Paul Meehl and his colleagues (e.g., Grove, 2004; Grove & Meehl, 1993; Meehl & Golden, 1982; Meehl & Yonce, 1994, 1996; Waller & Meehl, 1998). The name of this method reflects the process by which many of its procedures work: One determines whether predictable results are obtained (hence *coherent*) as a cutting point is moved through a distribution of indicator scores to create subsamples (hence *cut kinetics*). Our use of the synonymous terms *taxometric method* and *taxometrics* refers solely to the family of analytic procedures subsumed within Meehl's coherent cut kinetics approach, as well as one additional analytic tool (L-Mode; see chap. 5) that does not technically follow the coherent cut kinetics approach, but that was presented by Waller and Meehl (1998) in the context of the taxometric method.

The taxometric method is a hypothesis-testing (rather than exploratory) technique with several noteworthy features. First, taxometric procedures explore the relations among indicators, rather than the distributions of individual indicators, to infer something about the latent structure of the construct under investigation. Second, each procedure yields estimates of important parameters of the taxonic structural model as well as graphical results that can be visually inspected for structural information. Third, nonredundant lines of evidence are contributed by procedures that operate in mathematically different ways. Fourth, results are examined for converging evidence within a tolerable margin of error, rather than submitted to tests of statistical significance. Fifth, the final determination of latent structure is based partly on the degree of consistency among all obtained results.

In the remainder of this chapter, we briefly discuss each of these defining features of the method. Subsequent chapters delve more deeply into the types of data that are required for informative taxometric analyses (chap. 4), the individual procedures that comprise the taxometric method (chaps. 5 and 6), the wide range of consistency tests that have been employed in and proposed for taxometric investigations (chap. 7), and a number of factors to consider when interpreting taxometric results (chap. 8).

Feature 1: Examining Relations Among Indicators

Taxometric analysis addresses a bootstraps problem (Cronbach & Meehl, 1955) in that it uses fallible indicators to evaluate the latent structure of a construct in the absence of a gold standard criterion. Regardless of whether indicators of a construct are distributed categorically or continuously, the relationships *between* these indicators can provide valuable clues about the underlying structure of the target construct. Taxometric procedures capitalize on predictable differences in the way that indicators relate to one another when a taxonic boundary is either present or absent. This is somewhat different from the approaches taken by the alternative data-analytic techniques described earlier, which attempt to model the uni- or multivariate distributions of the indicator scores. However, taxometric procedures are similar to most of these approaches in their inclusion of multiple indicators within each analysis, allowing researchers to test the structure of constructs for which no single variable is a necessary and sufficient defining characteristic.

Feature 2: Providing Graphical Results and Estimates of Model Parameters

Taxometric procedures yield graphs that provide clues to latent structure and can be used to generate numerical estimates of important model parameters (e.g., the taxon base rate). When taxometric procedures are provided with appropriate data, they tend to produce different curve shapes for taxonic and dimensional latent structures. Examining taxometric curve shapes permits investigators to consider the effects of potentially problematic factors (e.g., indicator skew) that might otherwise be obscured by purely quantitative results, allowing the taxometric method to accommodate data that might otherwise violate unnecessarily strict assumptions. Finally, there is evidence that structural inferences can be drawn from taxometric results in an accurate and reliable manner. In two Monte Carlo studies (Meehl & Yonce, 1994, 1996), taxometric graphs were visually inspected with a high level of interrater agreement. Moreover, under the conditions studied in these investigations, the estimates of model parameters yielded by taxometric analyses were quite accurate. However, it is important to note that the data sets in these studies were somewhat idealized and may have produced taxometric curves that were easier to interpret than those that researchers will often obtain. Also, there is some evidence that more realistic data may bias the estimation of model parameters such as the taxon base rate (see chap. 7). Thus, although research supports the likely reliability and validity of conclusions arising from taxometric investigations, this research is still in its infancy. Consequently, we emphasize throughout this book the need for appropriate caution when interpreting the output of taxometric procedures and consistency tests, and we draw attention to those areas where additional research is especially needed.

Feature 3: Providing Nonredundant Lines of Evidence

The taxometric method encompasses a set of diverse analytic procedures that assess latent structure in varying ways. Each procedure is designed to distinguish the same structural models (i.e., taxonic structure as captured in the General Covariance Mixture Theorem, and dimensional structure as reflected in the common factor model; see chap. 6). However, each uses a unique approach to distinguish these models. Because taxometric procedures differ in their theoretical rationales and mathematical operations, they can contribute complementary pieces of evidence to a structural investigation. A bedrock principle of the taxometric method is that multiple analytic procedures should be employed to provide independent lines of evidence to inform a structural inference, thereby reducing the likelihood of reaching an incorrect conclusion based on a single aberrant result. This aspect of taxometric methodology can be contrasted with statistical techniques for which there are few or no nonredundant sources of evidence, or for which there is little or no expectation that different ways of implementing the available procedures should suggest the same underlying structure.

Feature 4: Requiring Converging Evidence Within a Tolerable Margin of Error

The results obtained from taxometric analyses (both graphs and model parameter estimates) are evaluated in terms of their consistency with taxonic or dimensional latent structure, not through tests of their statistical significance. There are two reasons that the method does not employ significance tests. The first is that *any* structural model is likely to only imperfectly correspond to the true latent structure of the target construct. Because the mathematical derivations of all statistical tests include simplifying approximations, tests that measure departures from perfect fit are confounded by the possibility that one structural model will appear to fit the data better than another model solely because it better satisfies the assumptions of the test.

For example, in latent class analysis, models containing a larger number of latent classes are more likely to meet the assumption of local independence than are models containing fewer latent classes. This is because as the number of classes in the model increases, each class becomes more homogeneous, and the restricted range of scores within each progressively homogeneous class drives down within-class variability (and hence within-class associations) among the indicators. Consequently, significance tests performed on competing latent class models using the assumption of local independence may tend to favor models that include more groups, regardless of their structural accuracy, simply because they better approximate this assumption.

A second reason that taxometric procedures do not rely on tests of statistical significance concerns the methodological paradox by which researchers can be penalized for working with large samples (Meehl, 1967). Fit indexes based on p values almost always reveal statistically significant departures from perfect fit in large samples, and large samples are required for informative classification research. Given that no model of latent structure is likely to achieve perfect fit, the penalty for working with appropriately large samples virtually guarantees the rejection of even exemplary models. Of course this drawback is lessened when one uses fit indexes that are not linked to p values, or when multiple competing models are compared, as is done in most applications of mixture modeling, cluster analysis, or latent class analysis. Nonetheless, the informational value of the model-fit statistics remains confounded with the degree to which simplifying assumptions are met by each model, making it impossible to disentangle the extent to which models truly fit the data from the extent to which important assumptions of these models are satisfied. To the degree that the comparative measures of fit are sensitive to sample size, large samples might exacerbate this fundamental confound between model fit and the satisfaction of model assumptions.

Feature 5: Requiring Consistency Checks

In place of an exclusive reliance on a null hypothesis significance test or a particular fit index that is calculated for competing structural models, the taxometric method relies on the convergence of evidence from as many nonredundant sources as possible, with increasing coherence of findings leading to increasing confidence in a structural inference. Because the method consists of several independently derived analytic procedures, each of which affords a judgment about latent structure, the use of multiple taxometric procedures in an investigation allows these techniques to serve as consistency tests for one another. This system of consistency checks, unique to the taxometric method, guards against the potential misinterpretation of isolated results that may be anomalous and misleading.

There are many additional ways to check the consistency of taxometric results. For example, each taxometric procedure may be conducted multiple times using different configurations of the available indicators. Likewise, multiple sets of indicators can be constructed from the available data using a variety of data combination approaches, or analyses may be performed in samples drawn from different populations. Assessing the convergence of graphical results and model parameter estimates across analyses performed with multiple sets and configurations of indicators provides many opportunities to compare results. In a well-conducted taxometric investigation, each analysis adds to the rigor with which latent structure is tested. Whereas any single finding may be ambiguous or misleading, confidence is bolstered when evidence from multiple sources consistently points to the same structural inference.

INFERENTIAL FRAMEWORKS FOR THE TAXOMETRIC METHOD

Each taxometric procedure or consistency test is expected to produce different results for taxonic and dimensional data. The interpretation of these results depends in part on the inferential framework one uses. There are at least three possibilities, and at present the consensus of opinion regarding the most appropriate choice is modest at best. We briefly outline all three potential frameworks and describe why we prefer the third.

Inferential Framework 1: Dimensional Structure as a Null Hypothesis

The first perspective that one might adopt identifies the dimensional structural model as a null hypothesis (H_0) that one attempts to reject with taxometric evidence. Rejecting this H_0 would increase support for an alternative hypothesis (H_A) of taxonic structure, although one must remember that rejecting the dimensional model does not logically entail support for the taxonic model; there are other nontaxonic models that may be even more appropriate (e.g., nonlinear or hierarchical models), hence one can make Type I errors. However, the failure to reject H_0 would place the researcher in an even more ambiguous position: The dimensional model may be correct (H_0 is true) or there may have been insufficient power to correctly reject this H_0 (e.g., the data were inadequate to detect taxonic structure, a taxometric procedure was implemented ineffectively), in which case one would make a Type II error. In either case, then, additional factors must be considered to support an inference of latent structure.

In the event of apparently taxonic results, one would need to evaluate the plausibility of more complex dimensional models before accepting the H_A of taxonic structure. In the event of apparently dimensional results, one would need to evaluate the appropriateness of the data and the procedural implementation to judge whether any conclusion is warranted. If the data appear highly amenable to the taxometric procedure, implemented in that particular way, then one might tentatively reach an inference of dimensional structure. However, if the adequacy of the data and procedural implementation is ambiguous, one might refrain from hazarding any structural conclusions. The logic of this situation should be familiar to any user of statistical significance tests. Beauchaine (2003) and Beauchaine and Beauchaine (2002) contended that this is the appropriate inferential framework for the taxometric method, and they urged extreme caution in reaching conclusions of dimensional structure.

Inferential Framework 2: Taxonic Structure as a Null Hypothesis

The second perspective that one might adopt identifies the taxonic structural model as the H_0, which one attempts to reject with taxometric evidence. Rejecting this H_0 would increase support for an H_A of dimensional structure, whereas the failure to reject H_0 would be ambiguous: The taxonic model may be correct, or there may have been insufficient power to correctly reject it. To our knowledge, nobody has suggested that taxonic structure be conceptualized as an H_0, so we will not consider this possibility further.

Inferential Framework 3: Taxonic and Dimensional Structure as Competing Hypotheses

This third and final perspective does not identify either structural model as an H_0. Rather than granting either model *privileged* status as the H_0, both models are treated as viable competing hypotheses. Likewise, rather than attempting to reject one H_0 and tentatively accept a specific H_A, this inferential framework dictates that one assess the taxometric evidence for its *relative* consistency with each model. To the extent that the evidence appears consistent with one model and not the other, a structural inference is reached. If the evidence is judged too ambiguous to distinguish between the two models or equally inconsistent with both, one would refrain from offering any structural conclusion. J. Ruscio and Ruscio (2004a) contended that this is the appropriate inferential framework for the taxometric method, and Meehl (2004) appeared to support this position when he stated:

> It is important to realize that the use of taxometrics does not require that the investigator entertains a taxonic conjecture. Taxometrics may be used in that way, analogously to confirmatory factor analysis. However, a researcher with no theoretical opinion can properly use taxometrics as a decision procedure, with one possible outcome being that the latent structure is nontaxonic. (p. 39)

It is clear in all of Meehl's writing that the term *nontaxonic* corresponds to the dimensional structural model that his method is designed to tease apart from the taxonic model. We share the view that the taxometric method is most appropriately used as a decision procedure, with neither potential outcome given privileged status. As one anonymous reviewer of this book aptly noted: "The subtitle of the Waller and Meehl [1998] book was 'Distinguishing types from continua'; it wasn't 'detecting types or failing to detect types'."

This inferential framework is analogous to signal detection theory in that one hopes to find that two distributions of results—those for taxonic and dimensional structural models—overlap as little as possible and that a particular research result is unambiguously more consistent with those in one distribution. These ideals correspond to a statistically powerful test between the competing taxonic and dimensional structural models, with a given set of research data yielding a clear conclusion. Of course idiosyncratic aspects of a particular sample and the procedural implementation that one employs can affect the power of such a test, and in any event there is no guarantee of easily interpretable results. Although differing as a matter of degree, three patterns of results can be distinguished:

1. The research data yield results consistent with those for one structure (taxonic or dimensional), but inconsistent with those for the other. One would reach a structural inference corresponding to the better-supported model.
2. The research data yield results equally consistent with those for both structures. Such ambiguous results might reflect inadequacies in the research data, with results for taxonic and dimensional structures virtually indistinguishable from one another, or an intermediate result that lies midway between those for the two competing structural models. In such

cases, one would refrain from reaching a taxonic or dimensional structural inference.

3. The research data yield results inconsistent with those for both structures. This type of ambiguity can obtain even when there exists a substantial difference in results for taxonic and dimensional models. The ambiguity would stem not from inadequate data or intermediate results, but the fact that neither the taxonic nor the dimensional model adequately characterizes the latent structure of the target construct, hence the results observed for the research data differ from those for both models. In the event of such results, one would refrain from reaching a taxonic or dimensional structural inference.

The logic of this inferential framework can be contrasted with that of the first: Rather than attempting to infer one structure (taxonic) only by demonstrating negative evidence for another (dimensional), one considers both the positive and negative evidence bearing on two competing hypotheses. This means that some data sets can afford a powerful test between taxonic and dimensional structure (i.e., the results for these models may differ substantially) and provide support for either one. Thus, under certain circumstances, one can muster evidence in support of a dimensional structural inference, rather than uniformly shying away from this possibility by equating it with a null hypothesis and invoking the taboo of its acceptance. Likewise, this inferential framework recognizes that the rejection of either a dimensional or taxonic structural model does not necessarily provide support for the other model. Instead a compelling case for a structural inference must be based on evidence that is both inconsistent with one model and consistent with another specific alternative.

Although we prefer this competing-hypothesis inferential framework, we acknowledge that our perspective is not universally shared. As noted earlier, some authors prefer the first inferential framework, in which dimensional structure is treated as a null hypothesis that one seeks to reject by finding evidence inconsistent with it. Until a stronger consensus of opinion is reached within the broader taxometrics community, researchers will have to judge for themselves which inferential framework appears more appropriate.

CONCLUSIONS

A number of analytic techniques have been developed in an effort to address the classification problem: determining whether a particular construct is taxonic or dimensional at the latent level. Whereas examining frequency distributions for evidence of bimodality was long a popular method of searching for taxa, this approach has significant shortcomings and has become outmoded as advances in computing power have facilitated the routine application of more sophisticated multivariate techniques. Unfortunately, there have been few direct comparisons of the validity with which finite mixture modeling, cluster analysis, latent class analysis, and the taxometric method distinguish taxonic from dimensional structure. Comparative research is sorely needed. Until such studies are performed, there remain several open questions about the extent to which, and the conditions under which, each of these techniques affords the most valid differentiations between taxonic and dimensional latent structures.

At the same time, it has been demonstrated that finite mixture modeling, cluster analysis, and latent class analysis can have difficulty determining the correct number of latent classes underlying a construct, with particular difficulty in distinguishing two-group (taxonic) from one-group (dimensional) models. The taxometric method, by contrast, has been shown to be capable of distinguishing taxa from dimensions in a reliable and valid fashion under the handful of conditions that have been studied. This method also has several valuable features that distinguish it from alternative procedures and can serve to increase confidence in structural inferences. For these reasons, we regard taxometrics as an especially promising tool for addressing the fundamental classification question. Although we adopt the competing-hypothesis inferential framework, most of what we describe throughout this book applies equally whether one agrees with our perspective or prefers to treat dimensional structure as a null hypothesis. Having made the argument that taxometric analyses can provide useful clues to latent structure, we now turn to the second part of the book, in which a more in-depth description of the taxometric method and recommendations for its successful application are presented.

PART TWO

TAXOMETRIC METHOD

CHAPTER FOUR

Data Requirements for Taxometrics

The previous chapter provided an overview of several approaches to evaluating latent structure, with particular emphasis on the taxometric method. Before introducing the procedures included in the taxometric method, however, it is important to consider what sort of data one must have to perform these procedures. Like other data-analytic tools, taxometric procedures require adequate data to yield informative results. Several features of a data set influence its appropriateness for taxometric analysis, including the nature and construction of the sample, the breadth and appropriateness of the indicator set, and important statistical characteristics of the data. We review the relevant conceptual and empirical data considerations, including the conventional guidelines derived from Monte Carlo studies that have been used to judge the adequacy of a data set for taxometric investigation.

Following this discussion, we introduce a technique that is intended to supplement these more general guidelines with direct empirical evaluation of the adequacy of a particular data set for a planned taxometric analysis. The technique involves the generation of empirical sampling distributions of results through analyses of simulated taxonic and dimensional comparison data that reproduce key characteristics of the research data (e.g., sample size, indicator distributions and correlations). These comparison data are generated using a bootstrap procedure in which data-based estimates of population parameters are used to draw random samples of data (Efron & Tibshirani, 1993). This is done first in a way that yields comparison data that conform to the taxonic structural model and then in a way that yields comparison data that conform to the dimensional structural model. All bootstrap samples of comparison data are then submitted to parallel taxometric analyses, yielding empirical sampling distributions of taxometric results representing expected patterns for each structure. To the extent that these distributions diverge, one gains confidence that the unique set of research data is likely to afford a genuine test between these two competing structural hypotheses. After describing the rationale behind this approach and highlighting the strengths and limitations of the technique, we discuss the nuts and bolts of how it is implemented and how its results can be used to inform a taxometric investigation.

SAMPLING CONSIDERATIONS

Sample Size

On the basis of Monte Carlo research (e.g., Meehl & Yonce, 1994, 1996), Meehl (1995a) recommended that data sets submitted to taxometric analyses include a minimum of 300 cases. Although taxometric procedures have been shown to successfully distinguish taxonic from dimensional structure in smaller samples, this has generally been found only when the data possess other highly favorable characteristics (e.g., equal-sized groups separated by a large amount on indicators that are uncorrelated within groups and normally distributed along continuous scales). Because actual research data are unlikely to possess all of these desirable properties, many researchers have adopted Meehl's rule of thumb as a reasonable standard.

Haslam and Kim (2002) reviewed 66 published and unpublished taxometric studies and found a median sample size of 585. This figure jumped to 809 among the 57 published taxometric investigations in the updated review we present in chapter 10. Moreover, sample size appears to be on the rise in published taxometric reports: The median sample size was 639 for the 21 studies prior to 2000 and 923 for the 36 studies published during or after 2000. As is apparent from these trends, large sample sizes are the norm—and the expectation—in taxometric investigations.

Size and Base Rate of the Taxon

To be appropriate for taxometric analysis, a sample must not only be large, but also contain a sufficient number of putative taxon members to permit their detection in the analysis. This characteristic of the data is conventionally discussed as the *taxon base rate* because the taxon is often the smaller of the two groups in the sample. However, this data requirement pertains more generally to the relative sizes of the two latent groups, indicating that neither the taxon base rate (P) nor the complement base rate (Q) should be so small that the distinction between these groups is missed by the taxometric analyses. The same Monte Carlo studies cited by Meehl (1995a) to support his recommended minimal sample size found that taxometric procedures correctly identified taxonic structure in samples with a taxon base rate as low as $P = .10$. Although this finding was based on a small number of data configurations and involved no analyses with $P < .10$, many researchers have adopted the $P \geq .10$ threshold as a reasonable base rate standard. However, there are several reasons to be flexible in the use of this guideline.

First, Monte Carlo studies seldom evaluated the performance of taxometric procedures in samples with taxon base rates smaller than .10. As a result, the lower limit of what these procedures can detect is presently unknown. Beauchaine and Beauchaine (2002) generated and analyzed taxonic data with base rates lower than .10 and suggested that these small taxa could be detected under some conditions, but their criterion for success was based on two consistency tests later shown to poorly identify taxonic structure (see chap. 7 for details). In another simulation study, J. Ruscio (2005) found that taxon base rates as low as .05 were sometimes estimated accurately. This finding provides only indirect evidence, but the ability to accurately estimate their size suggests that the procedures can detect small taxa under at least some data conditions.

Second, preliminary evidence suggests that the absolute number of taxon members in a sample may be at least as important as the proportion of the sample that they comprise. J. Ruscio and Ruscio (2004a) found that taxometric procedures continued to detect a taxon of a constant absolute size even when large numbers of complement members were added to the sample, causing the taxon base rate to fall well below .10. For this reason, we prefer to break with convention by referring to a *small taxon* rather than a *low base rate taxon*, underscoring the potential importance of the absolute size of the group *as well as* its base rate in the sample. Indeed, although the conventional rule of thumb of $P \geq .10$ may be appropriate in relatively small samples, it may be somewhat conservative in especially large samples, wherein even a very low base rate may correspond to a substantial number of taxon members that can be detected by taxometric procedures.

At present, it is not well understood how the taxon base rate and the absolute size of the taxon jointly influence the sensitivity of each taxometric procedure to detecting a taxonic boundary. Thus, we suggest that researchers consider both of these values when evaluating the appropriateness of their data for a taxometric analysis. For example, if the base rate of a putative taxon is expected to be near .10, one could look to the absolute size of this taxon for additional guidance on the likely capacity of the analysis to distinguish taxonic from dimensional structure. If the taxon base rate is estimated to be lower than .10, it would be especially important to demonstrate, before proceeding with the analysis, that there are enough members of the taxon in the sample to give credence to their detection. Although no additional general guidelines are presently available to help researchers make this judgment, we discuss an approach later in this chapter that researchers can use to empirically evaluate the adequacy of a data set (including the size of the taxon) for the intended taxometric analyses.

Third, the sensitivity of taxometric procedures to the existence of a small taxon almost certainly depends on many other characteristics of the data. The validity with which the indicators separate the groups is probably most important, but also relevant are factors such as the magnitude of within-group correlations and the degree of within-group indicator skew. When all or many of these data properties are strongly favorable for taxometrics, it may be easier to detect relatively small taxa. To the extent that these data properties are weak or questionable, a small taxon may be missed. In addition, sensitivity to a small taxon may vary across taxometric procedures. A recent demonstration using one data configuration found a substantial difference in the ability of three taxometric procedures to detect an increasingly small taxon (J. Ruscio & Ruscio, 2004a).

For all of these reasons, it may be misleading to set a single acceptable base rate threshold without regard for other characteristics of the data or the analysis plan. Until further Monte Carlo research is conducted to explore this issue, we recommend that researchers (a) use the $P \geq .10$ rule of thumb flexibly, (b) strive to collect data from a mixed-group population in which the putative taxon naturally occurs with a base rate closer to the ideal of .50, and (c) empirically evaluate the appropriateness of their data for each planned analysis as a more direct check of the adequate representation of taxon members.

The Population Sampled

The nature of the (mixed-group) population from which cases are sampled will influence the likely size of the taxon and the range of scores represented on the indicator variables. For example, samples drawn from relevant clinical populations will often contain a larger psychopathology taxon (as well as more intermediate or subthreshold cases) than will samples drawn from community or analogue populations, meaning that a far larger nonclinical sample will be required to amass the same number of taxon members found in a smaller but well-chosen clinical sample. In addition, the broader range of symptom severity present in an appropriate clinical sample will often result in a larger range of scores on the indicator variables, increasing the likelihood of detecting a taxon located at the upper end of the score distribution and facilitating the implementation of taxometric procedures.

It should be noted, however, that there may also be potential disadvantages to conducting taxometric investigations of psychopathology constructs using certain clinical samples. First, some clinical samples may contain too *many* cases belonging to the taxon. For instance, a clinic that specializes in the treatment of one particular disorder may have a client roster consisting almost entirely of individuals with that disorder, thereby including too few members of the complement to distinguish it from the pathological taxon. Similarly, if the demand for services exceeds the capacity of a particular clinic, the staff may choose to provide services only to those individuals exhibiting the most severe levels of distress or impairment. This may eliminate most or all members of the complement and artificially constrain the range of functioning within the sample to the point where the data are no longer appropriate for taxometric analysis. Second, just because a sample is drawn from a clinical population does not necessarily mean that it contains a high enough rate of a particular form of psychopathology to be studied using taxometric analysis. For example, some conditions may be too rare to be powerfully investigated in a general outpatient sample, requiring data to be collected in an inpatient facility or a specialty clinic to yield a large enough taxon for analysis. Other conditions or related constructs may be sufficiently prevalent in the general population to be appropriately studied in epidemiological samples (see Kessler, 2002a). In extremely large epidemiological samples, even members of rare taxa may be represented in sufficiently large numbers to afford the detection of taxonic structure.

Questionable Sampling Techniques

In addition to the sample considerations discussed thus far, three sampling approaches are worthy of special note because they can undermine the results of a taxometric analysis by introducing plausible alternative explanations for the observed results. One practice that has been used with some frequency is to combine patient and nonpatient samples (or other distinct samples) into a single sample for analysis. This technique is often based on the reasoning that admixing these anticipated taxon (patient) and complement (nonpatient) members allows the investigator to influence the taxon base rate, assembling a sample such that the taxon base rate will more closely approximate the ideal of .50 than might be possible in either subsample alone. There are two potential problems with this approach. First, depending on the nature of samples that are mixed together, this may systematically omit cases with intermediate or subclinical symptom levels and retain cases scoring near the two extremes. This might bias results toward a taxonic solution. Second, mixing samples drawn from separate populations that differ on numerous characteristics other than the target construct could also lead to spurious taxonic results. For example, Grove (1991a) showed that pseudotaxa can emerge when one of the samples is selected based on high scores on one dimension. When researchers obtain apparently taxonic results in analyses of admixed samples, they should consider whether threshold effects or the omission of intermediate scores might explain apparently taxonic results. The conditions under which admixed samples bias taxometric results toward taxonic—or dimensional—results warrant additional study.

A second potentially problematic practice consists of splitting the available sample into subsamples within which the taxometric analyses will be performed. Although replication is, in principle, a worthwhile undertaking, analyses performed in subsamples that are drawn from a single sample (and therefore from a single population) can examine only the influence of sampling error and do not address the more important question of whether results generalize to other populations or to different measures or conceptualizations of the target construct. Such an exact replication provides far less compelling evidence of the generalizability and consistency of findings than would a conceptual replication employing different measures of the construct or samples drawn from different populations. Because taxometric procedures are better able to differentiate taxonic from dimensional structure using large samples, the best approach may be to submit all available cases in a sample to a single series of maximally powerful taxometric analyses. Dividing a sample into subsamples may be most problematic when a small taxon is hypothesized because this practice reduces the number of taxon members included in each analysis, and thereby lowers the odds of correctly detecting the taxon in any analysis—let alone consistently detecting it in all of them.

A third questionable sampling practice involves discarding likely complement members from the sample to increase the base rate of the putative taxon. The appeal of this technique probably stems from an unstated assumption that the taxon base rate (and not the absolute number of taxon members) is the critical factor in determining whether a taxon can be detected. Earlier we noted that the number of taxon members in the sample may also be an important factor in taxon detection. Because the removal of likely complement members does not increase the number of taxon members, this practice may do little to increase analytic sensitivity to a small taxon. In fact, as J. Ruscio, Ruscio, and Keane (2004) demonstrated, eliminating cases with the lowest indicator scores (i.e., those who are most likely to belong to the complement class) may not affect the regions of taxometric graphs most important for interpretation. For example, if taxon members are overwhelmed by complement members even among high-scoring subsamples, removing cases with lower scores may not help.

An alternative method for driving up the taxon base rate can also be problematic. This method uses a fallible criterion, such as diagnostic status, to identify likely complement members and then eliminates a certain number or proportion of these cases at random from the sample. For example, one could eliminate the number of complement members that would equate the sizes of the two hypothesized groups, bringing the taxon base rate to .50. To illustrate the potentially distorting effect of this sampling approach, consider the case of a researcher who posits that 100 out of a sample of 1,000 individuals are taxon members and that complement members should be dropped to equate the base rates of the putative classes. The left panel of Fig. 4.1 shows the score distribution of an indicator of a latent dimension in the full sample of 1,000 cases, with the highest scoring 100 cases and lowest scoring 900 cases on this dimension designated by dotted lines. The right panel shows the score distribution of this indicator in a subsample formed by randomly dropping all but 100 of the complement members (i.e., removing 800 of the 900 lowest scoring cases). Following the random removal of complement members, the indicator distribution becomes bimodal; the dotted lines reflect artificial groups created by this sampling technique. Analysis performed using one or more such indicators in the modified sample is at risk of yielding pseudotaxonic results. Like other analytic procedures, taxometric analyses cannot be expected to distinguish this type of artificial bimodality from

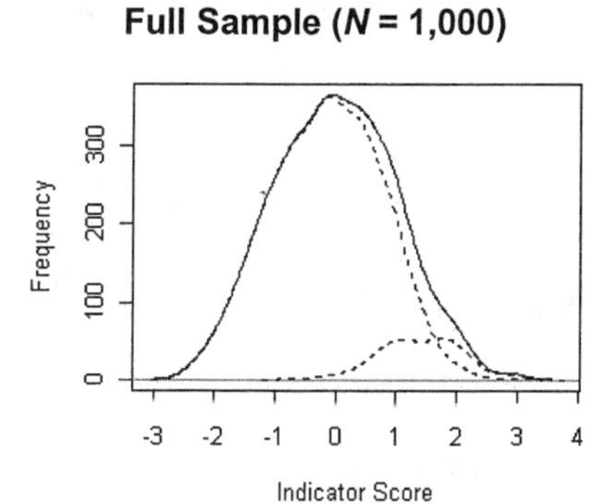

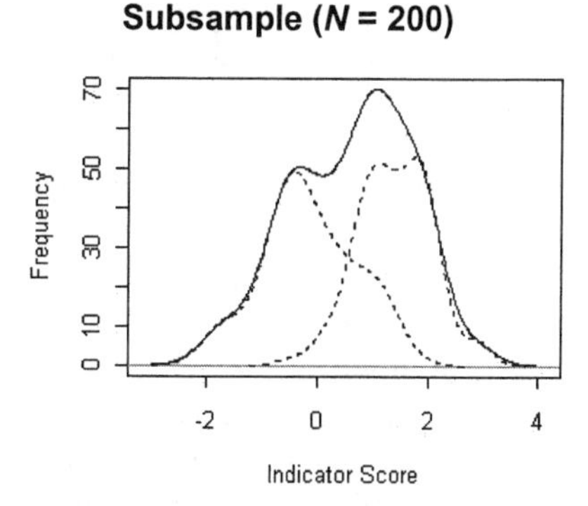

FIG. 4.1. Frequency distribution of an indicator of a latent dimension in a sample of 1,000 cases (left) and in a selected subsample of cases in which 800 of the lowest scoring 900 cases have been randomly discarded (right). Solid lines represent the distribution of all cases in the sample (or subsample), and dotted lines represent the distributions of the highest scoring 100 cases and the lower scoring cases.

the real thing caused by taxonic structure. Of course bimodal distributions do not necessarily yield taxonic results, and taxometric analyses of dimensional data sets modified by dropping low-scoring cases as outlined earlier yielded mixed results—sometimes erroneously taxonic in appearance, sometimes correctly dimensional, and sometimes ambiguous. The potential confound introduced by this sampling approach may outweigh any potential benefits that it appears to offer.

To minimize inaccurate structural conclusions, we urge researchers to draw their samples from populations that (a) provide sufficient representation of the hypothesized taxon and complement groups, and (b) include the full range of potential scores on the indicators. We further recommend that researchers use all available cases in each analysis and avoid sample selection and construction techniques that may increase the odds of pseudotaxonicity through sampling artifacts. When these guidelines are not followed and a taxon is observed, it may be difficult to confidently draw a taxonic inference and rule out the rival hypothesis of pseudotaxonicity. However, interpreting dimensional results obtained under these conditions would be less problematic because the previously mentioned sampling biases generally work against finding a dimensional solution, and thus may not introduce as compelling an alternative explanation for the results.

One final note with regard to sampling merits consideration. Due to ordinary sampling error or to strategies such as oversampling individuals with relatively rare characteristics, the base rate of a taxon in one's sample may not match that in the population from which it was drawn. This does not necessarily pose a problem when attempting to distinguish taxonic and dimensional structural models, but it does suggest that one must exercise caution in generalizing the taxon base rate estimated in a given sample to any particular population.

INDICATOR CONSIDERATIONS

Indicator Representation of the Target Construct

The ability to draw inferences about the latent structure of a construct requires careful consideration of how well that construct is represented by the indicators submitted to analysis (Widiger, 2001). This concern can be conceptualized using the familiar terms of content and discriminant validity (Cronbach & Meehl, 1955). For a taxometric analysis (or any other technique examining latent structure) to yield informative results, one must establish that the indicators (a) assess all relevant facets of the target construct, *and* (b) do not inadvertently triangulate on one or more other constructs. For example, a taxometric investigation of major depressive disorder (MDD) whose indicators exclusively assess the somatic symptoms of depression may raise questions about content validity because cognitive and affective symptoms are also important characteristics of this disorder. However, even if the indicators assess each of the primary features believed to characterize MDD, it is important that they not equally represent one or more other mood disorders (e.g., dysthymia, bipolar disorder), anxiety disorders (e.g., generalized anxiety disorder), or psychotic disorders (e.g., schizoaffective disorder) whose features overlap to varying but substantial degrees with those of MDD. If a set of indicators ambiguously represents multiple latent constructs, its poor discriminant validity would make it difficult to identify the construct whose structure is revealed by the taxometric results. As when determining the nature of factors yielded by a factor analysis, the identity of the target construct in a taxometric analysis is largely determined by the constellation of indicators that are used to study it. Thus, careful selection of indicators is essential for meaningful results.

An indicator may be a single item or rating selected from an assessment device or a composite of two or more items or ratings assessing a particular symptom, feature, or facet of the target construct. Using the available variables in a data set, an investigator can construct multiple indicator sets—each combining variables according to different criteria—that can each be submitted to taxometric analysis and used as consistency checks on the results. Different indicator sets may be constructed to reflect one or more theoretical conceptualizations of the construct, to capture the relations among variables within a particular data set, or to take joint consideration of both theoretical and empirical criteria. For example, to create theory-based indicators for testing the latent structure of major depression, a researcher might construct one indicator set containing a composite indicator for each symptom of MDD listed in the *DSM*, a second indicator set based on features of a cognitive theory of depression, a third indicator set based on elements of an interpersonal theory of depression, and so forth. To create indicators based on the empirical relations among symptoms of depression, a researcher might construct an indicator set by combining items that share higher pairwise correlations with one another than with other items in the sample. Finally, the researcher could attend to both theoretical and empirical considerations by forming one or more sets of composite indicators based on a priori theoretical conceptualizations of the construct as well as observed relations among items within the present sample.

As described later, regardless of the specific indicator construction strategy that is used, each candidate indicator set can and should be evaluated prior to analysis to ensure the adequacy of important data properties.

Each indicator set providing good content and discriminant validity vis-à-vis the target construct and judged to possess acceptable properties can then be submitted to taxometric analysis.

Indicator Validity and Within-Group Correlations

In addition to adequately representing the target construct, the indicators chosen for taxometric analysis must distinguish the putative groups with sufficient validity and be correlated at a tolerably low level within these groups. As noted in chapter 3, validity refers to the extent to which the conjectured latent classes are separated on each indicator. If indicator validity is too low (i.e., the indicator score distributions of the two groups overlap too much), taxometric procedures may be unable to detect a genuine taxonic boundary. For this reason, it is essential to demonstrate that a set of indicators possesses adequate validity to pose an informative test between taxonic and dimensional structural models. In the taxometric literature, indicator validity is usually expressed as the mean difference between the taxon and complement, standardized by the pooled within-groups variance, the metric known more familiarly as Cohen's d:

$$d = \frac{M_t - M_c}{\sqrt{\frac{(SD_t^2)(n_t - 1) + (SD_c^2)(n_c - 1)}{N - 2}}}, \qquad (4.1)$$

where terms subscripted with a t denote values calculated within the taxon and terms subscripted with a c denote values calculated within the complement. As a general rule, Meehl (1995a) suggested that an indicator for taxometric analysis should separate the taxon and complement with a minimum of $d = 1.25$.

A related concern involves the extent to which indicators are sensitive at trait levels that are appropriate for detecting a taxonic boundary. For example, consider the situation in which individuals differ on a common latent dimension, but there is also a taxonic boundary. In other words, latent structure can be considered taxonic, but with additional within-groups variation on a common latent trait. In the language of item-response theory (IRT; e.g., Embretson, 1996), an indicator can be understood as discriminating to varying degrees across different levels along a latent trait. (Multidimensional IRT models have been developed, but a single latent trait is presumed here for simplicity.) An indicator is seldom, if ever, equally sensitive to differences between individuals who are very low, moderate, and very high on a trait. Usually, sensitivity is maximized at a certain trait level and drops off as trait levels deviate from this value. If there is a mismatch between the trait levels at which the indicators are maximally sensitive and the point at which a taxonic boundary exists, one may fail to detect this boundary even with apparently valid indicators. For example, if a relatively rare and severe form of psychopathology (e.g., melancholic depression) is assessed in a community sample using self-report measures assessing subclinical symptom variation (e.g., sadness, dysphoria), even individuals falling well below the taxonic boundary may receive high scores on these measures, leading to poor discrimination in the severe region of interest. Ideally, indicators should be valid not only in the sense that taxon members score appreciably higher than complement members, but also in the sense that their discrimination power is maximized at trait levels near the location of the taxonic boundary.

In addition to the importance of indicator validity, taxometric procedures work under the simplifying approximation that within-group indicator correlations remain tolerably low. The reason for this approximation becomes clearer when specific taxometric procedures are introduced in chapters 5 and 6. Because taxometric procedures are somewhat robust to deviations from the idealization of $r = .00$ (Meehl & Yonce, 1994, 1996), within-group correlations need not literally be zero. Meehl (1995a) suggested the general guideline that indicators should be correlated no more than $r = .30$ within groups. In practice, a good rule of thumb is to ensure that within-group correlations are substantially smaller than full-sample correlations. However, because taxometric procedures grow more powerful as within-group correlations approach the ideal of zero, and because larger within-group correlations can preclude an informative taxometric analysis, it is important to minimize these as much as possible.

One way to reduce within-group correlations is to make use of multimethod assessments (Meehl, 1995a). For example, drawing all indicators exclusively from self-report questionnaires may generate larger within-group correlations than using self-report measures, interview ratings, behavioral observations, physiological measures, and other relevant forms of data as indicators of conceptually distinct facets of the target construct. Multimethod data collection has the advantage of not only minimizing within-group correlations stemming from a shared method, but also enhancing breadth of coverage of the target construct. A second way to reduce within-group correlations is to select and construct indicators with this goal in mind. When variables assessing similar aspects of the target construct are combined rather than used as separate indicators, and when decisions about which variables to combine are informed by consideration of the resulting within-group correlations, they are more likely to be acceptably low. By contrast, some taxometric investigators have taken the approach of first identifying a desired number of indicators and then assigning items to composite indicators either at random or through a scheme that is independent of item content (e.g., Items 1, 4, and 7 are combined to form Indicator 1, Items 2, 5, and 8 are combined to form Indicator 2, and Items 3, 6, and 9 are combined to form Indicator 3). This strategy can be problematic because it virtually guarantees high indicator redundancy and, in turn, large within-group correlations.

How can a researcher determine whether the available indicators fall within conventionally acceptable limits of validity and within-group correlations? A firm determination would require both the certainty that the target construct is taxonic and an infallible criterion to classify cases into the taxon and complement groups, both absent when latent structure is the focus of study. Lacking this information, it can be challenging to estimate these important model parameters, let alone evaluate their adequacy for a planned series of taxometric analyses. Nevertheless, there are ways to gauge the likely validity and within-group correlations of a candidate indicator set. For example, one can assign cases to groups based on a criterion measure such as diagnostic status (e.g., diagnosed cases are presumed to be taxon members, nondiagnosed cases are presumed to be complement members) or apply a conventional threshold to scores on a well-validated assessment instrument. Alternatively, one can estimate the likely base rate of taxon members in the sample, then assign the highest scoring cases on all available indicators to the taxon and the remaining cases to its complement (this base rate classification method is described in greater detail in chap. 6). Once cases are classified, it is a simple matter to calculate the indicator correlations within groups as well as to estimate indicator validity. A program on this book's companion Web site can be used for this purpose.

Although Meehl's (1995a) recommended indicator validity and within-group correlation values of $d \geq 1.25$ and $r \leq .30$, respectively, may provide useful rules of thumb, we caution researchers to evaluate their data not by the adequacy of each parameter in isolation, but by their *joint* sufficiency for taxometric analysis. For example, although the analysis of four indicators with an average validity of $d = 1.25$ may detect a taxon when $N = 600$, $P = .50$, and within-group indicator correlations are close to zero, greater validity may be required if there are fewer indicators, the sample is smaller, the taxon is substantially smaller (or larger), or there are nontrivial indicator correlations within groups. Just as we suggested that there is probably no universally applicable range of acceptable base rates (or absolute sizes) of taxa that can be detected using taxometrics, we believe that indicator validity and within-group correlations should also be considered in the context of other relevant factors. Shortly, we discuss a procedure that can be used to empirically evaluate the adequacy of a set of indicators for taxometric analyses—one that takes into account the unique configuration of data characteristics in that indicator set as well as the specific way that each planned taxometric procedure and consistency test is implemented.

In the process of attempting to select or construct a set of indicators with sufficiently high validity and small within-group correlations for informative taxometric analyses, it is possible that a researcher will be unable to meet these goals. If a concerted effort to identify adequate indicators fails, there are at least two possible explanations. First, it could be that the data are inadequate for taxometric analysis. For example, many measures that are sufficiently valid for other purposes may fail to distinguish groups with the large effect sizes required by the taxometric method. Second, it could be that latent structure is dimensional. For example, there may be no assignment of cases to groups that yields sufficiently large between-group differences and small within-group correlations because there is no taxonic boundary to be detected, and indicators correlate due to shared loadings on one or more latent dimensions. To the extent the researchers with appropriate samples of data exhaust the available supply of reliable and valid measures yet cannot construct a set of indicators that appears adequate for taxometric analyses, this provides indirect evidence that the structure of the target construct is unlikely to be taxonic.

Indicator Distributions

Although taxometric analyses involve no assumptions regarding normality or continuity, deviations from either of these properties can exert an influence on taxometric curve shapes and estimates of model parameters. For example, many taxometric procedures tend to produce rising curves when positively skewed indicators are submitted for analysis (A. M. Ruscio & Ruscio, 2002), which can make it more challenging to distinguish results for taxonic and dimensional structures (J. Ruscio, Ruscio, & Keane, 2004). As noted in chapter 3, psychological measures frequently yield skewed distributions (Micceri, 1989). This is not a concern when indicator skew arises solely from the mixture of taxon and complement groups (e.g., a small taxon mixed with a larger complement usually generates positive skew in the mixed-group sample). Rather, taxometric results can become more difficult to interpret when indicators of dimensional structure are skewed or indicators of taxonic structure are skewed within groups. Although indicators can be used even when they are skewed in these ways, it is important to recognize that such skew can complicate interpretation.

Likewise, when indicators vary across ordered categories (e.g., dichotomous items, polytomous Likert scales), rather than continuous scales (e.g., reaction time), this can also influence curve shapes (J. Ruscio, 2000) and may require special accommodations to perform taxometric analyses. For example, most taxometric procedures require at least one indicator whose score distribution varies across a large range of values, either to allow multiple cutting scores to be used or to form ordered subsamples for analysis. If the available variables vary only as ordered categories, it may be necessary to aggregate them into composite indicators whose distributions better approximate continuity. Alternatively, as we discuss in chapters 5 and 6, some taxometric procedures can be modified to accommodate ordered categorical indicators by forming composites as needed to run the analysis (Gangestad & Snyder, 1985; J. Ruscio, 2000). Although accommodations can be made for ordered categorical data, it is preferable to use continuous measures whenever possible (Grove, 2004).

Although taxometric procedures do not make distributional assumptions, results are often easier to interpret to the extent that indicators approximate continuous, normal distributions. Moreover, Monte Carlo studies have relied almost exclusively on simulated data sets containing continuous indicators that are normally distributed (in the full sample for dimensional data sets, within-groups for taxonic data sets; but see Cleland & Haslam, 1996; Haslam & Cleland, 1996, who began with the data sets from the Meehl & Yonce [1994, 1996] Monte Carlo studies and skewed them for reanalysis). As a result, many researchers are only familiar with the taxometric curve shapes yielded by such highly idealized data. In chapter 8, we consider interpretational issues raised by factors that can influence taxometric results. For now we simply emphasize that the distributions of indicators submitted to taxometric analysis can exert important effects on curve shape.

Number of Indicators

The number of indicators included in a taxometric analysis has implications for analytic flexibility and, under certain conditions, statistical power. Although some taxometric procedures can be performed with as few as two indicators, the range of analyses that can be conducted—and the statistical power of some procedures—increases as the number of valid indicators increases.

That being said, it is important to emphasize that we do not recommend simply submitting as many indicators as possible to taxometric analyses. As noted earlier, taxometric procedures require not only that indicators be positively correlated in the full sample, but that each be relatively independent of the others within groups. Careful construction of conceptually distinct indicators that assess unique facets of the target construct using different methods can help keep within-group correlations low. However, no matter how much care is used, it may be the rare psychological or behavioral construct that can be represented by more than a half-dozen indicators without introducing substantial conceptual redundancy and within-group correlations. Although using all available items as separate indicators may appear desirable by maximizing the total number of indicators, the potential downside is that within-group correlations may reach intolerable levels because many of the indicators are redundant with one another.

We recommend an alternative approach in which researchers begin with a list of theoretically relevant facets of the target construct and then form composite indicators by joining together those variables that most validly assess each facet. The adequacy of this candidate indicator set can then be evaluated and the indicators refined as necessary. Thus, rather than beginning with a desired number of indicators and forcing the available variables into this mold, we recommend that one consider how many relatively distinct facets of the target construct exist and then select or construct the best possible set of indicators to represent each of these facets. In many cases, the ideal indicator set would include one valid indicator per relevant facet.

Finally, it is worth noting that statistical power does not necessarily increase with the addition of extra indicators into an analysis. Rather, gains in statistical power are only likely to be achieved when the added indicators are of comparable or superior validity to the existing indicators *and* when within-group correlations are not increased to an extent that offsets the validity of the entire indicator set. Thus, taxometric analyses using a small number of valid and nonredundant indicators may better distinguish taxonic from dimensional structure than analyses using a larger number of less valid or more redundant indicators.

Having read about many aspects of the data that can influence taxometric results, the reader may wonder just how to tell whether a particular data set is appropriate for taxometric analysis, especially when—as often happens with real research data—some characteristics easily surpass the suggested threshold values while others fall very close to (or even below) the values recommended for informative taxometrics. Our reading of the literature suggests that investigators have largely based this determination on the Monte Carlo studies with idealized data sets that led to Meehl's (1995a) rules of thumb for acceptable data parameters. Fortunately, a newly developed procedure allows researchers to empirically examine the adequacy of a unique data set. It is to this procedure that we turn in the next section.

EVALUATING THE DATA BY GENERATING EMPIRICAL SAMPLING DISTRIBUTIONS

The remainder of this chapter describes a technique by which one can determine whether a particular set of indicators is appropriate for a planned taxometric procedure or consistency test implemented in a particular way. Our approach involves simulating and then analyzing multiple sets of taxonic and dimensional comparison data that hold constant as many characteristics of the research data as possible while differing only in their underlying structural model (J. Ruscio, Ruscio, & Keane, 2004; J. Ruscio, Ruscio, & Meron, 2005). In this approach, taxonic data are simulated using a two-group taxonic model that allows for between-group separation on each indicator as well as within-group indicator correlations. Dimensional data are simulated using the common factor model in which indicator correlations stem only from variation along one or more latent factors. Comparison data are simulated by using the common factor model to reproduce the observed indicator correlation matrix (in the full sample for dimensional data and within-groups for taxonic data) and bootstrapping each indicator's score distribution (Efron & Tibshirani, 1993). The bootstrap works by using an observed distribution as an unbiased estimate of the population distribution, from which random samples can be drawn. Specifically, a bootstrapped score distribution is randomly resampled (with replacement) from each indicator's observed distribution.

Each comparison data set is then subjected to the taxometric analyses planned for the research data. These analyses generate empirical sampling distributions of taxometric results, with *empirical* referring to the fact that no distributional assumptions are made to derive hypothetical sampling distributions analytically. Our approach to evaluating the adequacy of data involves the ability of the unique attributes of the research data to produce discernibly different distributions of taxometric results in the presence of taxonic versus dimensional latent structure. To the extent that the empirical sampling distributions overlap little or not at all, the data appear to afford a powerful test between the two competing structural hypotheses. When these distributions cannot be distinguished, the data appear inadequate to provide informative results, at least when analyzed in this particular way. As Efron and Tibshirani (1993) explained, an approach such as this, which is grounded in the bootstrap, "is used not to learn about the general properties of a statistical procedure, as in most statistical simulations, but rather to assess its properties for *the data at hand*" (p. 395; italics original).

Strengthening the Support for Structural Inferences

One advantage to evaluating the data in this way is that it can provide stronger support for the structural inference drawn from the obtained results. It does this by helping to rule out the possibility that the latent structure revealed by taxometric analysis represents a pseudotaxonic or pseudodimensional result. Earlier we discussed some of the factors that can lead to pseudotaxonic findings, such as the influence of indicator skew. Successfully defending an inference of taxonic structure made on the basis of seemingly taxonic results requires that such alternative explanations be convincingly eliminated.

Likewise, numerous factors can lead to pseudodimensional taxometric findings. For example, a taxon may exist, but the indicators submitted to analysis may possess insufficient validity to detect it, yielding results that appear consistent with dimensional structure. Other data characteristics that may increase the probability of pseudodimensional inferences include an overly small sample, too few taxon members, substantial within-group correlations, or poorly chosen analytic strategies for implementing a taxometric procedure or consistency test. In instances where results appear more dimensional than taxonic, successfully defending an inference of dimensional latent structure requires that one rule out a host of potential alternative explanations for these results.

By generating empirical sampling distributions of taxonic and dimensional results, one can assess the extent to which inadequate data or poorly made implementation decisions may have led to misleading results. If the data or the procedural implementation are especially problematic, the analysis would not be sufficiently powerful to accurately distinguish latent structure and the sampling distributions for taxonic and dimensional comparison data sets would overlap to a substantial extent. In such a case, results obtained for the research data would be regarded as ambiguous and would not be interpreted. However, if the data and procedural implementation are not problematic, one would expect to observe discernible differences in the empirical sampling distributions. In this case, results obtained for the research data could not easily be attributed to unacceptable data or a poorly implemented analysis, lending increased support to the structural solution suggested by the research data.

Establishing that one's data are appropriate for a particular analysis can help to mitigate the force of one potential criticism raised by some taxometric researchers: That dimensional structure cannot be inferred from a taxometric analysis because this is equivalent to accepting a null hypothesis. This criticism arises in traditional null hypothesis significance testing because it is always possible that the alternative hypothesis was not supported due to an overly small sample, insufficient reliability or validity of measured variables, or other factors that weakened the test. However, when a researcher can persuasively counter each of the major alternative explanations (e.g., by demonstrating that statistical power was high and that variables were measured with good precision), support builds for the conclusion that there actually is no effect. Consistent with our preferred inferential framework (see chap. 3), we believe that the same considerations apply to structural inferences drawn from taxometric results. Whether one concludes that the results support taxonic or dimensional latent structure, there will always be a number of alternative explanations to challenge this conclusion. To the extent that these alternatives can be met with evidence-based counterarguments, one increases support for the structural inference revealed by the analysis. Because it may be more likely that inappropriate data will give rise to pseudodimensional than pseudotaxonic results, it is particularly important to marshal strong evidence that the data were adequate—and the taxometric analysis implemented satisfactorily—to detect a taxon had one existed. Although establishing the appropriateness of a given data set for a specific taxometric procedure or consistency test does not altogether eliminate the risk of pseudodimensional findings, we believe it goes a long way toward preventing their potential misintepretation.

What Empirical Sampling Distributions Do and Do Not Achieve

Several aspects of our proposed technique to evaluate the adequacy of data and help interpret results are noteworthy, and it is important to understand what analyses of taxonic and dimensional comparison data do and do not accomplish in a taxometric investigation.

First, empirical sampling distributions are designed to supplement, but not replace, the findings of Monte Carlo studies. Large-scale, systematic studies of the performance of taxometric procedures and consistency tests under a wide range of data and analytic conditions can provide useful guidelines to suggest whether a planned analysis is likely to be appropriate for a particular set of data. However, there are many ways in which the idealizations of Monte Carlo studies limit the practical guidelines that they can offer to the taxometric investigator. For example, as noted earlier, most Monte Carlo studies have been performed with data simulated by drawing values from continuous, normal distributions. Actual research data deviate to some degree from this idealization because most psychological measurements are not strictly continuous and often yield nonnormal score distributions (Micceri, 1989). Likewise, Monte Carlo studies typically involve data that are simulated so that all indicators possess similar validity, exhibit similar within-group correlations, and are distributed similarly to one another. Although this is often necessary to make such studies feasible, it does not address the fact that actual research data sets typically contain indicators that differ from one another along these and other factors.

Monte Carlo studies also have not systematically varied the ways in which each taxometric procedure can be implemented. In chapters 5 and 6, we discuss many important decisions that must be made when performing each taxometric procedure, some of which may have a dramatic influence on the obtained results. At present, almost no Monte Carlo evidence can be brought to bear at many of the choice points involved in conducting a taxometric analysis. An understandable result is that most researchers appear to follow established conventions even when such conventions may result in less powerful or appropriate analyses.

Perhaps most important, no published Monte Carlo study of a taxometric procedure or consistency test has been performed in an exhaustive, crossed factorial manner that realistically represents the full range of conditions that researchers are likely to encounter. To include all relevant data characteristics and implementation options as individual factors—as well as a wide range of values to serve as levels within each factor—would result in an unmanageable number of cells in any study design. This is especially the case given the graphical nature of some of the most important output of taxometric procedures, making it remarkably difficult to present (or even to summarize) curve shapes for a large number of replications in each of a large number of cells. To avoid this combinatorial explosion and yield a feasible design, researchers performing Monte Carlo studies must select a relatively small number of potentially important factors, vary these across a relatively small number of reasonable levels, implement the procedure(s) under study in only one way, and/or study the impact of one or two factors at a time, rather than varying them in all factorial combinations. The result is that large regions of the parameter space have not yet been explored. Hence, when using Monte Carlo results to guide decision making in a taxometric investigation, the problem that researchers face is not usually one of interpolation (e.g., determining what the results might look like for $P = .40$ when the simulations involved $P = .10$, .25, and .50), but one of extrapolation (e.g., determining what the results might look like for a specific configuration of data characteristics that was not studied, analyzed using a procedural variant that was also not studied), which poses a much more formidable challenge.

This is where empirical sampling distributions can be most useful. By tailoring the simulation of taxonic and dimensional comparison data to the estimated population parameters of one set of research data, and examining the results of an analysis implemented in the manner intended for those data, one can determine how well the proposed analysis can distinguish these two latent structures. The technique takes into account the unique configuration of characteristics in the research data (e.g., sample size, number and distributions of the indicators, indicator validity, within-group indicator correlations) as well as the specific variant of each analytic procedure or consistency test under consideration. Analyses of taxonic and dimensional comparison data sets yield sampling distributions of taxometric results for a very small region of the parameter space—one that has almost certainly not been directly assessed in any Monte Carlo study. In cases where the adequacy of one or more aspects of the data—

or, more important, their joint acceptability—is questionable, or where a procedure is going to be implemented in a way that has not received Monte Carlo study, study-specific empirical sampling distributions can be highly informative.

Second, when empirical sampling distributions are generated, one simultaneously assesses the adequacy of one's data and the manner in which a specific procedure is implemented. It is important to recognize that a data set may be adequate for one analytic procedure but not another, or for an analysis that is performed in some ways but not others. For this reason, we recommend that researchers evaluate each indicator set for each planned taxometric procedure or consistency test, drawing structural inferences on the basis of only those results that appear informative when interpreted in the context of the empirical sampling distributions.

Third, even if analyses of comparison data are performed and the empirical sampling distributions are quite distinct from one another, it is possible that results for the research data will not be interpretable. The comparison of empirical sampling distributions allows a researcher to ask whether a planned analysis is capable of differentiating taxonic from dimensional structure. Under some circumstances, however, neither structural model may correspond to the true structure of the target construct. For example, the results for the research data may be ambiguously intermediate, falling midway between those for the taxonic and dimensional comparison data. Or the results for the research data may be equally inconsistent with both structures, suggesting that neither of these models is even approximately correct. Thus, generating empirical sampling distributions does not guarantee that the planned analysis will provide interpretable or informative results. We believe, however, that researchers are less likely to draw unwarranted structural inferences when confronted with evidence that neither the taxonic nor dimensional model fits appreciably well, in an absolute sense, or appreciably better than the other, in a relative sense.

Fourth, because empirical sampling distributions are generated through analyses of comparison data tailored to population parameters estimated from a particular sample, it cannot address potential problems arising from the sampling strategy that produced the research data, nor can it rule out the possibility that the research sample—due to normal sampling error—deviates in important ways from the target population. For example, if an investigator begins with a data set in which cases were selectively sampled from the extremes of one or more pertinent distributions, problematic characteristics of these data will be reproduced in the comparison data sets, making pseudotaxonic results a potential concern in each sample. Likewise, if sampling error produces a research sample that is unrepresentative of its population, empirical sampling distributions cannot reveal this problem. Thus, although the adequacy of one's data and procedural implementation can be usefully assessed, the influence of sampling considerations must be addressed through careful sampling practices, reasoned arguments, and—ideally—replications in new samples of data.

Having described the major strengths and limitations of the use of empirical sampling distributions of taxometric results, we refer interested readers to Appendix A for technical details on how to simulate the requisite taxonic and dimensional comparison data. We have illustrated the utility of empirical sampling distributions in a number of recent articles (e.g., J. Ruscio & Ruscio, 2004a, 2004b, 2004c; J. Ruscio, Ruscio, & Keane, 2004), and further empirical illustrations appear in chapters 7 and 8 in the context of consistency testing and interpretational issues. For now we close by reiterating that we believe this to be the most rigorous technique currently available for evaluating the adequacy of data for taxometric analysis. Users of traditional analytic tools are applying the bootstrap to generate empirical sampling distributions for purposes such as estimating standard errors or constructing confidence intervals when analytic assumptions may be violated (e.g., Efron & Tibshirani, 1993). Users of more advanced analytic tools—such as IRT or latent variable mixture modeling—can use a bootstrapping approach to generate empirical sampling distributions for a variety of purposes (e.g., Muthén, 2001; Stone, 2000). Taxometric researchers can likewise apply the bootstrap to supplement the results of Monte Carlo studies by taking into consideration the complex array of data characteristics in their unique sample of research data and the particular way in which they intend to implement a given analysis. By focusing on the localized region of the vast parameter space that is most directly relevant to their investigation, researchers can make an informed decision about the appropriateness of their data for the planned analyses.

CONCLUSIONS

As with any approach to the study of latent structure, the taxometric method requires a large sample of data collected with an eye toward avoiding methodological artifacts. Other important sample characteristics include sufficient representation of a taxon, high indicator validity, low within-group indicator correlations, and indicators whose number and score distributions are appropriate for the procedures and consistency tests that will be performed. Although decisions about the adequacy of a particular data set for a particular analysis have traditionally been based solely on general guidelines arising from Monte Carlo studies, we suggest that such decisions be informed by generating empirical sampling distributions through analyses of bootstrap samples of taxonic and dimensional comparison data.

It is important to keep in mind the capabilities and limitations of our proposed technique. In particular, although it does test the joint sufficiency of the data and the planned analysis, it does not detect or rule out sampling artifacts. However, when used with appropriate caution, empirical sampling distributions can fruitfully supplement the rules of thumb derived from Monte Carlo studies by taking into account the unique configuration of data characteristics and procedural implementation in a taxometric investigation. We strongly urge researchers to bootstrap taxonic and dimensional comparison data for each candidate indicator set, and to submit each set of comparison data to every procedure and consistency test in the analysis plan. When the empirical sampling distributions of results for the taxonic and dimensional data sets are distinguishable, one can judge an indicator set adequate for this analysis with added confidence. When the empirical sampling distributions are indistinguishable, we recommend that the corresponding research indicators be used with caution, if at all, for that analysis. Following these guidelines can help to muster empirical support for the appropriateness of indicators and procedural implementations by ruling out the inappropriateness of the data or analytic strategy as viable alternative explanations for the observed results. This can be especially important when one reaches an inference of dimensional structure.

Researchers have several options when empirical sampling distributions suggest that taxonic and dimensional structural models cannot be distinguished with adequate statistical power. For example, they can estimate key model parameters (e.g., indicator validity, within-group correlations) to try to determine what problematic aspect(s) of the data may be present, combine items in new ways that may yield more valid indicators or lower within-group correlations, or attempt different taxometric procedures (or more appropriate ways of implementing the procedures) that may better differentiate taxonic from dimensional structure. Further refinement of the indicator set and/or the analysis plan may ultimately yield a more powerful distinction between structural models. However, it is important to acknowledge that, in some cases, the available data will simply be inadequate for an informative taxometric analysis, regardless of the efforts made. Although certainly disappointing, this outcome must be preferred to drawing unwarranted structural inferences when the data are incapable of distinguishing between these two competing latent structures. In this way, the rigorous generation and use of empirical sampling distributions may ultimately lessen the risk of reaching inaccurate structural conclusions on the basis of misleading results.

CHAPTER FIVE

Taxometric Procedures I: MAXSLOPE, MAMBAC, and L-Mode

The taxometric method comprises a number of analytic procedures that can be used to evaluate the latent structure of the construct under investigation. Because many of these procedures are derived in mathematically distinct ways, they provide nonredundant evidence that can be used to check the consistency of results and thereby build confidence in an inference of latent structure. The present chapter introduces three of these taxometric procedures: MAXSLOPE (*MAX*imum *SLOPE*; Grove, 2004; Grove & Meehl, 1993), MAMBAC (*M*ean *A*bove *M*inus *B*elow *A C*ut; Meehl & Yonce, 1994), and L-Mode (*L*atent *Mode*; Waller & Meehl, 1998). Two additional procedures that are closely related to one another—MAXCOV (*MAX*imum *COV*ariance) and MAXEIG (*MAX*imum *EIG*envalue)—are introduced in chapter 6. Chapters 5 and 6 are intended to be read and used together. We have divided this material between two chapters to make it easier to digest and have chosen to begin with the procedures that require fewer implementation decisions and are therefore somewhat simpler to perform. This order of presentation is not meant to imply that MAXSLOPE, MAMBAC, and L-Mode are in any way preferable to MAXCOV or MAXEIG or vice versa. After the five procedures have been described and illustrated, chapter 7 provides suggestions for choosing complementary taxometric procedures for an investigation with an eye toward the goal of consistency testing.

In chapters 5 and 6, we review the logic underlying each taxometric procedure, describing its conceptual rationale and mathematical underpinnings. We then proceed to the mechanics of how the procedure is implemented, including the critical decisions that must be made when the procedure is applied in a particular investigation. Within this context, we describe some recent methodological innovations aimed at enhancing the analytic flexibility of several procedures. These points are illustrated through analyses of the taxonic and dimensional data sets described in chapter 3. Throughout both chapters, we emphasize the importance of making sound decisions at the many choice points encountered when applying each procedure to a unique set of research data. We also review evidence from systematic investigations of the procedures, where such evidence is available, and suggest issues that remain to be studied in future Monte Carlo work.

A handful of taxometric procedures that were introduced relatively early in the development of the method are not discussed in this book. For example, the *consistency hurdles* procedure (Golden, 1982; Golden, Tyan, & Meehl, 1974; Meehl & Golden, 1982) is seldom used in contemporary taxometric studies perhaps because it does not produce graphical output or because it shares some important similarities to more recent procedures. Likewise, the normal minimum chi-square procedure (Golden & Meehl, 1973; Meehl, 1968; Meehl, Lykken, Burdick, & Schoener, 1969), which is performed with only a single indicator variable, has more in common with finite mixture models than with coherent cut kinetics taxometric procedures. The five procedures included in this book are all used, to varying degrees, in present-day taxometric investigations.

Before embarking on a tour of taxometric procedures, we want to clarify a couple of points about our style of presentation and the guidelines that we offer. First, our aim in presenting analyses of the eight data sets introduced in chapter 3—as well as additional data sets in some places—is only to illustrate the basic nature of taxometric analyses and the decisions that must be made to implement them. As noted in chapter 3, six of our sample data sets (TS#1 to TS#3 and DS#1 to DS#3) are idealized in many ways; the remaining two data sets (TS#4 and DS#4) contain several of the complicating factors that are frequently encountered in real research data. Given our instructional focus, we chose to analyze a relatively small number of data sets rather than performing more wide-ranging analyses that would be cumbersome to summarize in a thorough manner and that may distract readers from the primary tutorial purpose at hand. It is important to emphasize at the outset that these analyses by no means comprehensively cover the range of possible outcomes that may be observed in actual taxometric research. Instead, our description of the results expected for taxonic and dimensional structure—as well as many results of our illustrative analyses—are intended to represent prototypical outcomes to help readers gain a general sense for the output generated by taxometric analyses.

Second, the advice that we offer is based on the broader taxometric literature, not simply the analyses presented in this book. We draw on Monte Carlo studies of taxometric procedures as well as lessons learned through our own hands-on experience in performing taxometric investigations of many psychological constructs using a wide range of research data and conducting both informal and formal Monte Carlo simulations. We emphasize Monte Carlo evidence where it is available; where it is not available, we present the logic underlying our advice so that researchers can evaluate its applicability to specific situations. Finally, we identify topics in need of systematic investigation, both to highlight questions that are presently unresolved and to stimulate research that will lead to more generally applicable, empirically driven guidelines. Because taxometric methodology is at a relatively early stage of development and evaluation, our methodological recommendations should be considered tentative rather than authoritative. Indeed, a certain degree of informed flexibility may be necessary for conducting appropriate and informative taxometric analyses. In summary, our goal in presenting these procedures is not to prescribe a rigid blueprint for their application, but to raise awareness of important aspects of the taxometric method and encourage investigators to make thoughtful, reasoned decisions about how to apply the method to their own research questions.

MAXSLOPE

The MAXSLOPE procedure is conceptually the simplest of the taxometric procedures in part because it requires only two indicators of the target construct and in part because it involves a straightforward graphical analysis.

Logic of the Procedure

A MAXSLOPE analysis begins with a scatterplot that displays the relationship between two indicators: One indicator (the *input*) is placed on the x axis and the other indicator (the *output*) is placed on the y axis. When latent structure is taxonic and the two indicators distinguish the taxon from the complement with sufficiently strong validity and tolerably low within-group correlations, two distinct clouds of points will be visible. A constellation of points representing taxon members is located toward the upper right region of the scatterplot because these individuals tend to score relatively high on both indicators, and a constellation of points representing complement members is located toward the lower left region of the scatterplot because these individuals tend to score relatively low on both indicators. Rather than using simple linear regression to calculate the best-fitting line through this scatterplot, a local regression curve is generated using any of a number of scatterplot smoothing techniques that estimate the regression slope within restricted regions of the scatterplot. This approach allows for a curved regression function, rather than forcing the function to be linear across the entire range of x values. Grove and Meehl (1993) used the LOWESS technique (LOcally WEighted Scatterplot Smoother; Cleveland, 1979) to smooth scatterplots in their MAXSLOPE analyses, and we do the same.

When the indicators of taxonic structure are relatively independent *within* the taxon and complement classes (i.e., there are negligible within-group correlations), the local regression curve will take on an S-shaped appearance: The curve will be fairly flat toward both ends and steepest in the intermediate region where the taxon and complement members intermingle. In contrast, if the association between the two indicators stems from dimensional latent structure, the scatterplot will take the form of a single homogeneous cloud of points. The local regression curve drawn through this cloud will slope upward, but will remain fairly straight across the full span of the graph. Thus, dimensional structure produces a comparatively linear MAXSLOPE plot that is distinct from the S-shaped plot indicative of taxonic structure.

Consider a concrete example. Suppose that measures of auditory hallucinations and bizarre delusions are administered to a large sample of psychiatric inpatients and the two scale scores are correlated. If one half of the individuals belong to a schizophrenia taxon, the remainder of these individuals constitute a heterogenous mixture of other mental disorders, and the hallucination and delusion measures validly distinguish these groups but are negligibly correlated within them, then a MAXSLOPE analysis should reveal taxonic structure. The scatterplot might look like the graph on the left in Fig. 5.1. Two distinct clouds of points are evident: Patients diagnosed with schizophrenia cluster together in the upper right region, and other patients cluster together in the lower left region. The dashed lines show within-group regressions, which are fairly linear because the indicators are uncorrelated within groups, and the solid line shows the MAXSLOPE curve (the local regression curve calculated in the full sample), which is S-shaped due to the mixture of groups. If, however, schizophrenia is not taxonic, then the scatterplot might look more like the graph on the right in Fig. 5.1. A single cloud of points is observed, and the MAXSLOPE curve is fairly linear throughout its range because there are no groups to distinguish. Please note that both of these curves were constructed using highly idealized data parameters so that we could obtain unambiguously taxonic and dimensional results. We present these graphs to illustrate the logic of the MAXSLOPE procedure, not to suggest that

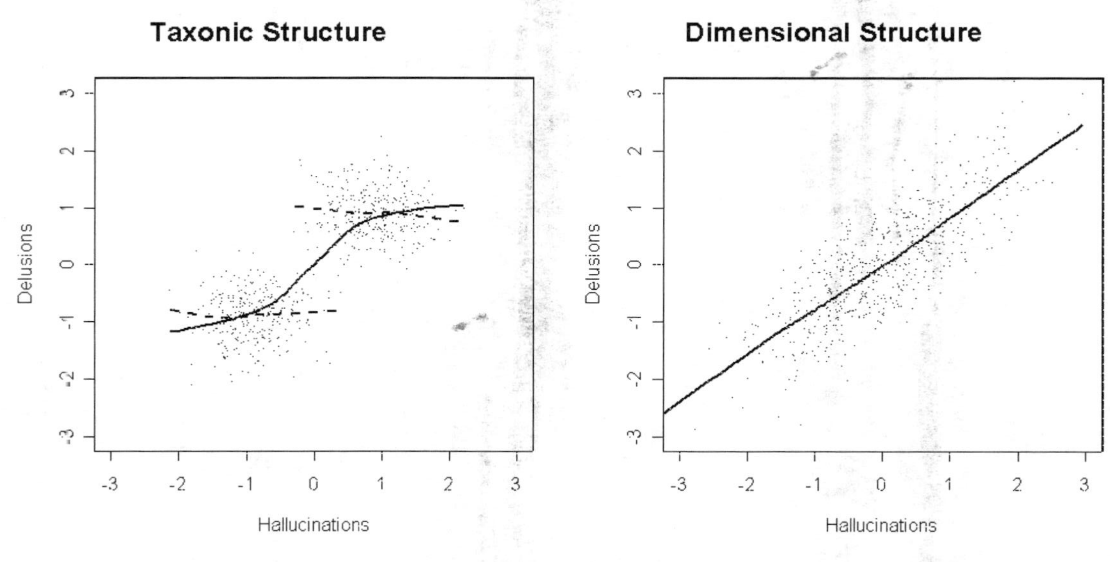

FIG. 5.1. Illustrative MAXSLOPE graphs for taxonic (left) and dimensional (right) structures. Solid lines are MAXSLOPE curves (local regressions calculated using the LOWESS scatterplot smoother); in the graph for taxonic structure, dashed lines are within-group LOWESS curves.

analyses of actual research data will yield curves that can be interpreted as easily as these prototypes.

Estimating the Taxon Base Rate

The location of the steepest point along MAXSLOPE's local regression curve can be used to estimate the relative sizes of the taxon and complement. With respect to the input indicator (i.e., the x axis of the scatterplot), taxon and complement members are best discriminated near the place where the regression takes on its maximum slope. Depending on the nature of the within-group distributions, this point of maximum slope may correspond to the *hitmax* value for that indicator or the cutoff score that would maximize the hit rate when classifying cases (Grove, 2004). For example, when within-group indicator distributions are normal, the hitmax is expected to coincide with the point of maximum slope. For other distributions, they may diverge to some extent. For groups of equal size, the point of maximum slope is located near the center of the scatterplot because taxon and complement members overlap maximally in that region. For a small taxon this point is deflected toward the right, whereas for a large taxon it is deflected toward the left.

Grove and Meehl (1993) recommended estimating the taxon base rate by locating the point of maximum slope, using this as the best estimate of the hitmax value, and calculating the proportion of cases scoring above this point on the input indicator. They also suggested smoothing the slopes before locating the maximum value. Grove (2004) provided a more complex technique for estimating the taxon base rate that does not presume that the point of maximum slope will correspond to the hitmax value—an error in the original technique that yielded biased estimates for non-normal within-group distributions. Presumably, the unbiased estimator will provide more accurate base rate estimates. At present, however, neither the accuracy of taxon base rate estimates obtained through MAXSLOPE analyses of taxonic data nor the consistency of such estimates for taxonic or dimensional data has been systematically studied.

Implementation Decisions

There is only one significant implementation decision required when MAXSLOPE is performed: deciding how to assign the available variables to the roles of input and output indicators. Because MAXSLOPE requires just one input and one output indicator (whose values form the x and y axes of the graph, respectively), the most straightforward way to conduct MAXSLOPE analyses is to assign variables to the input and output roles in all possible pairwise combinations. Because two different scatterplots and local regression curves can be obtained by reversing the input/output indicator assignments for a pair of variables, this technique yields $k(k - 1)$ MAXSLOPE plots for k indicators. Thus, if only two indicators are available, two MAXSLOPE plots are generated.

Alternatively, if more than two indicators are available, one can assign a single variable to serve as the output indicator and combine all remaining variables into a single composite input indicator (e.g., by summing them), yielding k curves (one per indicator). For example, assigning each of four variables to all possible pairwise input/output indicator combinations generates 12 curves, whereas using each of the four variables as an output indicator in turn, and summing the remaining three to form a composite input indicator, generates 4 curves. One benefit of forming composites is that the resulting input indicator may contain a larger range of values that provides a more reliable rank ordering of cases; this can be especially helpful when individual indicators vary across ordered categorical response scales. Using composite input indicators also allows all of the data to be included in each analysis, which can increase the clarity of the results—provided that each of the indicators does in fact validly distinguish a taxon and its complement. Aggregating valid and invalid indicators can weaken, rather than strengthen, results.

Although it is not typically treated as a decision point in MAXSLOPE analysis, one could consider alternative procedures for calculating the local regression curve and its required parameters (e.g., the proportion of cases used in estimating each new point along the curve). As noted earlier, the handful of researchers who have included MAXSLOPE in their taxometric investigations have followed the lead of Grove and Meehl (1993) by using the LOWESS function to generate local regression curves, and our MAXSLOPE program does likewise. Alternative techniques for smoothing the scatterplot—as well as variations in the proportion of the sample used to estimate each new point along the LOWESS curve—have not been studied.

Empirical Illustrations

To illustrate the nature of the results yielded by a MAXSLOPE analysis, we began by conducting MAXSLOPE using the data sets introduced in chapter 3, whose characteristics are briefly reviewed here. All data sets contained four indicators in a sample of 1,000 cases. TS#1 contained indicators that separated the taxon (base rate $P = .50$) and complement by $d = 1.00$, TS#2 contained indicators that separated the taxon ($P = .50$) and complement by $d = 2.00$, and TS#3 contained indicators that separated the taxon ($P = .25$) and complement by $d = 2.00$. The within-group correlations in each of these three taxonic samples were set to zero; like the specified indicator validities, this was subject to normal sampling error arising in the creation of these particular samples. The correlations among indicators in DS#1, DS#2, and DS#3 reproduced those in their corresponding taxonic data sets (mean *rs* = .20, .50, and .43, respectively); correlations were also subject to sampling error. The indicators in each taxonic data set were continuous and normally distributed within each group; the distributions of indicators in the dimensional data sets matched those in the corresponding taxonic data sets.

Unlike its previous three counterparts, TS#4 contained indicators that separated the taxon ($P = .30$) and complement with varying validity and were characterized by variable and nontrivial within-group correlations. The distributions of indicators in TS#4 and DS#4 were identical to one another, with each indicator positively skewed to a different extent and taking on a different number of ordered categorical values. Thus, TS#4 and DS#4 were probably more representative of real research data than the other six data sets, which were comparatively idealized in several ways. The key properties of all eight data sets are summarized in Table 5.1; additional information about their indicator distributions and correlations is provided in chapter 3 and summarized in Fig. 3.1 and Tables 3.1 and 3.2.

For each of the first three sets of taxonic and dimensional data, we implemented MAXSLOPE by assigning the four variables to input/output indicator roles in all pairwise combinations. Because this technique yielded 12 curves in each panel, we present only the first 4 curves to conserve space. Figure 5.2 shows the results for the dimensional data sets, and Fig. 5.3 shows the results for the taxonic data sets. For each of the dimensional

TABLE 5.1
Summary of the Data Sets Used in Illustrative Analyses

Data Set	*Taxon Base Rate (P)*	*Indicator Validity (Cohen's d)*[a]	*Within-Group Correlations (r)*[a]	*Within-Group Distributions*
TS#1	.50	1.00	.00	Continuous, normal
TS#2	.50	2.00	.00	Continuous, normal
TS#3	.25	2.00	.00	Continuous, normal
TS#4	.30	> 2.00, but variable across the four indicators	About .15, but varying widely between pairs of indicators within groups	Ordered categorical (6, 8, 10, and 12 values for the four indicators, respectively), positively skewed (to a variable extent for each indicator)

Note. $N = 1{,}000$ for each data set. Each dimensional data set reproduced the full-sample indicator distributions and correlations in the corresponding taxonic data set.

[a]Population parameters used to generate each data set; actual values in the sample differ due to normal sampling error.

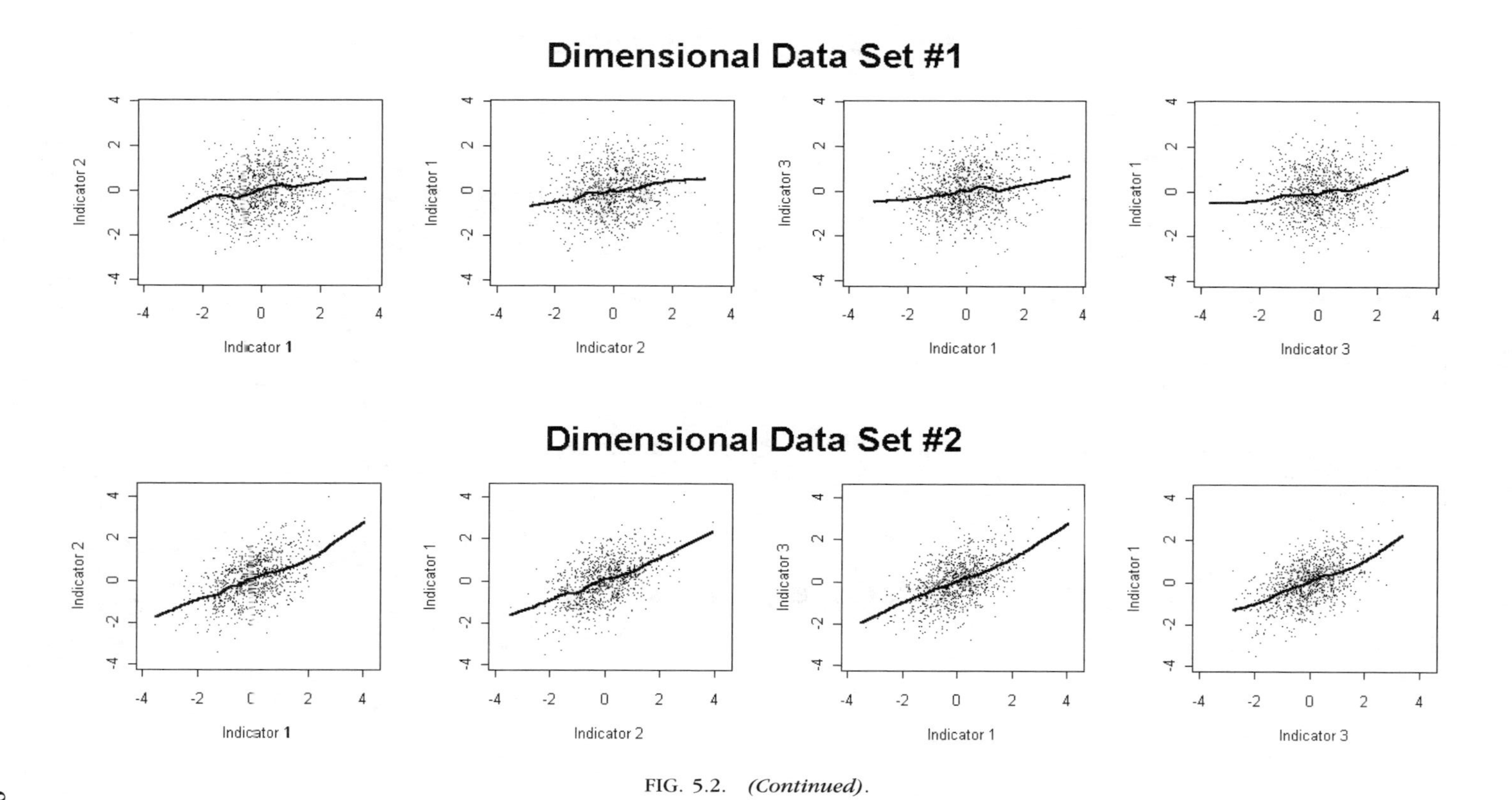

FIG. 5.2. *(Continued).*

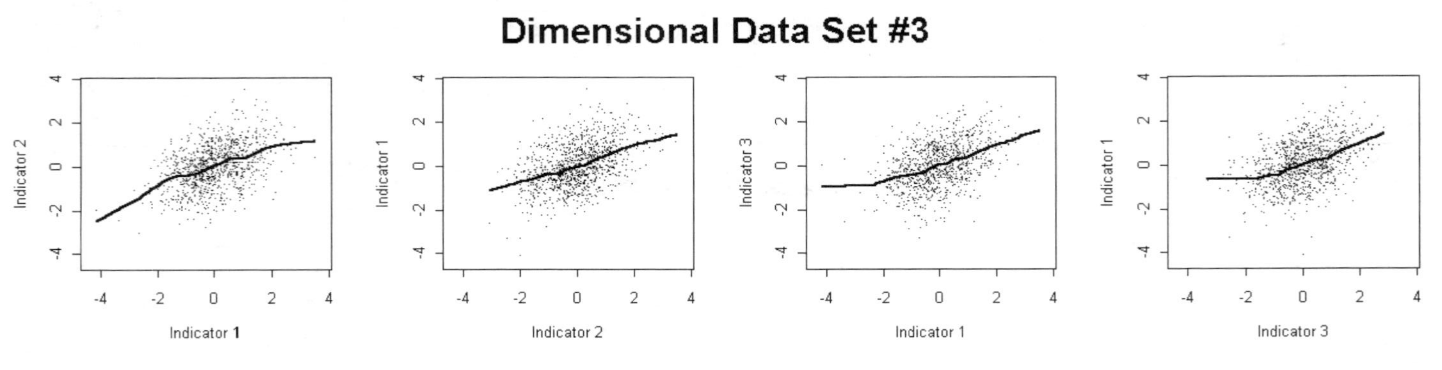

FIG. 5.2. Representative MAXSLOPE graphs for the first three dimensional data sets. Variables were assigned to the input indicator (x axis) and output indicator (y axis) roles in all pairwise configurations, and the first four graphs from each panel of 12 are presented.

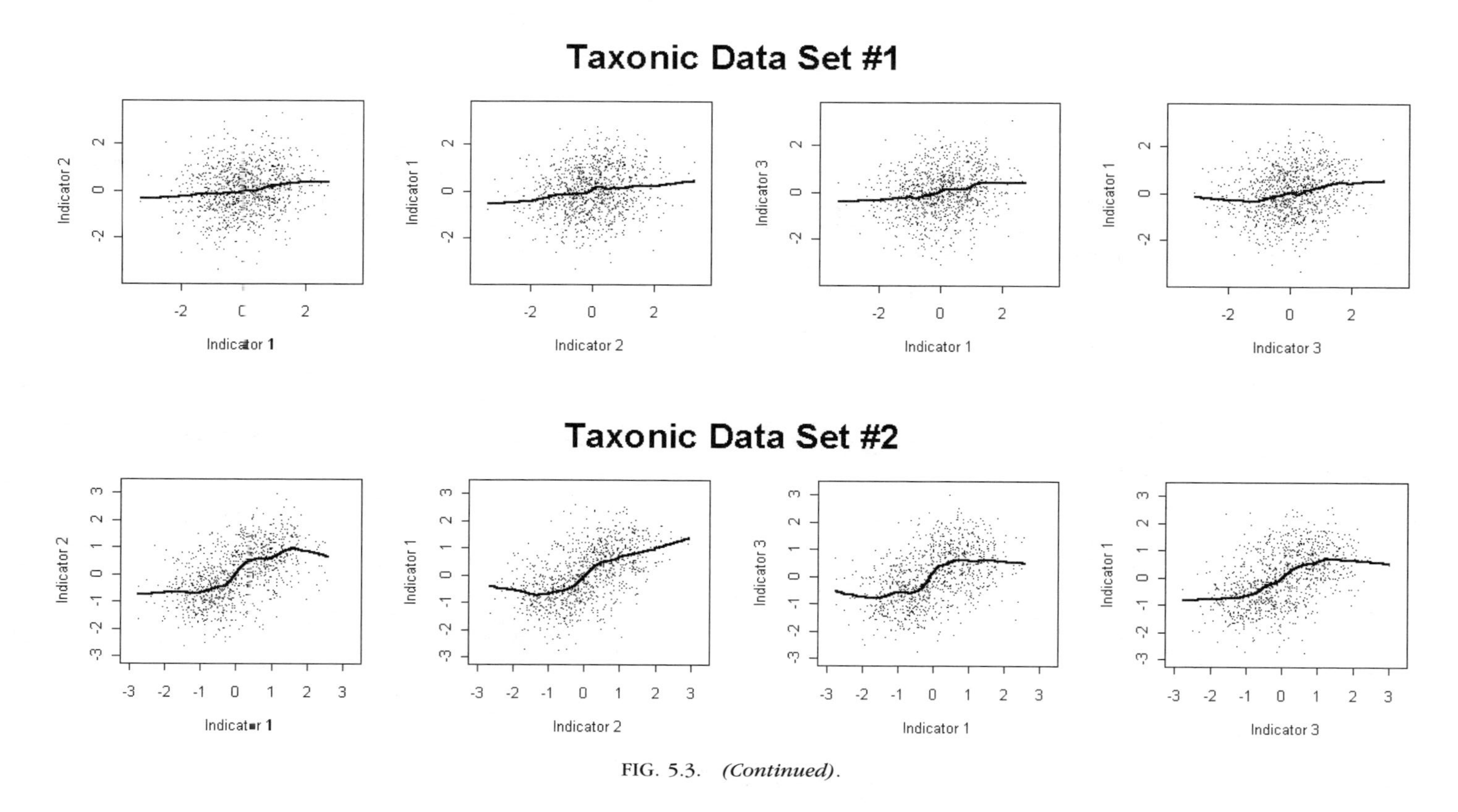

FIG. 5.3. *(Continued)*.

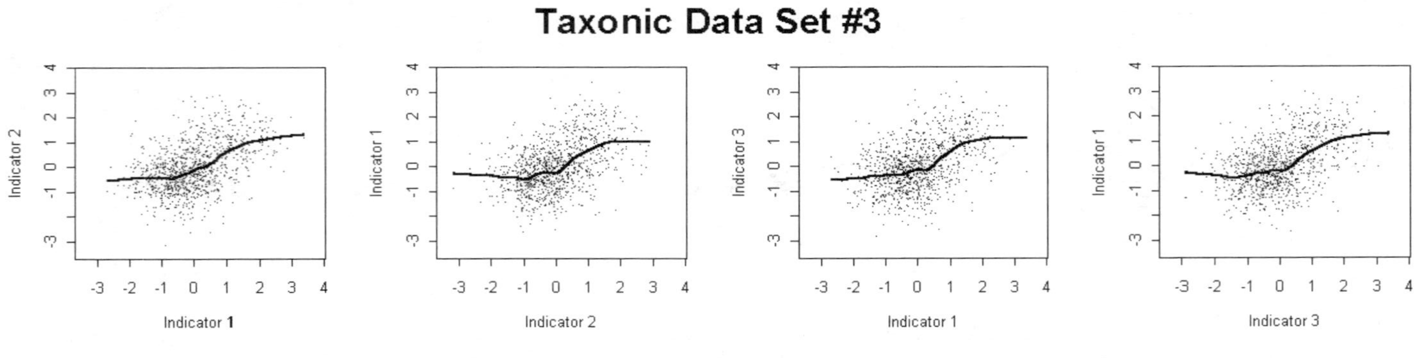

FIG. 5.3. Representative MAXSLOPE graphs for the first three taxonic data sets. Variables were assigned to the input indicator (x axis) and output indicator (y axis) roles in all pairwise configurations, and the first four graphs from each panel of 12 are presented.

data sets, the local regression curves in the MAXSLOPE plots were linear as expected. For TS#1, the curves were also linear, suggesting that the indicators were not sufficiently valid to reveal taxonic structure. For TS#2 and TS#3, however, the local regression curves were S-shaped, with the point of maximum slope located in the region where taxon and complement members overlapped. Estimates of the taxon base rate, appearing in Table 5.2, corroborated the visual inspection of curve shapes. These estimates varied widely for analyses of dimensional data and of TS#1, but were accurate (note the *M* values) and highly consistent (note the *SD* values) for analyses of TS#2 and TS#3.

Because the ordered categorical indicators in TS#4 and DS#4 varied across too few values to generate interpretable scatterplots, an alternative technique was needed to assign these variables to the required input/output indicator roles. Thus, for these data sets, we conducted MAXSLOPE by using one variable at a time as the output indicator and summing the remaining three variables to form a composite input indicator. This yielded four plots for each data set, which appear in the top half of Fig. 5.4. Although some of the local regression curves appeared slightly S-shaped, the curves produced by TS#4 were difficult to distinguish from those of DS#4. Estimates of the taxon base rate also varied considerably for both data sets (see Table 5.2). Thus, MAXSLOPE could not clearly differentiate taxonic

TABLE 5.2
Taxon Base Rate Estimates for MAXSLOPE Analyses

Estimates for Fig. 5.2		
Data Set	*# Estimates*	*M (SD)*
DS#1	12	.52 (.35)
DS#2	12	.39 (.41)
DS#3	12	.39 (.18)
Estimates for Fig. 5.3		
Data Set	*# Estimates*	*M (SD)*
TS#1	12	.43 (.23)
TS#2	12	.50 (.03)
TS#3	12	.25 (.06)
Estimates for Fig. 5.4		
Data Set	*# Estimates*	*M (SD)*
TS#4	4	.43 (.23)
DS#4	4	.43 (.14)
TS#3	4	.32 (.02)
DS#3	4	.37 (.13)

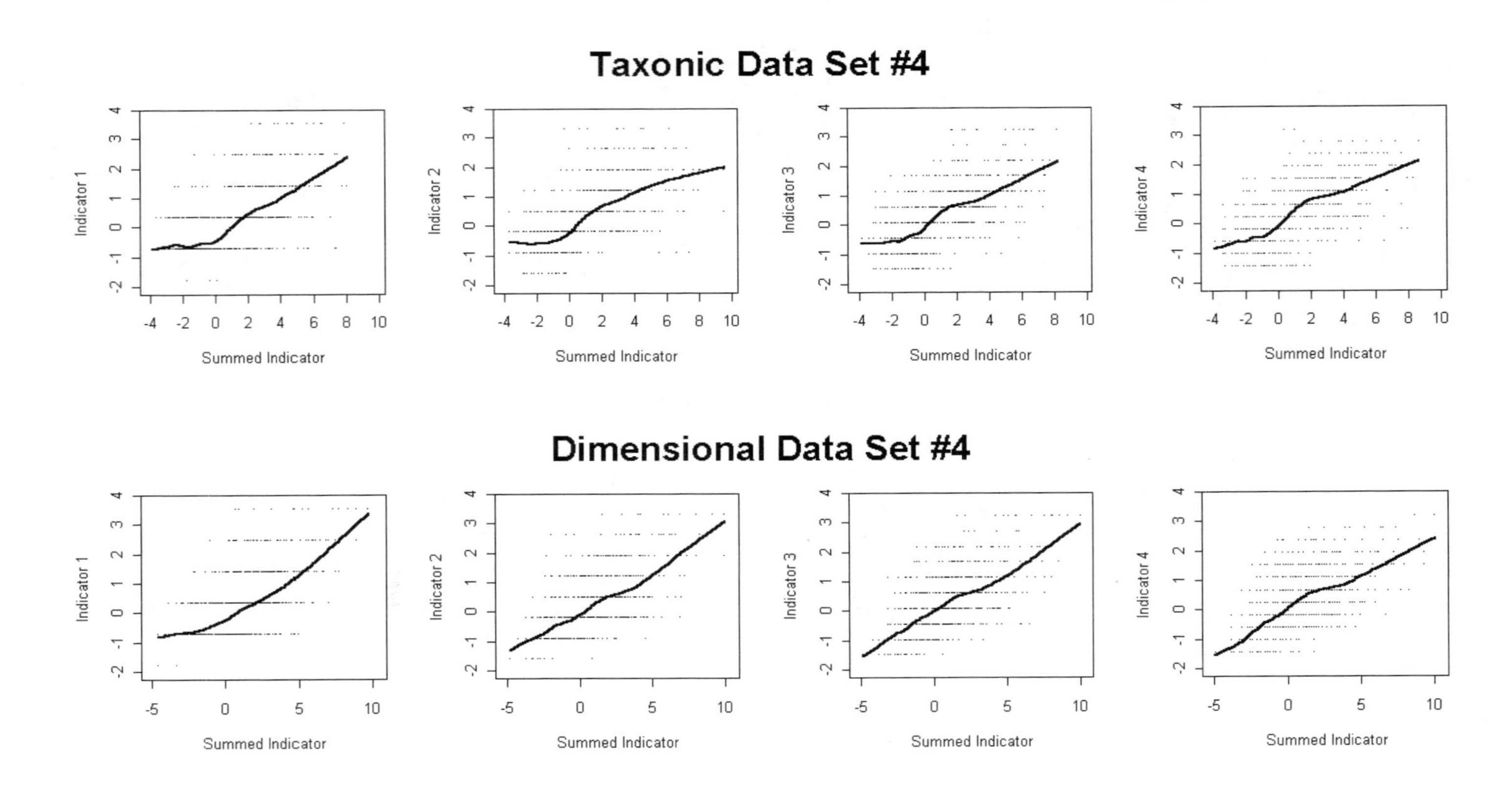

Taxonomic Data Set #4
Indicator 1
Indicator 2
Indicator 3
Indicator 4
Summed Indicator
Dimensional Data Set #4
Indicator 1
Indicator 2
Indicator 3
Indicator 4
Summed Indicator

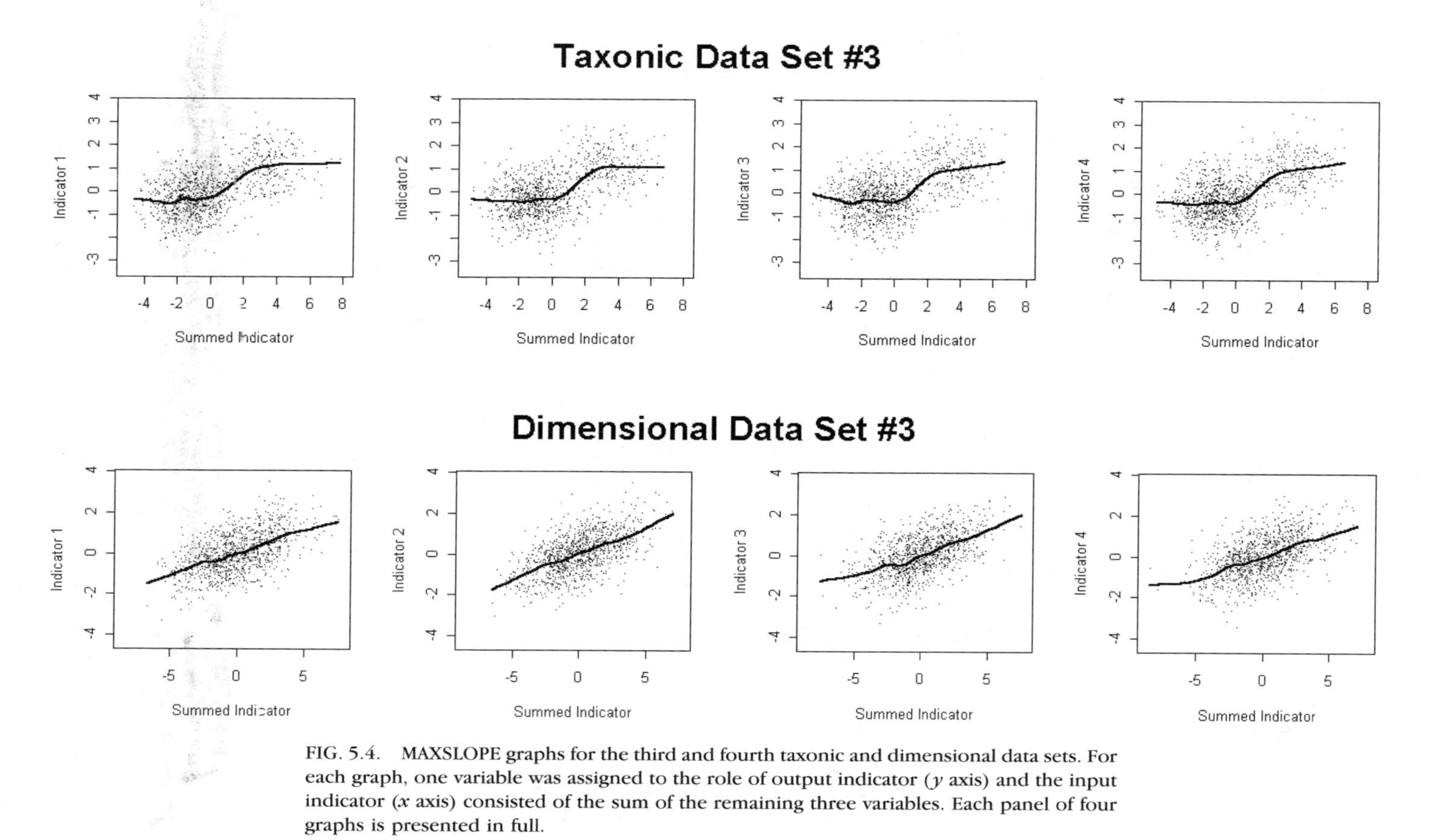

FIG. 5.4. MAXSLOPE graphs for the third and fourth taxonic and dimensional data sets. For each graph, one variable was assigned to the role of output indicator (y axis) and the input indicator (x axis) consisted of the sum of the remaining three variables. Each panel of four graphs is presented in full.

from dimensional structure given the properties of these two data sets; in particular, positive indicator skew and within-group correlations may have served to obscure the structural differences underlying these data.

To provide an example in which the composite input indicator *did* facilitate interpretation, the same technique was repeated in MAXSLOPE analyses of TS#3 and DS#3; results appear in the bottom half of Fig. 5.4. In these analyses, the technique resulted in clearer distinctions between taxonic and dimensional structure than was provided by earlier analyses in which individual indicators served as the input. That is, because the composite input indicators more validly distinguished the taxon and complement than did any individual indicators, the two clouds of points were even more distinct than they were in the corresponding plots shown in Fig. 5.3. Thus, although the composite input indicator technique did not overcome the limitations of other unfavorable data characteristics (as in TS#4 and DS#4), it did appear to improve the interpretability of some MAXSLOPE curves (as in TS#3 and DS#3). The relative performance of MAXSLOPE with pairwise input–output indicators versus composite input indicators awaits study across a wide range of data conditions.

Systematic Study of the Procedure

The MAXSLOPE procedure has yet to be evaluated in Monte Carlo studies. At present, virtually nothing is known about the conditions under which it can distinguish taxonic and dimensional structure or about the accuracy or coherence with which taxon base rates (or other model parameters) are estimated in analyses of taxonic and dimensional data.

MAMBAC

Like MAXSLOPE, the MAMBAC procedure also requires only two indicators, but can be performed with composite input indicators that use all available data for each curve, thereby helping to increase the clarity of results under certain conditions.

Logic of the Procedure

The MAMBAC procedure takes advantage of the fact that if two groups exist there must be an optimal cutting score for distinguishing them. That is, if an indicator validly differentiates a taxon and complement, there must be a particular value on this indicator that will classify cases into groups with a minimum number of false positives and false negatives. In the absence of taxonic structure, there are no groups to be distinguished, and hence no optimal cutting score for doing so. Thus, MAMBAC is based on the search for an optimal cutting score. If such a score can be found, this suggests taxonic structure; if it cannot, this suggests dimensional structure.

MAMBAC uses two indicators to search for an optimal cutting score. As in MAXSLOPE, one of these is treated as the input indicator and placed along the *x* axis of a graph. Cases in the data set are sorted according to their scores on this input indicator, which is used to create a series of cutting scores. Beginning with a cutting score toward the low end of the input indicator, the mean score on the other (output) indicator for all cases falling below the cut is subtracted from the mean score of all cases falling above the cut. This mean difference is plotted as the *y* value for the cutting score (*x* value) that was used. The subtraction is repeated for each cutting score along the *x* axis, and each mean difference is plotted as the corresponding *y* value. Thus, MAMBAC involves plotting mean differences on the output indicator for cases scoring above each cutting score on the input indicator minus cases scoring below that cutting score.

The shape of the resulting MAMBAC curve allows an inference to be drawn about latent structure. If an optimal cutting score exists, mean differences should be largest near the location of this score and should decline as lower or higher cutting scores are used. Thus, taxonic structure tends to yield peaked MAMBAC curves, with the peak suggesting the region of the input indicator in which the optimal cutting score lies. In contrast, dimensional structure tends to yield concave curves without a discernible peak, often bowing upward at one or both ends. The absence of a peak is the critical clue that the latent structure is dimensional.

Why do dimensional MAMBAC curves often bow upward at one or both ends rather than simply appearing flat? This upward bowing stems from the fact that, as cutting scores approach the ends of the input indicator, they isolate increasingly small samples of extremely high- or low-scoring cases (Meehl & Yonce, 1994). When the subtraction is performed to compare the mean score of these cases with the more moderate mean score of the remaining sample, this produces a large difference. In contrast, cuts made near the middle of the input indicator divide the sample roughly in half. Differences between these two halves of the sample—each containing some moderate-scoring cases that are quite similar to one another—will be considerably smaller than the differences obtained at the extremes. These small differences at cutting scores near the middle of the input indicator range, along with large differences at more extreme cutting scores, produce the prototypically concave dimensional MAMBAC curve.

Let us return to the hypothetical example used earlier for MAXSLOPE: a large sample of psychiatric inpatients assessed for hallucinations and delusions. To perform a MAMBAC analysis, cases are sorted along one indicator and mean differences are calculated using another. For example, cases can be sorted according to scores on the hallucination measure and a series of cutting scores created (for this example, cutting scores were located at every *n*th case in the data, beginning and ending near the extremes; this and other techniques for locating cutting scores for MAMBAC analyses are discussed shortly). The mean score on the delusion measure is calculated for cases scoring above a cut and below that cut; the mean difference is plotted for each cutting score. If one half of these individuals belong to a schizophrenia taxon and the remaining individuals belong to its complement group (i.e., nonschizophrenic mental disorders), the indicators validly distinguish these groups, and the indicators correlated at tolerably low levels within each group, the resulting MAMBAC graph might look like the one in the left side of Fig. 5.5. Mean differences on delusions reach a clear peak when cases are cut into approximately equal-sized groups (i.e., when the cut is located near the 300th of the 600 cases in this illustrative data set) and taper off as the location of the cut moves to the left or to the right. The emergence of such a peak in a MAMBAC curve suggests that there are in fact two groups mixed in the full sample of data; a peaked curve is expected for taxonic structure. However, if there are no groups being mixed, then the MAMBAC graph might look more like the one on the right side of Fig. 5.5. In this curve, no particular cut between cases sorted by their scores on hallucinations stands out as maximizing the mean difference on delusions. Instead the curve takes on the prototypically concave or bowed shape expected for dimensional structure. Again, note that these illustrative curves were constructed using highly idealized data parameters to reveal unambiguously taxonic and dimensional results. We present these graphs only to illustrate the logic of the MAMBAC procedure; curves for actual research data are unlikely to match these curve shapes.

Estimating the Taxon Base Rate

In addition to inspecting the shapes of MAMBAC curves, one can calculate an estimate of the taxon base rate from each curve. For taxonic data, the location of the peak suggests the relative sizes of the taxon and complement. Equal-sized groups produce a peak toward the center of the curve because the optimal cutting score lies near the middle of the input indicator range. For a small taxon, the optimal cutting score is higher, and thus the peak is deflected toward the right. Similarly, for a large taxon, the optimal cutting score is lower and the peak is deflected toward the left. Meehl and Yonce (1994, Appendix A) derived the following formula for estimating the taxon base rate from a MAMBAC curve:

$$\hat{P} = \frac{1}{R+1}, \qquad (5.1)$$

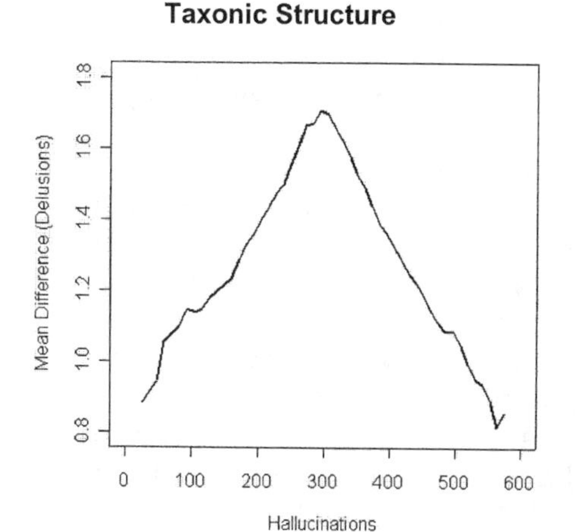

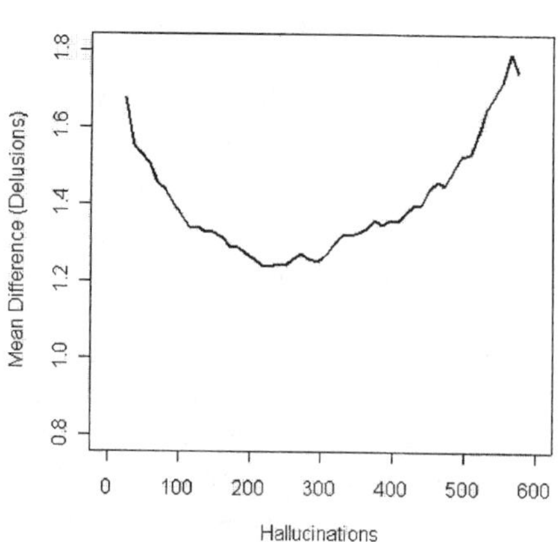

FIG. 5.5. Illustrative MAMBAC graphs for taxonic (left) and dimensional (right) structures.

where $R = d_{upper} / d_{lower}$, with d_{upper} the mean difference for the highest cutting score (the y value of the right endpoint on the MAMBAC curve) and d_{lower} the mean difference for the lowest cutting score (the y value of the left endpoint on the MAMBAC curve). Using this formula requires the assumption that cases above the highest cutting score are exclusively taxon members and cases below the lowest cutting score are exclusively complement members. Meehl and Yonce showed that when this is true, $d_{upper} = Q \times D_y$ and $d_{lower} = P \times D_y$, where P is the taxon base rate, Q is the complement base rate $(1 - P)$, and D_y is the raw validity of output indicator y (e.g., $\bar{y}_t - \bar{y}_c$). By forming the ratio $R = d_{upper} / d_{lower}$, indicator validity cancels out to leave $R = Q / P$. Equation 5.1 transforms R to estimate the taxon base rate in the sample. When the number of cases above the highest cutting score does not equal the number of cases below the lowest cutting score, an adjustment to this formula is necessary (see Meehl & Yonce, 1994, pp. 1118–1119). As discussed shortly, we recommend the use of equally spaced cutting scores that begin and end with equal numbers of cases beyond the most extreme cutting scores, therefore we will not discuss this adjustment further.

Although Meehl and Yonce's (1994) Monte Carlo results suggest that this technique provides good estimates of the taxon base rate, it is important to note that the estimate depends on only two data points from the MAMBAC curve. Meehl and Yonce recommended taking steps to ensure that the two endpoints of a MAMBAC curve are actually representative of the curve shape, rather than deviant values arising from sampling error. To remove or reduce the influence of sampling error on the calculation of mean differences at the endpoints of a MAMBAC curve, one can make the first and last cuts at less extreme locations along the input indicator, or one can smooth the MAMBAC curve prior to estimating the taxon base rate, as discussed further later.

There are at least two ways in which MAMBAC base rate estimates can be biased. Nonzero within-group correlations and indicator skew can influence the mean differences obtained at the ends of a MAMBAC curve independent of the actual taxon base rate, and thereby bias its estimation. Because within-group correlations tend to increase mean differences, base rate estimates can be biased toward .50 when indicators are correlated within both groups. For example, increasing both d_{upper} and d_{lower} by a similar amount causes their ratio R to approach 1.0. When R = 1.0, the taxon base rate estimate will be 1 / (1 + 1) = .50. When indicators are correlated at different levels across groups, base rate estimates can be biased in other ways (e.g., large indicator correlations only within the taxon can biased estimates downward, with the reverse true for large indicator correlations only within the complement). When indicators are skewed within groups, the effect is to tilt a MAMBAC curve, with positive skew yielding a rising curve (J. Ruscio, Ruscio, & Keane, 2004). This often increases the ratio R of d_{upper} to d_{lower}, thereby reducing the taxon base rate estimate. The opposite would be expected for negative skew.

Implementation Decisions

Three decisions must be made when implementing the MAMBAC procedure, each of which can have a significant influence on results. The first decision concerns how to assign variables to the roles of input and output indicators, with the available choices identical to those for MAXSLOPE analyses. One can use variables in all possible input–output indicator pairs, generating $k(k - 1)$ MAMBAC curves for k indicators. Alternatively, one can remove a single variable at a time to serve as the output indicator and sum all remaining variables to form a composite input indicator, yielding k curves.

Having determined how to assign variables to input–output indicator roles, the second decision concerns where to place cuts along the input indicator. Cuts may be made between each successive case, yielding the maximal number of points on the curve for interpretation. However, a smaller number of points usually suffices for interpretational clarity and may dramatically reduce the computational demands of the analysis (when a data set contains hundreds or thousands of cases and a large number of MAMBAC curves are being generated, this is not a trivial issue). Methods of selecting a more limited number of cut points include cutting at fixed SD intervals (e.g., place a cut every .25 SD along the input indicator, which is usually standardized for this purpose; Meehl & Yonce, 1994), at intact scale values (e.g., place cuts at particular scores on the input indicator), or at every *n*th case along the input indicator (e.g., determine the desired number of points to appear on the MAMBAC curve and place evenly spaced cuts to achieve this number). When the input indicator varies across a small number of ordered categories, the use of either intact scale values or *SD* units may yield too few data points to clearly reveal the shape of the MAMBAC curve.

Locating cutting scores at every *n*th case (or between each case) in the data set often means that one or more cuts will fall between equal-scoring cases on the input indicator. In such instances, the mean difference will be influenced to some extent by the arbitrary partitioning of these equal-scoring cases above and below the cut. Therefore, it is useful to perform internal replications by reshuffling the equal-scoring cases after performing the analysis and then recalculating the MAMBAC values. The average values across multiple replications can be plotted to produce a curve for a given input–output indicator configuration (J. Ruscio & Ruscio, 2004b, 2004c). With increasing numbers of internal replications, the obfuscating effect of cutting between equal-scoring cases is reduced.

As noted earlier in the context of base rate estimation, when cuts are placed at every *n*th case or between every case in the sample, it is also important to place the first and last cut points far enough from the ends of the input indicator range to stabilize the endpoints of the curve. For example, many researchers have used a minimum *n* of 25 cases beyond which no further cuts are made. If sample size allows, one might wish to use a larger minimum *n*, especially given the importance of these endpoints in estimating the taxon base rate. At the same time, because the optimal cutoff score for small taxa is located at a very high value along the input indicator, reserving too many cases above the uppermost cut may eliminate interpretationally important information at the right end of a curve and make a small taxon more difficult to detect.

The third implementation decision concerns how the *x* and *y* axes should be scaled when the MAMBAC curve is plotted. Regardless of the number and placement of cuts that are used to obtain data points for MAMBAC, one has a choice about how to display these points on a graph for interpretation. The *x* axis of this graph may be constructed using either case numbers or the actual values of the indicator response scale. The latter approach has some intuitive appeal and is typically employed when cutting scores are located according to *SD* units or intact scale values. Plotting mean differences (the *y* values on a MAMBAC graph) above *SD* units or intact scale values shows the input score at which the maximum mean difference is achieved—a value that can be informative. However, constructing an *x* axis on the basis of case numbers may be preferable when a specified number of equally spaced cutting scores are used. Particularly when input scores vary across ordered categories, cutting at every *n*th case often yields two or more adjacent cuts at equal-scoring cases. This can produce a choppy, difficult-to-interpret curve when the number of cutting scores is larger than the number of distinct values on the input indicator, yielding multiple y values for a single x value.

Consider, for example, an input indicator ranging from 0 to 7, such as a composite input indicator produced by summing seven dichotomous items. When scaling the x axis for the MAMBAC curve, one could either (a) graph points at each of seven distinct scale values, yielding a 7-point curve (eight scale values yield seven points because each point represents a difference between scores), or (b) graph points according to the case numbers that gave rise to them, resulting in a single y value for each x value (case number). Although the latter approach will artificially create the impression of equidistant cutting scores (i.e., values will be equally spaced along the x axis even when the actual cuts on the input indicator are unequally spaced), we have found that this typically yields smoother, more interpretable curves in part because it produces a far larger number of points for graphing. The potential utility of plotting a large number of points on a MAMBAC graph even when there are few distinct values on the input indicator is explored in the empirical illustrations presented in the next section.

The interpretation of MAMBAC curves is usually facilitated by using the narrowest range of y values that includes all of the mean differences obtained in the analysis. Scaling the y axis from the smallest to the largest mean difference observed will draw out the shape of the curve, be it peaked or concave, most clearly. Although one might wish to hold constant the scaling of the y axes across a full panel of MAMBAC curves to facilitate their comparison, the intuitive appeal of this strategy is at least partly undermined by the fact that the mean differences (y values) observed for each curve depend in part on the units of measurement of the output indicator as well as the validity of the input and output indicators. Thus, one can reasonably expect to observe varying mean differences across a panel of MAMBAC curves generated by analyses of different indicators, and holding the y axes constant may make it more difficult to interpret some or all of these individual curves.

However, there is one situation in which it is arguably useful to apply consistent scaling of y axes across multiple MAMBAC graphs: when comparing the results of a particular analysis in the research data to those in empirical sampling distributions for taxonic and dimensional comparison data (see chap. 8 for illustrations). The goal here is to determine whether the results for the research data are more similar to those for simulated data of one structure than another. In this context, the absolute value of

MAMBAC mean differences across parallel analyses can be quite valuable in facilitating the visual inspection and comparison of curves produced by these comparison data sets.

Empirical Illustrations

Our first series of MAMBAC analyses was performed on all eight data sets in a traditional manner by assigning variables to input–output indicator roles in all pairwise combinations and placing cuts every .25 *SD* along the input indicator, with a minimum sample size of 25 cases beyond the most extreme cut at each end of the curve. Because this yielded 12 curves for each data set, we present a single averaged curve to conserve space; the results appear in Fig. 5.6. TS#1 yielded an ambiguously concave curve with a right-end peak that did not appear clearly dimensional or taxonic, suggesting that these indicators were not sufficiently valid to reveal the optimal cutting score that distinguished taxon from complement members. As can be seen in Table 5.3, these data also produced estimates of the taxon base rate that were highly variable. TS#2 and TS#3 yielded peaked curves indicative of taxonic structure, with the locations of the peaks and the taxon base rate estimates correctly revealing the actual taxon base rates of .50 and .25, respectively. TS#4 yielded an ambiguously rising curve that could reflect either a small taxon or a latent dimension; the fairly consistent taxon base rate estimates were close to the true value of .30. DS#1, DS#2, and DS#3 each yielded the concave curves expected of dimensional structure; base rate estimates varied considerably for DS#1, whereas they were relatively consistent for DS#2 and DS#3. DS#4 yielded a curve and base rate estimates that were highly similar—and equally ambiguous—to those generated by TS#4. In summary, this series of MAMBAC analyses produced results of variable utility.

A second series of MAMBAC analyses was performed by placing 50 evenly spaced cuts along the input indicator, beginning and ending 25 cases from the extremes; the results appear in Fig. 5.7. Although these analyses generated many more points on each curve, the structural conclusions that they suggested were generally similar to those for the previous series of analyses. Perhaps the only noteworthy exception was that the distinction between TS#4 and DS#4 became clearer in these analyses, with the curve yielded by DS#4 more likely to be correctly interpreted in support of dimensional structure.

A third series of MAMBAC analyses was performed with composite input indicators in an effort to obtain still clearer results. In addition to using composite indicators for all eight data sets, analyses of TS#4 and DS#4 incorporated 10 internal replications to help reduce the distorting effect of partitioning tied scores for these ordered categorical indicators.

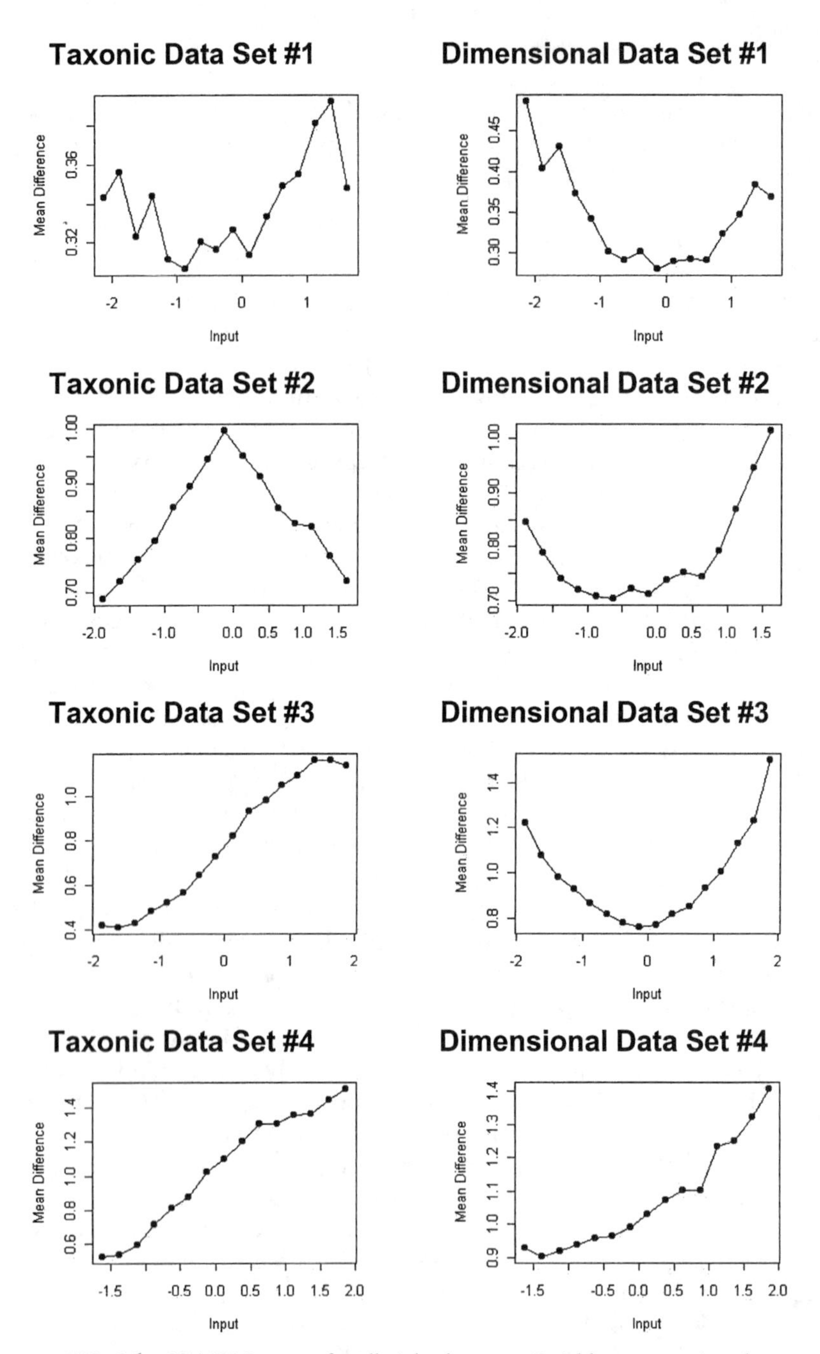

FIG. 5.6. MAMBAC curves for all eight data sets. Variables were assigned to the input–output indicator roles in all possible pairwise configurations, cuts were placed every .25 *SD* units along the input indicator (with a minimum of 25 cases beyond the most extreme cutting scores), and each panel of 12 curves was averaged for presentation.

TABLE 5.3
Taxon Base Rate Estimates for MAMBAC Analyses

	Estimates for Fig. 5.6		*Estimates for Fig. 5.7*		*Estimates for Fig. 5.8*	
Data Set	*# Estimates*	*M (SD)*	*# Estimates*	*M (SD)*	*# Estimates*	*M (SD)*
TS#1	12	.47 (.19)	12	.53 (.22)	4	.48 (.10)
TS#2	12	.49 (.05)	12	.50 (.09)	4	.40 (.08)
TS#3	12	.26 (.05)	12	.27 (.07)	4	.23 (.06)
TS#4	12	.33 (.07)	12	.25 (.06)	4	.29 (.06)
DS#1	12	.52 (.14)	12	.47 (.19)	4	.55 (.08)
DS#2	12	.49 (.03)	12	.46 (.03)	4	.48 (.01)
DS#3	12	.45 (.06)	12	.44 (.07)	4	.46 (.05)
DS#4	12	.40 (.04)	12	.35 (.02)	4	.35 (.06)

The results, shown in Fig. 5.8, were substantially easier to interpret than the curves of the earlier MAMBAC analyses. First, each of the dimensional data sets yielded concave curves, although results for DS#4 continued to appear ambiguous. Second, each of the taxonic data sets yielded a peaked curve. Results were especially impressive for TS#1, whose indicators did not clearly reveal taxonic structure in any prior MAXSLOPE or MAMBAC analyses, and for TS#4, in which a distinct peak emerged relative to the generally rising curve. In addition, the curve for TS#3 was more clearly peaked than in previous MAMBAC analyses. Thus, it appears that the use of composite input indicators may sometimes be preferable to the more traditional MAMBAC implementation approach, in which variables are typically assigned to input–output indicator roles in all pairwise combinations. As with MAXSLOPE, however, the aggregation of valid and invalid indicators could weaken results.

A final series of MAMBAC analyses was performed to illustrate the feasibility and utility of placing a large number of cuts along an input indicator that has few distinct values. The data for these analyses consisted of eight dichotomous items drawn from Scale 5 (Masculinity–Femininity) of the Minnesota Multiphasic Personality Inventory (MMPI) to distinguish the latent groups of females (n = 593) and males (n = 407). These eight items were selected because they most validly distinguished men from women and were correlated at relatively low levels within groups; for more information about the Hathaway Data Bank from which this subsample of 1,000 MMPIs was taken, see J. Ruscio and Ruscio (2000, Study 2). Because dichotomous items cannot serve as input indicators, MAMBAC was performed using composite input indicators. Figure 5.9 displays the averaged MAMBAC curves for analyses using cuts made only at intact scale values (top) and analyses using 50 evenly spaced cuts beginning and ending 25

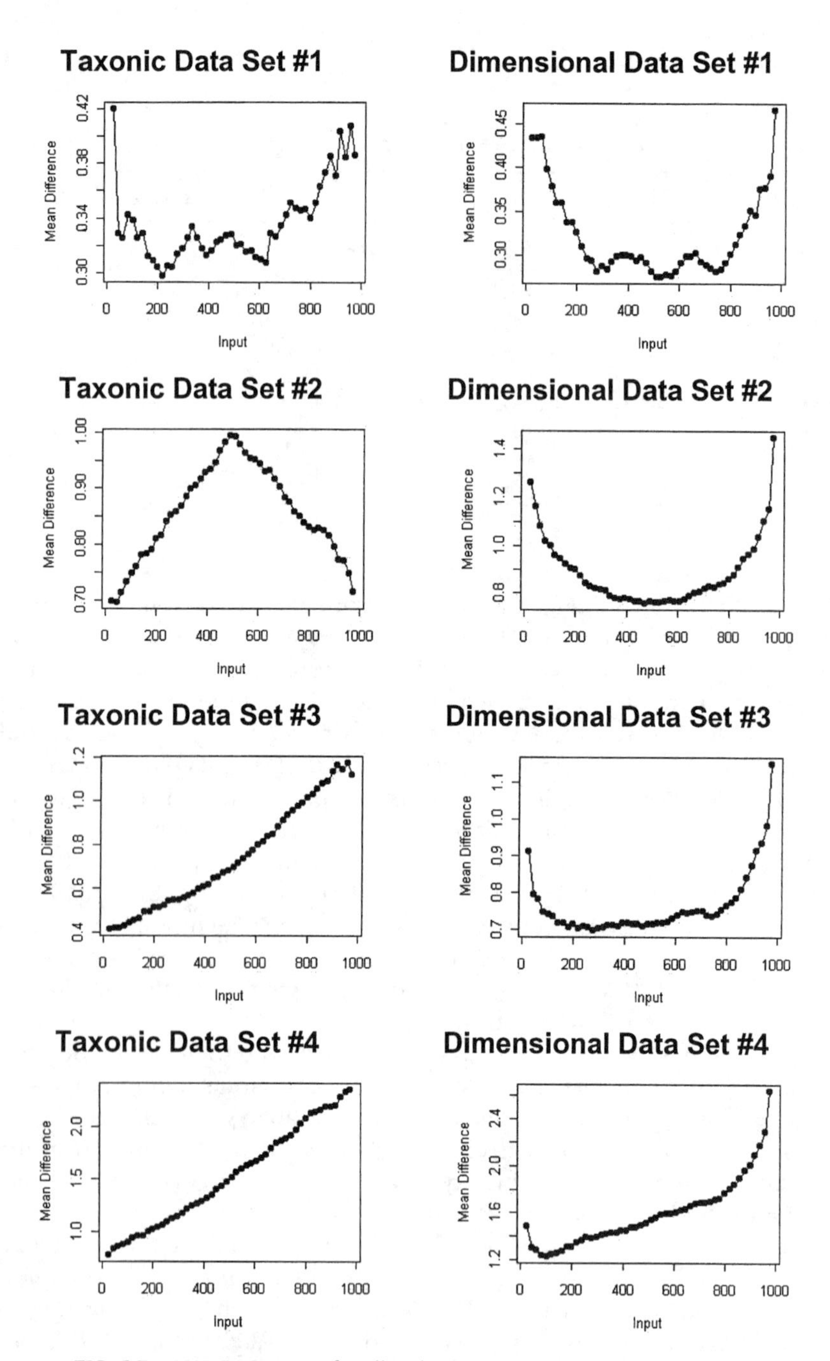

FIG. 5.7. MAMBAC curves for all eight data sets. Variables were assigned to the input–output indicator roles in all possible pairwise configurations, 50 evenly spaced cuts were used (beginning and ending 25 cases from each end of the input indicator), and each panel of 12 curves was averaged for presentation.

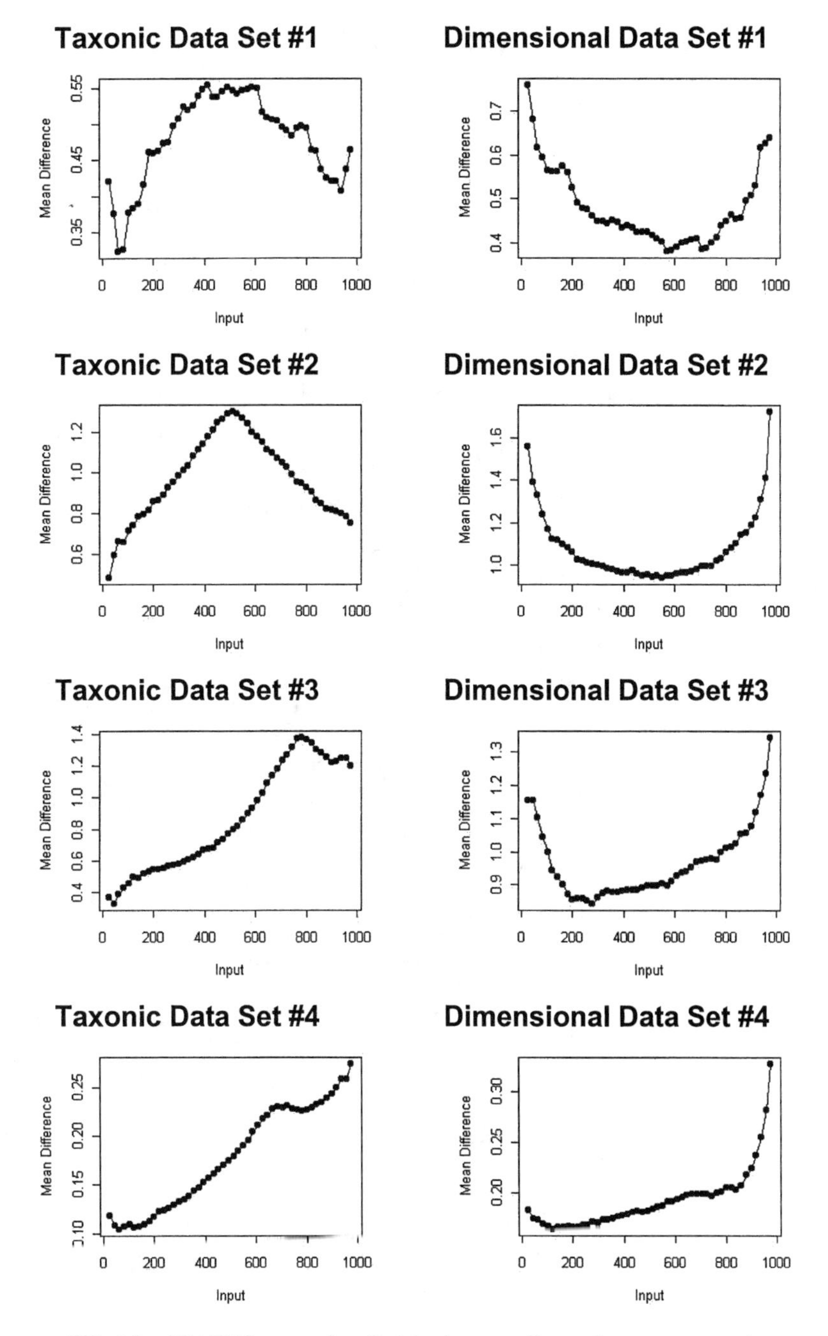

FIG. 5.8. MAMBAC curves for all eight data sets. For each curve, one variable was assigned to the role of output indicator and the input indicator consisted of the sum of the remaining three variables. Fifty evenly spaced cuts were used (beginning and ending 25 cases from each end of the input indicator), and each panel of four curves was averaged for presentation.

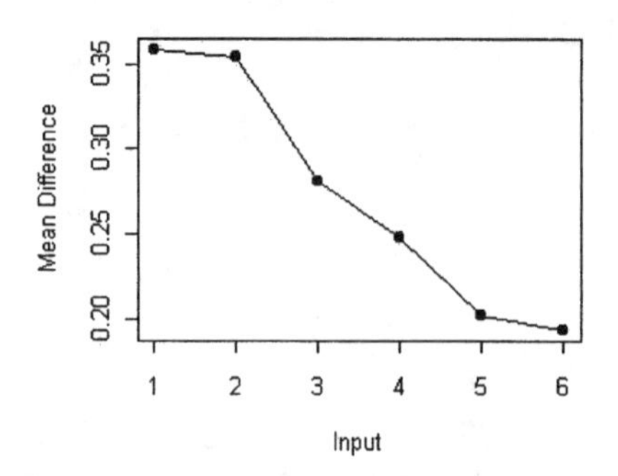

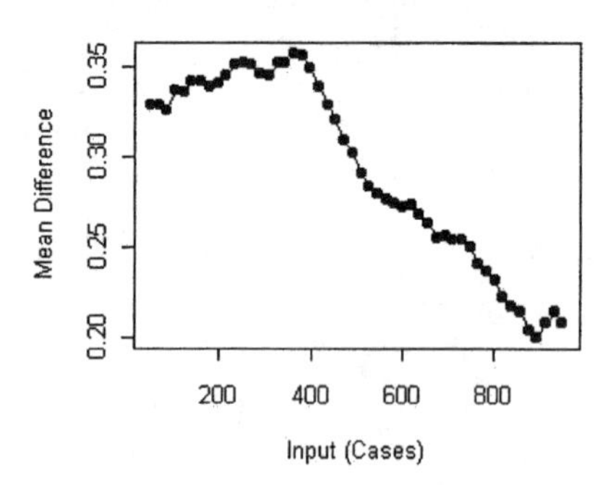

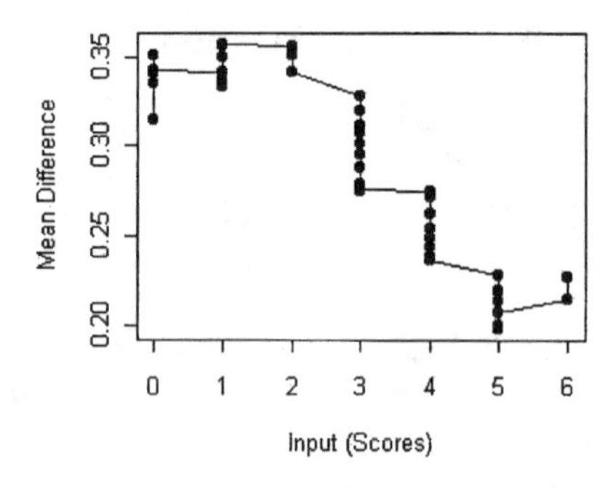

FIG. 5.9. MAMBAC curves for analyses of eight MMPI items. For each curve, one variable was assigned to the role of output indicator, and the input indicator consisted of the sum of the remaining three variables. Three different techniques were used to locate cutting scores and construct the x axis. Each panel of eight curves was averaged for presentation.

cases from the extremes of the input indicator (middle). A third MAMBAC curve (bottom) was generated using 50 evenly spaced cuts, but was plotted using intact indicator scale values—rather than case numbers—along the *x* axis. Two points can be made on the basis of these results. First, the use of (far) more cutting scores than there are distinct values on the input indicator can produce a more interpretable MAMBAC curve than the use of intact scale values alone. Moreover, in this example, the taxon base rate estimates were slightly more accurate (relative to the true taxon rate of .59) when 50 cutting scores were used (M = .60) than when intact scale values were used as cutting scores (M = .63). Second, when data points for MAMBAC have been obtained using a larger number of cutting scores than there are intact values on the input, the *x* axis may be more usefully constructed according to case numbers than intact values. The choppy curve yielded by the latter approach (reflecting a large amount of information plotted in a very crude way) was considerably more difficult to interpret than the well-defined curve yielded by the former approach.

Systematic Study of the Procedure

Meehl and Yonce (1994) provided detailed output for MAMBAC analyses performed on 700 taxonic and dimensional data sets created using 14 data configurations. Their results illustrate prototypical MAMBAC curve shapes for taxonic and dimensional data. However, it is important to note that Meehl and Yonce did not include indicator skew in any of their analyses, nor did they analyze data sets in which multiple characteristics deviated from fairly ideal values. Thus, the curve shapes presented in their monograph may reflect somewhat clearer results than researchers might reasonably expect to obtain in actual taxometric investigations. In a small-scale reanalysis of five of the Meehl–Yonce data sets, Cleland and Haslam (1996) transformed the indicators to take on positive skew and found that MAMBAC curves were robust to the influence of this skew. However, analyses of other data sets with more pronounced skew (A. M. Ruscio & Ruscio, 2002; J. Ruscio, Ruscio, & Keane, 2004) have suggested that MAMBAC curves—and parameter estimates derived from these curves—can be affected by indicator skew.

A Monte Carlo study of taxon base rate estimates (J. Ruscio, 2005), in which a large number of data parameters were fully crossed in the design, revealed biases in MAMBAC estimates under many conditions. For example, MAMBAC analyses often substantially underestimated the size of moderate taxa (P = .50) when indicators were positively skewed and substantially overestimated the size of relatively small taxa (P = .25, .10, or .05) in the presence of nontrivial within-group correlations. To some extent, positively skewed indicators were found to offset the influence of within-group correlations, but it would be a most fortunate coincidence if these two influences perfectly canceled one another in any particular analysis. The overall accuracy of MAMBAC base rate estimates was discovered to be poorer than that of MAXCOV or MAXEIG base rate estimates (see chap. 6).

A considerable amount of research is needed to better understand the conditions under which MAMBAC curve shapes are most informative and the techniques for implementing the procedure that yield the greatest utility. For example, even the large-scale Monte Carlo study of base rate estimates noted earlier implemented MAMBAC in a single way: Variables were assigned to input–output indicator roles in all pairwise combinations, and 50 evenly spaced cuts were located beginning and ending 25 cases from the extremes of each input indicator. It may not be safe to generalize the results to other techniques for implementing MAMBAC.

L-MODE

The final taxometric procedure that we discuss in this chapter is quite different from MAXSLOPE and MAMBAC, as well as from the two procedures to be described in the next chapter (MAXCOV and MAXEIG). Whereas all of these other procedures are based on the coherent cut kinetics approach to taxometrics that was outlined in chapter 3 (Meehl, 1995a), wherein the relations among indicators are examined relative to sliding cut points, L-Mode involves no such sliding cut. Instead it is based on Thurstone's (1935, 1947) belief that latent factors do not necessarily represent continuous variables and can provide useful information about categorical variables. Waller and Meehl (1998) took advantage of this observation to develop a factor-analytic taxometric procedure.

Logic of the Procedure

Traditionally, factor analysis has been used almost exclusively to explore the dimensionality of psychological constructs, usually to resolve the number of latent dimensions underlying the variation in responses to a large number of items. In contrast, the L-Mode taxometric procedure attempts to differentiate taxonic and dimensional structure. It does this by graphing the distribution of cases' estimated scores on a single latent factor calculated using Bartlett's (1937) method of factor score estimation. Like any other composite of variables that are each a valid measure of a common latent construct, such factor scores should separate a taxon and complement more validly than the individual indicators (Waller & Meehl, 1998). With enough indicators of sufficient validity, taxonic structure should yield a bimodal distribution of factor scores, whereas dimensional structure should yield a unimodal distribution of factor scores. Hence, L-Mode results should become increasingly informative with larger numbers of valid indicators (see Waller & Meehl, 1998, pp. 71–72).

Estimating the Taxon Base Rate

The L-Mode procedure provides several estimates of the taxon base rate (Waller & Meehl, 1998). The first two estimates are derived from the location of each mode in the distribution of estimated factor scores, one positive and one negative. Waller and Meehl (1998) recommended that when these two base rate estimates are close to .50, their average should be used. An analytic examination of the expected accuracy of the estimates calculated using the location of each mode led to an additional recommendation: When the base rate is less than .50, the estimate from the upper mode should be used; when the base rate is greater than .50, the estimate from the lower mode should be used. Waller and Meehl did not discuss how to determine whether the base rate is less than or greater than .50. Presumably, this decision is made on the basis of the estimates obtained from the locations of the two modes: When these are both below .50 or above .50, the taxon base rate is less than or greater than .50, respectively. However, it is possible that these estimates will diverge substantially and lie on either side of .50 (see the analysis of TS#4 in the empirical illustrations discussed shortly), in which case it is unclear which estimate to use or whether their average is to be preferred.

An additional base rate estimate can be obtained by classifying each case as a likely taxon or complement member (Waller & Meehl, 1998, p. 67). Cases whose estimated factor scores are closer (in terms of absolute distance) to the factor score beneath the upper mode of the L-Mode graph are assigned to the taxon, and cases whose estimated factor scores are closer to the factor score beneath the lower mode are assigned to the complement. The base rate is then estimated as the proportion of cases classified as taxon members. Waller and Meehl provided no guidelines about when this estimate is to be used.

Implementation Decisions

Each L-Mode plot is graphed with factor scores along the x axis and relative frequencies along the y axis. Because L-Mode automatically includes all supplied indicators in the calculation of one factor-score frequency distribution, there are fewer decisions involved when implementing the procedure than for MAXSLOPE or MAMBAC. However, the question of which indicators to include in the analysis must be addressed. Just as one can perform MAXSLOPE or MAMBAC using indicators in all possible pairwise configurations, one can perform a series of L-Mode analyses using all possible combinations of three or more indicators (because at least three indicators are required to uniquely identify factor loadings). Just as one can perform MAXSLOPE or MAMBAC using all available indicators in the calculations for each curve (e.g., with one variable serving as the output indicator and the sum of the remaining variables serving as the input indicator), one can perform a single L-Mode analysis using all available indicators.

Another issue arises when L-Mode is performed with substantially different taxon and complement base rates. By default, L-Mode programs search for latent modes by moving outward in each direction from a factor score of $x = 0$. One consequence of this is that if a discernible mode exists, but its peak value does not reach the height of the curve at $x = 0$, the program will fail to locate this mode correctly. For example, suppose that a small taxon produces a mode on the right side of the graph, whereas the larger complement produces a much taller mode on the left side of the graph. The region of the distribution corresponding to complement members may remain taller at $x = 0$ than the tallest point in the region corresponding to taxon members. In this situation, L-Mode program would correctly locate the lower mode, but would incorrectly locate the upper mode very near $x = 0$. The solution to this problem is to manually set the starting position of the search to a value in the visible trough between $x = 0$ and the score beneath the mode whose correct position was not initially detected. It should be noted, however, that this decision requires some subjective judgment. That is, because it is sometimes difficult to determine whether a bump in the curve reflects a true mode or is a product of the lumpiness of chance, it must be subjectively determined whether and where to manually start the search for an apparent mode.

Empirical Illustrations

Results of the L-Mode analyses of all eight data sets appear in Fig. 5.10; base rate estimates appear in Table 5.4. Each of the dimensional data sets yielded a unimodal distribution of factor scores; even the curve produced by DS#4, although lumpy, was not clearly bimodal. Base rate estimates generated from the location of the lower and upper modes were particularly inconsistent because of the absence of two distinct modes. This caused the search for both modes to terminate close to the starting point of $x = 0$, leading the lower mode to estimate an extremely small taxon and the upper mode to estimate an extremely large taxon.

Taxonic Data Set #1

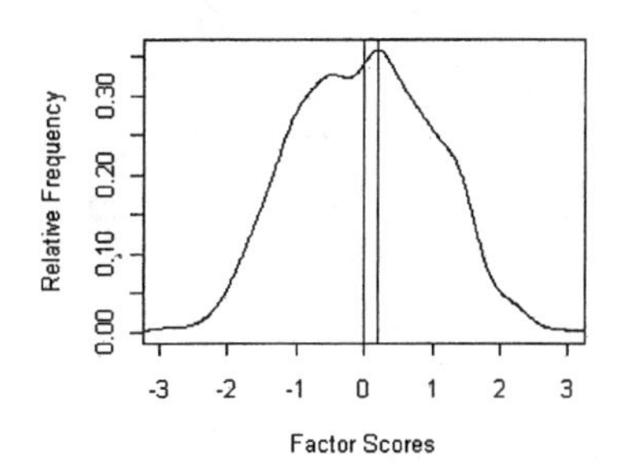

Dimensional Data Set #1

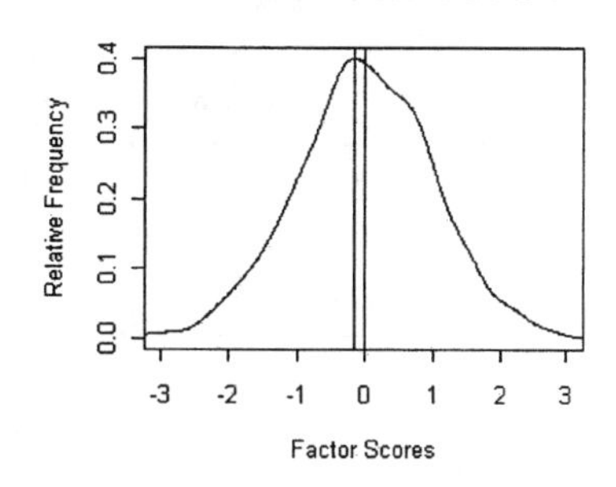

Taxonic Data Set #2

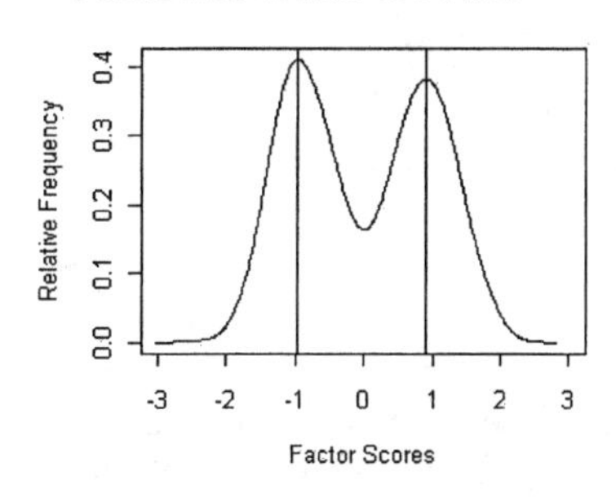

Dimensional Data Set #2

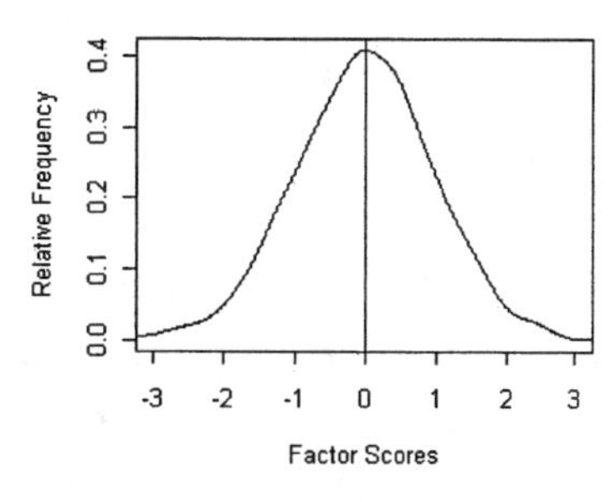

Taxonic Data Set #3

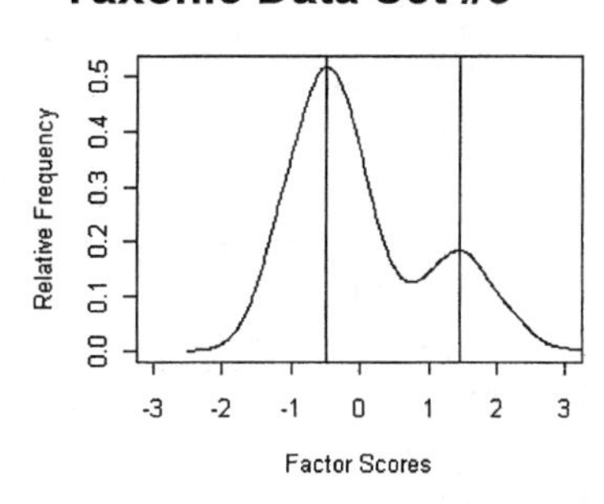

Dimensional Data Set #3

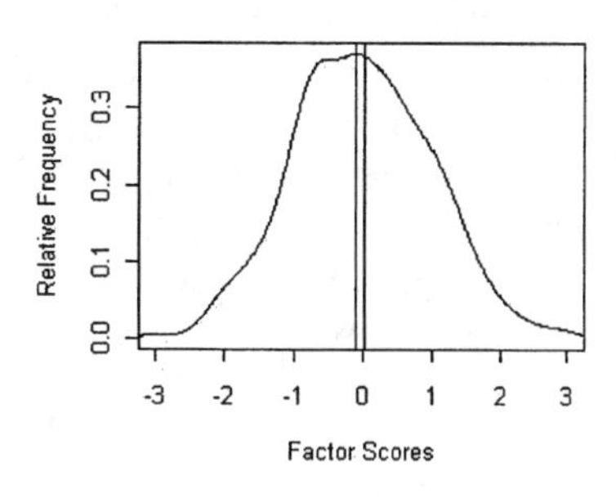

Taxonic Data Set #4

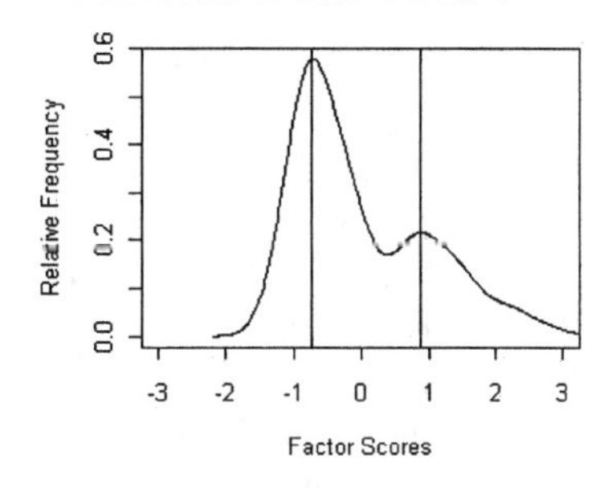

Dimensional Data Set #4

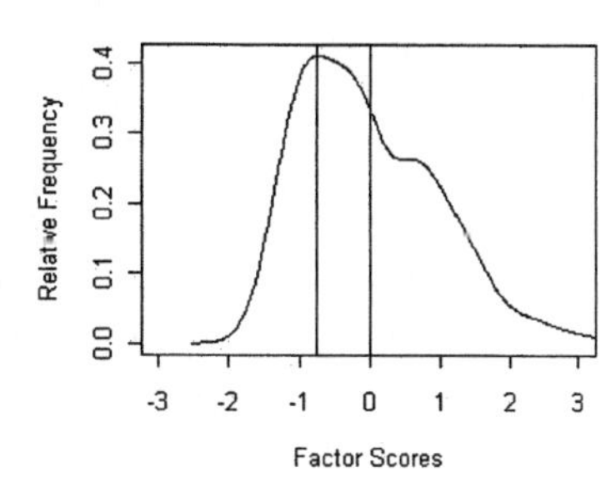

FIG. 5.10. L-Mode graphs for all eight data sets. For the third and fourth taxonic data sets, the search for the upper mode was set to begin at $x = .5$, near the visible trough in the distribution of factor scores.

TABLE 5.4
Taxon Base Rate Estimates for L-Mode Analyses

	Base Rate Estimation Technique			
Data Set	*Location of Lower Mode*	*Location of Upper Mode*	*Average of* P_{lower}, P_{upper}	*Empirical Classification*
TS#1	.00	.96	.48	.48
TS#2	.49	.55	.52	.50
TS#3	.20	.32	.26	.27
TS#4	.35	.57	.46	.37
DS#1	.03	1.00	.51	.53
DS#2	.00	1.00	.50	1.00
DS#3	.01	1.00	.51	.50
DS#4	.36	1.00	.68	.56

Like its dimensional counterpart, TS#1 yielded a lumpy, but essentially unimodal L-Mode graph and discrepant estimates of the taxon base rate, suggesting that its indicators were not sufficiently valid to detect the underlying groups using L-Mode. TS#2, TS#3, and TS#4 each yielded bimodal distributions and far more consistent estimates of the taxon base rate, although base rate estimates were not especially accurate for TS#4. As recommended in the preceding section, the starting point for the search for the upper mode was set manually (to $x = .5$) for TS#3 and TS#4 because the height of the distribution at $x = 0$ was greater than the height at the upper mode. The multivariate power of L-Mode can be seen by comparing the manifest distributions of the indicators in TS#2, TS#3, and TS#4 (shown in Fig. 3.1), with the corresponding distributions of factor scores estimated using L-Mode. Although the indicators were unimodally distributed, the L-Mode analyses produced bimodal curves that clearly distinguished the taxon and complement.

Systematic Study of the Procedure

The shape of L-Mode graphs has yet to be studied in Monte Carlo investigations, so relatively little is known about the conditions under which the procedure can distinguish latent structures. Waller and Meehl (1998) showed that L-Mode can be used to estimate taxon base rates and assign cases to groups quite accurately under the 14 data configurations examined by Meehl and Yonce (1994, 1996). It would be useful to know how L-Mode performs across a broader range of more realistic conditions.

CONCLUSIONS

The present chapter reviewed the logic and implementation of three taxometric procedures: MAXSLOPE, MAMBAC, and L-Mode. Each procedure was illustrated using the eight simulated taxonic and dimensional data sets first introduced in chapter 3, and available Monte Carlo findings for each procedure were reviewed. We reserve general conclusions until the end of the next chapter, following our description of the two remaining taxometric procedures (MAXCOV and MAXEIG) whose application predominates in the taxometric literature.

CHAPTER SIX

Taxometric Procedures II: MAXCOV and MAXEIG

The present chapter rounds out the review of taxometric procedures begun in the previous chapter. Here we introduce two procedures—MAXCOV (*MAX*imum *COV*ariance; Meehl, 1973; Meehl & Yonce, 1996) and MAXEIG (*MAX*imum *EIG*envalue; Waller & Meehl, 1998)—that are conceptually and mathematically similar. A significant number of implementation decisions must be carefully considered when applying either of these procedures. Therefore, we review the available options, highlight their implications, and describe what is known about the performance of each procedure under various data conditions. We also present the results of MAXCOV and MAXEIG analyses performed on the eight data sets introduced in chapter 3 and analyzed in chapter 5. Finally, we end by drawing general conclusions about all five of the taxometric procedures described in chapters 5 and 6 and suggesting how some methodological innovations can be used to best advantage in a taxometric investigation.

Among the family of analytic procedures that constitutes the taxometric method, MAXCOV has been used far more extensively than any other, performed in more than 80% of the taxometric investigations reviewed by Haslam and Kim (2002) and 77% of the published studies reviewed in chapter 10. By contrast, the next most popular procedure (MAMBAC) was performed in about half (58%) of the published studies. MAXEIG, a multivariate extension of MAXCOV, has been employed in 11 (19%) of the published studies, 10 of which have appeared in print since 2000. Although MAXCOV and MAXEIG have traditionally been implemented in different ways, they are based on the same underlying logic. One reason that we have grouped these procedures together in a single chapter is to highlight their conceptual similarities.

THE GENERAL COVARIANCE MIXTURE THEOREM

The way in which MAXCOV and MAXEIG distinguish taxonic from dimensional structure can be expressed by an equation that is fundamental to taxometrics, the General Covariance Mixture Theorem (GCMT; Meehl, 1973, 1995a):

$$\text{cov}(xy) = P\,\text{cov}_t(xy) + Q\,\text{cov}_c(xy) + PQD_xD_y \tag{6.1}$$

This algebraic identity expresses the covariance between two indicators x and y as a function of: (a) the covariance within the taxon—$\text{cov}_t(xy)$, (b) the covariance within the complement—$\text{cov}_c(xy)$, and (c) the validity with which each indicator separates the taxon and complement—D_x and D_y, representing the (unstandardized) mean differences between the taxon and complement on indicators x and y. Each term is weighted by the base rate(s) of the relevant class(es): the taxon base rate P and the complement base rate $Q = 1 - P$. If the indicators do not covary within these groups (see Meehl, 1995b, for a more complex technique that does not make this assumption), the first two terms can be dropped, and Eq. 6.1 simplifies to:

$$\text{cov}(xy) = PQD_xD_y \tag{6.2}$$

In other words, indicator covariance is a function of the taxon and complement base rates and the validity of the two indicators x and y.

MAXCOV

Logic of the Procedure

The MAXCOV procedure is made possible by the fact that the GCMT is applicable not only within a full sample of data, but also within any subset of cases that might be drawn from this sample (Meehl, 1973, 1995a; Meehl & Yonce, 1996). Equation 6.2 shows that indicator covariance should vary as a function of the taxon and complement base rates. Because indicator validities are assumed not to vary across subsamples, they can be replaced with a constant, further simplifying the GCMT:

$$\text{cov}_i(xy) = p_i q_i K, \tag{6.3}$$

where K represents the product of the indicator validities and p_i and q_i represent the taxon and complement base rates within a subsample.

Therefore, one can test latent structure by examining the covariance of two indicators within a series of subsamples, each likely to contain a different proportion of taxon and complement members. In subsamples consisting entirely of taxon members, $p_i = 1$, $q_i = 0$, $p_i q_i = 0$, and $\text{cov}_i(xy) = 0$. Likewise, in subsamples consisting entirely of complement members, $p_i = 0$, $q_i = 1$, $p_i q_i = 0$, and again $\text{cov}_i(xy) = 0$. However, in subsamples containing a mixture of taxon and complement members, $p_i > 0$, $q_i > 0$, $p_i q_i > 0$, and $\text{cov}_i(xy) > 0$. Covariance will reach a maximum value in a subsample consisting of an equal mixture of taxon and complement members, as $p_i = q_i = .5$, $p_i q_i = .25$, and $\text{cov}_{\text{max}}(xy) = .25K$. Rearranging the terms in the latter formula, one can estimate the product of the indicator validities to be $K = 4\text{cov}_{\text{max}}(xy)$.

It may now be apparent that the formalism expressed in Eqs. 6.1–6.3 conveys the same expectations that underlie a MAXSLOPE analysis in the presence of taxonic structure. Within regions containing relatively homogeneous subsamples of taxon or complement members, the slope of the local regression curve should be flat; this is analogous to zero within-group covariance. However, within regions containing a mixture of taxon and complement members, the slope of the local regression curve should be positive, reaching its maximum value near the point where the groups most evenly intermingle.

However, when latent structure is dimensional, the covariance between indicators should remain relatively constant across subsamples, rather than systematically rising to a maximal value. Envision, for example, a scatterplot formed by crossing two indicators of a latent dimension. If one were to divide this scatterplot using a series of vertical slices, the regression slopes across all slices would be essentially the same. This, too, is analogous to expectations in a MAXSLOPE analysis: Dimensional structure should produce a local regression curve whose full span is linear.

MAXSLOPE examines the local regression curve in a scatterplot of the association between two indicators, whereas MAXCOV examines the conditional covariance between two indicators, traditionally within nonoverlapping subsamples (or *intervals*) that have been ordered along a third indicator. This third indicator is referred to as the *input indicator* because its rank ordering of cases leads to the intervals that form the x axis of a MAXCOV graph. The other two indicators are referred to as the *output indicators* because they are used to calculate the data points that are plotted as y values for each successive interval.

To illustrate how MAXCOV works, suppose we wished to test the latent structure of human biological sex using the indicators of height, voice pitch, and nonverbal sensitivity (J. Ruscio & Ruscio, 2002). Within a subsample consisting exclusively of women (arbitrarily designated here as the taxon), two of these indicators—say voice pitch and nonverbal sensitivity—should be negligibly correlated; in other words, women with higher pitched voices should be no more sensitive to nonverbal cues than women with lower pitched voices. The same negligible correlation would be expected within a group of men (the complement). Thus, the presumption of negligible within-group correlations is satisfied. However, in a mixed-sex sample, women would tend to score higher on both voice pitch and nonverbal sensitivity than would men, causing the indicators to covary. Without infallible knowledge of group membership, how can pure and mixed groups be obtained to test for this difference in covariance? Here is where the third input indicator—in this case, height—comes into play. By forming intervals according to height, one can create subsamples that vary in their group composition. Intervals containing the shortest individuals will consist mostly of women, and the covariance between voice pitch and nonverbal sensitivity in these subsamples will approach zero. Likewise, intervals containing the tallest individuals will consist mostly of men, and the covariance in these subsamples will also approach zero. However, intervals containing individuals of intermediate heights will include a mixture of women and men, causing voice pitch and nonverbal sensitivity to covary. This covariance will peak in the interval comprising the most equal proportions of women and men. Thus, indicator covariances vary in a systematic, predictable way when the latent structure of a construct is taxonic.

When latent structure is dimensional, however, there is no reason for covariances to differ across ordered subsamples. In the absence of taxa, indicators will covary at similar levels within successive intervals, although at a level lower than in the total sample due to the predictable influence of range restriction within each subsample. Given appropriate data (see chap. 4), the absence of a clear covariance peak is taken as evidence against taxonic structure and in favor of dimensional structure.

Estimating the Taxon Base Rate

Because the location of a MAXCOV peak suggests the interval in which taxon and complement members are mixed in approximately equal proportions, it also suggests the location of an optimal cutting score (or *hitmax* cut) on the input indicator to separate the two groups. Likewise, the GCMT can be used to estimate the proportion of taxon members within each interval following a procedure introduced by Meehl (1973) and presented in detail in Appendix B.

Class Assignment and Estimation of Other Latent Parameters

Having estimated the proportion of taxon members in the sample, as well as within each input interval, one can then estimate the valid positive (*VP*) and false positive (*FP*) rates achieved by the hitmax cut on each input indicator. The *VP* rate represents the proportion of individuals scoring above the hitmax cut that belong to the taxon, whereas the *FP* rate represents the proportion of individuals scoring above the hitmax cut that belong to the complement. Combining this information with the overall taxon base rate estimate, one can then use Bayes' Theorem to assign each case to the taxon or complement class. Appendix B presents the step-by-step procedure for estimating *VP* and *FP* rates and assigning cases to groups.

Once all cases are assigned, one can then estimate the distribution of taxon and complement members' scores on each indicator; summary statistics such as the *M*, *SD*, skew, and kurtosis of the taxon and complement classes on each indicator; the indicator correlations within each group; and the validity of each indicator. Validity can be expressed either as the raw mean difference between groups (e.g., $D = M_t - M_c$) or, in the more familiar metric of Cohen's *d*, the standardized mean difference (see Eq. 4.1).

Implementation Decisions

The first implementation decision for MAXCOV concerns how variables will be assigned to the required input–output indicator roles. The choices here are similar to those available for MAXSLOPE and MAMBAC. One can form all possible triplets of indicators, allowing each variable in the triplet to serve as the input indicator for one MAXCOV graph, yielding $k(k - 1)(k - 2)/2$ curves for k indicators. Alternatively, one can remove two variables to serve as output indicators and sum the remaining variables to form a single composite input indicator, yielding $k(k - 1)/2$ curves. The latter technique allows MAXCOV to be conducted with indicators that vary across too few values to reliably rank order cases for subsample formation along the input indicator. For example, dichotomous items cannot serve individually as input indicators because this would yield only two intervals on the MAXCOV graph that could not reveal the full shape of the curve. However, one can remove two such items to serve as output indicators and use the sum of several other dichotomous items as a composite input indicator (Gangestad & Snyder, 1985; J. Ruscio, 2000). The use of composite input indicators not only accommodates ordered categorical data, but also can sometimes surpass the ability of individual input indicators to distinguish taxonic from dimensional structure.

The second implementation decision concerns how cases should be divided into intervals along the input indicator. Such intervals have been constructed on the basis of *SD* units (e.g., divide the sample every .25 *SD* along the input indicator, which is usually standardized for this purpose; Meehl & Yonce, 1996), intact scale values (e.g., divide the sample according to particular scores or ranges of scores on the input indicator), or fixed-size subsamples (e.g., divide the sample into deciles). If the latter method is used, internal replications can offset the obfuscating effect of drawing subsample divisions between equal-scoring cases, as was the case for MAMBAC analyses.

A third implementation decision concerns how the *x* axis should be constructed for each MAXCOV graph. When subsamples are formed along the input indicator according to *SD* units, the *x* axis is typically constructed such that points lie above the *SD* values (e.g., values of –1.50, –1.25, –1.00, . . . , 1.00, 1.25, 1.50 when .25 *SD* units are used, with the outermost *x* values determined by the most extreme values at which a sufficient number of cases exist to form subsamples). Another possibility is to compute the mean score of each subsample on the input indicator and use this as its *x* value, although in this case the results would probably differ little from the preceding approach. When subsamples are formed using intact scale values (e.g., values of 1, 2, 3, . . . along an input indicator that consists exclusively of positive integer values), the *x* axis typically consists of these scale values.

However, when subsamples are formed on the basis of equal numbers of cases (e.g., decile intervals), either ordinal numbers or subsample means can be used as the *x* values for the series of data points. For example, in a MAXCOV analysis using 10 intervals, one could construct the *x* axis using the numbers 1–10, plotting data points equidistant from one another along the *x* axis. Alternatively, one could construct the *x* axis such that the *x* value of each data point is the subsample mean on the input indicator. We view the latter strategy as preferable because it reveals the actual differences between data points along the *x* axis. To consider why this might be important, consider a MAXCOV analysis of indicators possessing high positive skew. The mean scores of low-scoring subsamples will differ relatively little from one another, whereas the mean scores of high-scoring subsamples will differ to a much greater degree. If these data points were plotted at equal distances along the *x* axis, the relative flatness toward the left side of the curve—and the relative steepness toward the right side of the curve—would be exaggerated. In contrast, plotting these data points according to subsample means along the input indicator will yield a less steeply rising curve because points toward the left side of the curve will lie close together and points toward the right side of the curve will be spread farther apart. In our experience, this graphing approach yields MAXCOV curves that more clearly differentiate results for a small taxon from results for skewed indicators of a latent dimension (J. Ruscio, Ruscio, & Keane, 2004). It is worth noting that the method used to construct the *x* axis for MAXCOV curves will not influence estimates of latent parameters such as the taxon base rate because these are based only on the covariance values of each data point and the number of cases within each subsample.

One final point is particularly important: Because the *x* axis can be constructed in different ways, it should always be clearly labeled or described so that readers can identify the construction approach that was used, evaluate its appropriateness, and inspect the resulting curves with an informed eye.

A fourth implementation decision concerns how the *y* axis should be scaled for each MAXCOV graph. Whereas the *y* range of MAMBAC graphs is often set at the minimum and maximum observed mean differences to better distinguish peaked from concave curves, a *y* range in MAXCOV graphs that is too narrow or too wide could artificially flatten a genuine taxonic peak or create the appearance of a spurious peak. Thus, it is important to override the default settings of computerized graphing programs—which typically scale graphs according to the observed minimum and maximum values—and instead consider what would constitute an appropriate range of *y* values for the graph.

No consensus of opinion has yet been reached on the most appropriate technique for scaling the *y* axis of MAXCOV graphs. The primary point of concern can be summarized as follows: If too narrow a range of *y* values is used, the influence of normal sampling error may be amplified such that results for even dimensional data appear to contain peaks. However, if too wide a range of *y* values is used, genuine taxonic peaks may appear to be flattened. Unfortunately, what constitutes "too narrow" or "too wide" a range of *y* values is unknown and almost certainly varies from study to study due to a number of data properties that influence the covariances. For example, more valid indicators yield taller taxonic peaks, and within-group correlations can elevate *y* values across an entire MAXCOV graph or within selected regions of this graph (Meehl, 1995b). The influence of indicator validity and within-group correlations is discussed further next and illustrated in chapter 8.

All computerized graphing programs have default procedures to scale *y* axes, but these procedures often involve little more than beginning and ending at *y* values just beyond the most extreme observed values. It is possible that such program defaults will tend to produce graphs with too narrow a range of *y* values. Holding the *y* scale constant across a full panel of curves may alleviate this problem, although it may be unreasonable to expect even genuine taxonic peaks to achieve similar heights across graphs due to differences in the metrics and validities of the indicators.

Our MAXCOV program employs an algorithm to scale the *y* axis that, under some circumstances, can help to distinguish taxonic peaks from nonpeaked dimensional curves. The rationale underlying this algorithm is based on an admittedly idealized conception of taxonic and dimensional MAXCOV graphs. Specifically, taxonic data are expected to yield curves that begin and end near the *x* axis (i.e., covariance ≈ 0) and rise to relatively large values in between, particularly at the peak; in contrast, dimensional data are expected to yield curves that vary little around a fairly constant covariance (Meehl & Yonce, 1996). When these expectations are met, one can expect a relatively small ratio of the *M* to the *SD* of all covariances for taxonic data and a relatively large ratio for dimensional data. That is, dimensional data produce fairly constant (i.e., low *SD*) covariances relative to the more variable (i.e., high *SD*) covariances for taxonic data. This observation led us to develop the following algorithm for determining the upper limit *y* value for a MAXCOV graph:

$$y_{\text{upper}} = \frac{y_M}{y_{SD}} \times \frac{y_{\max}}{15}, \tag{6.4}$$

where the *y* in y_M, y_{SD}, and $y_{\max}$ refers to covariance values on the MAXCOV curve. The lower limit *y* value is set to 0 unless negative values are observed (a rare occurrence, given that indicators must be positively correlated in the full sample to perform meaningful taxometric analyses); in such cases, the lowest observed *y* value defines the lower end of the axis.

For dimensional data that meet the previously mentioned expectations, this algorithm accentuates the flatness of the curve by using a large upper limit *y* value. For taxonic data that meet the previously mentioned expectations, the peaked nature of the curve is accentuated by using an upper limit *y* value that is only slightly larger than the peak. This algorithm was used to scale the *y* axes for all MAXCOV (and MAXEIG) graphs presented in this book, and graphs presented throughout the present chapter illustrate the potential utility of this algorithm.

There are several circumstances, however, in which this algorithm may prove less useful. For example, positive indicator skew (in the full sample for dimensional data, within groups for taxonic data) tends to produce rising curves regardless of latent structure, causing the *M*:*SD* ratio of the *y* values to be small even for dimensional data. Likewise, as illustrated in chapter 8, low indicator validity or substantial within-group correlations can flatten genuinely taxonic curves, causing the *M*:*SD* ratio of the *y* values to be large for such taxonic data. Thus, our algorithm is by no means a panacea for the challenge of scaling the *y* axes of MAXCOV curves, and we offer it only as a stopgap measure. More generally, we encourage researchers to consider the appropriateness of the *y* scaling employed in their own graphs and those of other taxometric studies. Although no optimal strategy has been demonstrated, researchers should be aware of the relevance of scaling for interpretation and be able to recognize choices that may be especially poor.

Two final notes can be offered with regard to scaling the *y* axes of MAXCOV graphs. First, it can be helpful to apply a common *y* scale to all graphs included in a panel of MAXCOV analyses for a full set of indicators. Each graph can be scaled from zero (or a constant negative value if negative covariances are observed in some plots) to the maximum covariance observed across the entire series of curves. This has the advantage of preventing the differential flattening or exaggeration of peaks in individual curves and may also facilitate comparison and interpretation of the full panel of curves. If this approach is taken, it may be helpful to standardize each indicator prior to analysis to remove any differences in metric that might yield meaningful differences in covariance values. For example, the covariance of two output indicators whose values range along 7-point Likert scales can be expected to be much lower than that of two indicators whose values range along percentile scales, making a common *y* scale across their graphs inappropriate. It is important to note that standardizing the indicators equates their variance but does not equate their validities (which also influence the covariance values), so one cannot expect all peaks for taxonic graphs to reach the same absolute height even after standardization. Second, when comparing MAXCOV curves for one's research data to those in empirical sampling distributions for simulated taxonic and dimensional comparison data, the particular *y* scale that is used may be less important than maintaining a consistent scale across these curves.

Empirical Illustrations

Our first series of MAXCOV analyses was performed on all eight data sets in a traditional manner by assigning variables to input–output indicator roles in all possible triplets and dividing cases into subsamples by placing cuts every .25 *SD* along the input indicator, with a minimum interval size of 25 cases (Meehl & Yonce, 1996). Because this yielded 12 curves for every data set, we present a single averaged curve for each data set to conserve space; these results appear in Fig. 6.1. Estimates of the taxon base rate calculated for all MAXCOV curves are summarized in Table 6.1. TS#1 yielded an ambiguous MAXCOV curve and highly variable estimates of the taxon base rate, TS#2 and TS#3 yielded clearly peaked curves and accurate, consistent base rate estimates, and TS#4 yielded a curve that was discernibly peaked as well as base rate estimates that were reasonably accu-

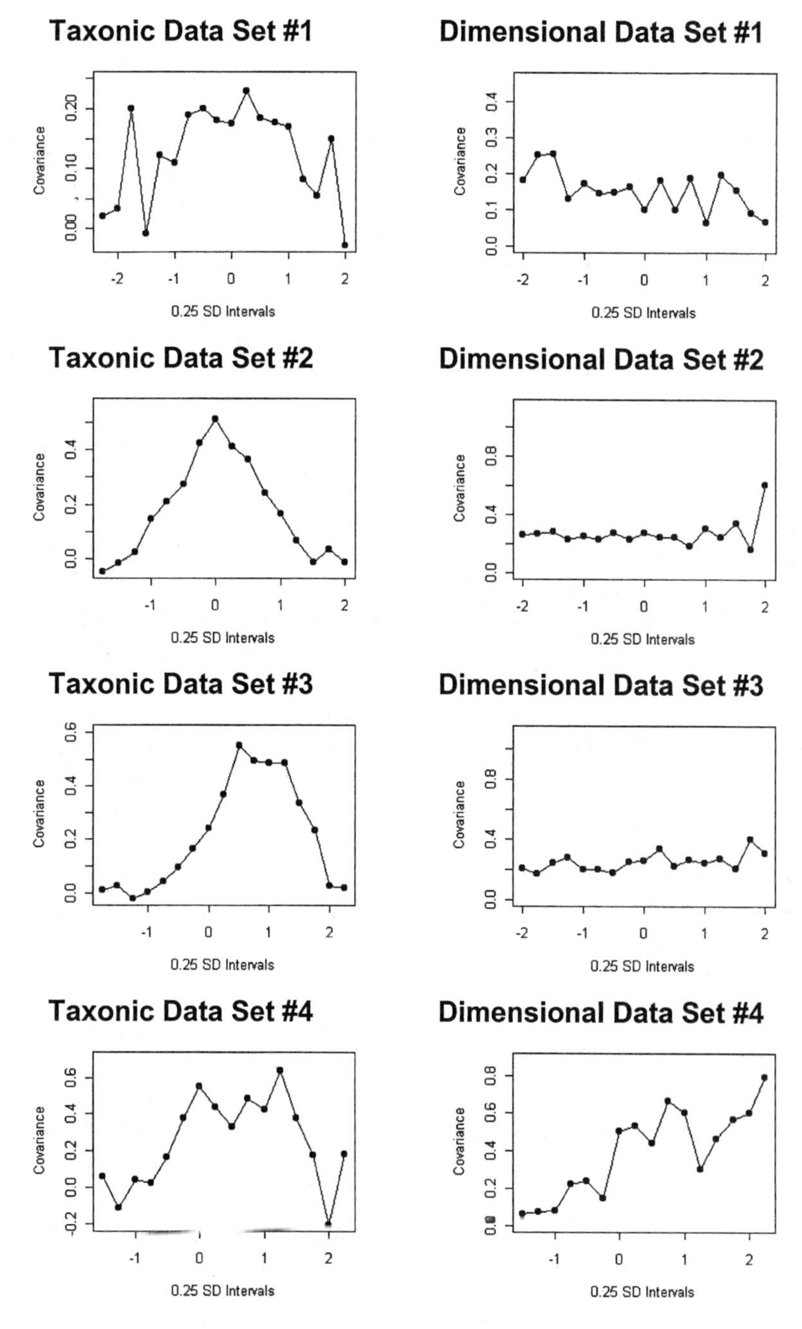

FIG. 6.1. MAXCOV curves for all eight data sets. Variables were assigned to the input–output indicator roles in all possible configurations of triplets, intervals were formed at .25 *SD* units along the input indicator (with a minimum of 25 cases beyond the most extreme cutting scores), and each panel of 12 curves was averaged for presentation.

TABLE 6.1
Taxon Base Rate Estimates for MAXCOV Analyses

Data Set	*Estimates for Fig. 6.1*		*Estimates for Fig. 6.2*		*Estimates for Fig. 6.3*	
	# Estimates	*M (SD)*	*# Estimates*	*M (SD)*	*# Estimates*	*M (SD)*
TS#1	12	.59 (.25)	12	.51 (.15)	4	.48 (.10)
TS#2	12	.48 (.03)	12	.49 (.03)	4	.49 (.02)
TS#3	12	.25 (.05)	12	.26 (.04)	4	.27 (.02)
TS#4	12	.25 (.10)	12	.34 (.04)	4	.26 (.10)
DS#1	12	.72 (.30)	12	.56 (.19)	4	.59 (.22)
DS#2	12	.19 (.19)	12	.33 (.21)	4	.25 (.17)
DS#3	12	.32 (.24)	12	.40 (.16)	4	.48 (.15)
DS#4	12	.18 (.09)	12	.18 (.05)	4	.16 (.08)

rate. DS#1, DS#2, and DS#3 each yielded relatively flat curves and variable estimates of the taxon base rate. In contrast, DS#4 yielded an ambiguously rising curve that—when combined with the consistently low taxon base rate estimates obtained for this data set—might be mistaken for evidence of a small taxon.

A second series of MAXCOV analyses was performed by dividing cases into intervals using deciles rather than *SD* units; the results appear in Fig. 6.2. On the whole, these curves were more clearly peaked for taxonic data sets and more clearly flat for dimensional data sets than in the first analysis series, with one noteworthy exception: DS#4 again yielded a rising curve and consistently low base rate estimates that could be mistaken for evidence of a small taxon.

A final series of MAXCOV analyses was performed using composite input indicators in an attempt to better distinguish taxonic and dimensional structures. Intervals along the composite input indicator were once again formed using deciles. This technique yielded six curves per data set; the averaged curves are shown in Fig. 6.3. In most cases, these results were even easier to interpret than those of either of the prior series of MAXCOV analyses, suggesting that composite input indicators may at times yield more interpretable curves than the more traditional MAXCOV implementation approach of assigning variables to input–output indicator roles in all possible triplets. However, DS#4 continued to produce curve shapes and taxon base rate estimates suggestive of a small taxon. Additionally, TS#1 yielded a peak divided into two apparent peaks, perhaps due to the influence of sampling error in the intermediate interval. Unfortunately, judging whether a curve is peaked remains a subjective process; there are no accepted guidelines for determining whether the results for TS#1 should be interpreted as peaked, consistent with taxonic structure, or sufficiently erratic (e.g., multiple peaks and valleys) that they are more con-

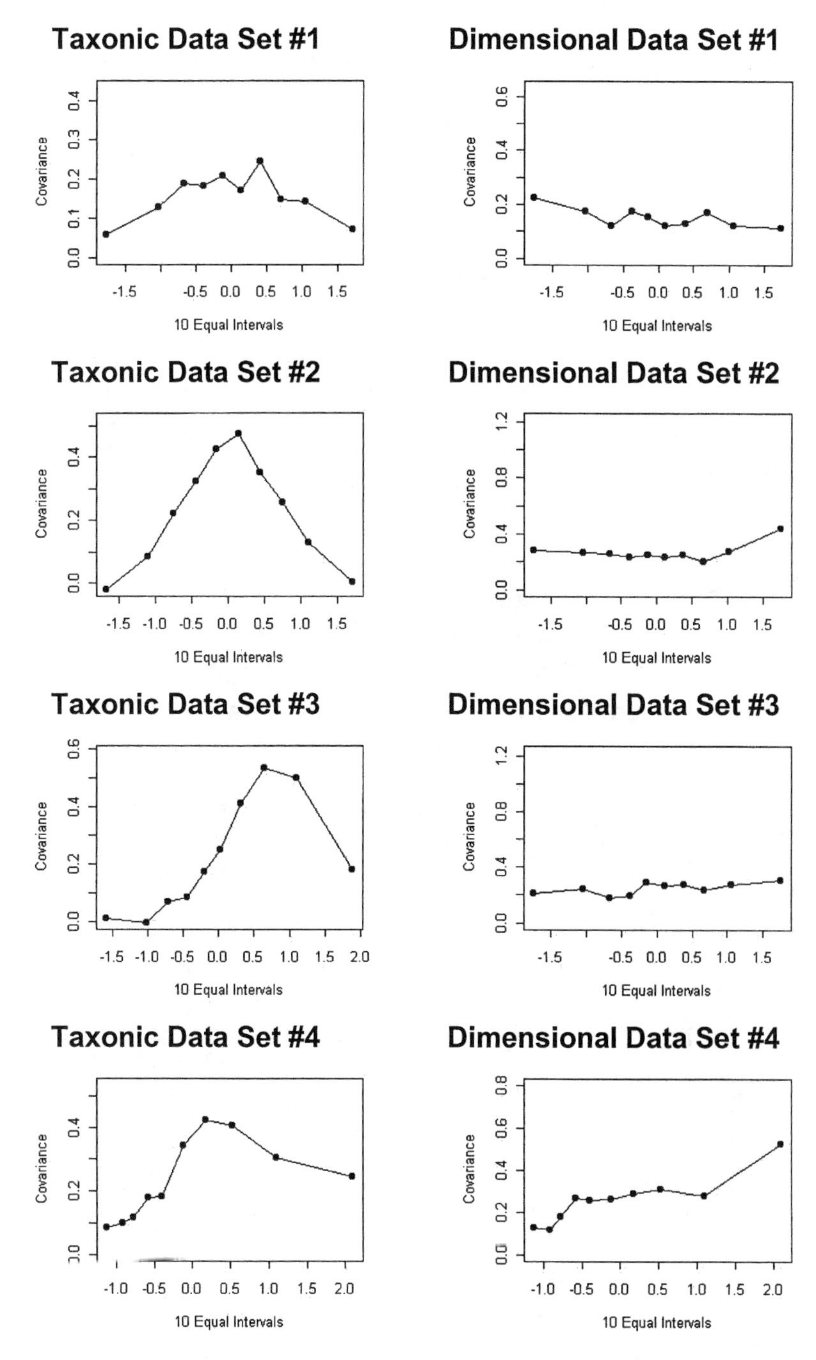

FIG. 6.2. MAXCOV curves for all eight data sets. Variables were assigned to the input–output indicator roles in all possible configurations of triplets, 10 equal-sized intervals (i.e., deciles) were formed along the input indicator, and each panel of 12 curves was averaged for presentation.

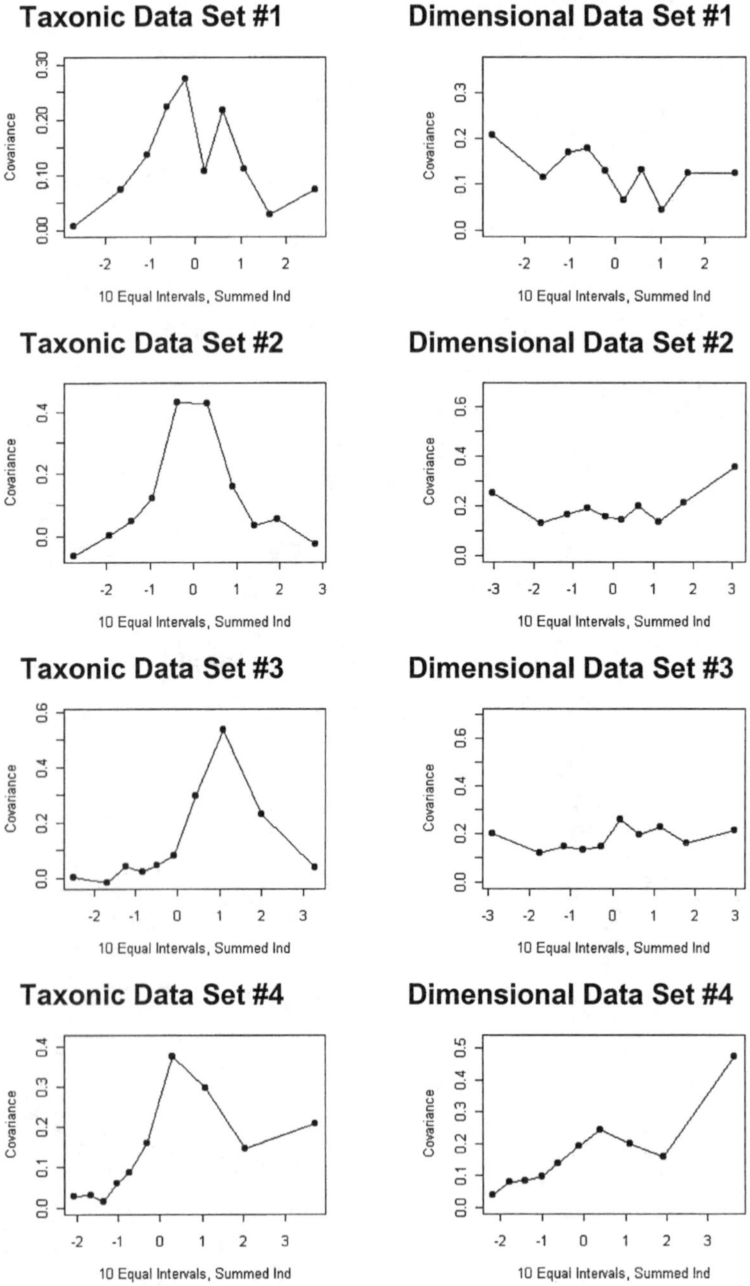

Taxonic Data Set #1

Dimensional Data Set #1

Taxonic Data Set #2

Dimensional Data Set #2

Taxonic Data Set #3

Dimensional Data Set #3

Taxonic Data Set #4

Dimensional Data Set #4

FIG. 6.3. MAXCOV curves for all eight data sets. Two variables were assigned to the role of output indicators, and the input indicator consisted of the sum of the remaining two variables. Ten equal-sized intervals (i.e., deciles) were formed along the input indicator, and each panel of six curves was averaged for presentation.

sistent with dimensional structure. Thus, even when all indicators validly distinguish taxon and complement members, using composite input indicators is not a panacea for the challenges posed when interpreting MAXCOV curves.

Systematic Study of the Procedure

MAXCOV has been the subject of more Monte Carlo research than any other taxometric procedure. Meehl and Yonce (1996) provided detailed results for MAXCOV analyses of the 700 data sets that they had created to study the MAMBAC procedure (Meehl & Yonce, 1994). As noted in chapter 5, these analyses may have produced clearer results than researchers will ordinarily obtain because each of the 14 data configurations was generally favorable to taxometric analysis. For example, none of the indicators was skewed, and all but one data parameter in each analysis was held constant in a highly favorable way. When Haslam and Cleland (1996) imparted positive skew to five of the Meehl–Yonce data sets and reanalyzed the data, they concluded that MAXCOV was robust to indicator skew. However, subsequent analyses of data sets with larger amounts of skew (A. M. Ruscio & Ruscio, 2002; J. Ruscio, Ruscio, & Keane, 2004), as well as a factorial Monte Carlo study (J. Ruscio, 2005), suggest that this conclusion may have been premature. It appears that indicator skew can exert a substantial influence on the shape of MAXCOV curves, with larger amounts of positive skew producing curves that rise more steeply. This effect occurs regardless of the taxonic or dimensional nature of the construct being studied, making it more difficult to distinguish a small taxon from a latent dimension in the presence of positively skewed indicators (J. Ruscio, Ruscio, & Keane, 2004). Of course the same effect would be expected in the presence of negative skew, but in the opposite direction (i.e., curves slanting steeply downward from the left rather than sloping upward toward the right, mimicking the effects of a large taxon). The interpretationally complicating effects involve skewed indicators of a dimensional construct or within-group skew in taxonic data. Even when within-groups indicator skew is absent, indicators of taxonic data are usually skewed in the full sample due to the mixture of groups, and skew arising for this reason should not cause any difficulties.

Beauchaine and Beauchaine (2002) performed an extensive evaluation of the accuracy with which cases can be classified into taxon and complement classes using the results of MAXCOV analyses. They classified cases using Bayes' Theorem, utilizing estimates of the taxon base rate and the *VP* and *FP* rates achieved by cutting each indicator at its hitmax value. Results suggest that such case classifications were usually quite accurate, often surpassing the classification accuracy of a *k*-means cluster analysis.

Beauchaine and Beauchaine also examined the sensitivity of MAXCOV results to taxonic structure across a number of data parameters. Applying thresholds to two of the consistency tests described in chapter 7 (GFI values and distributions of Bayesian probabilities of taxon membership), they scored each data set as correctly classified as taxonic or incorrectly classified as nontaxonic by these consistency tests. Unfortunately, two limitations of this study make it difficult to draw conclusions from its results. First, because all data sets in the study were taxonic, it is impossible to determine conditions under which MAXCOV could distinguish taxonic from dimensional structure. Second, as discussed and demonstrated in chapter 7, neither of the consistency tests used in this study has received enough empirical support to be viewed as a sufficiently valid criterion for the successful detection of taxa. In other words, it is impossible to disentangle the performance of the MAXCOV procedure (e.g., the extent to which curve shape interpretations afford valid structural inferences) from the performance of the consistency tests that served as criterion measures in the study. Thus, we are reluctant to draw conclusions about the utility of MAXCOV curve shapes from this study.

The factorial Monte Carlo study of taxon base rate estimates described earlier (J. Ruscio, 2005) suggested a few tentative conclusions about MAXCOV. First, base rate estimates were found to be more accurate when indicator validities were larger. Second, MAXCOV usually overestimated the size of small taxa (P = .10 or .05). However, with increasing amounts of indicator skew, MAXCOV base rate estimates shrank, often leading to dramatic *underestimates* of the size of moderate (P = .50) and small taxa. Third, whereas within-group correlations increased the size of MAXCOV base rate estimates for nonskewed data, it decreased the size of these estimates for skewed data. Although MAXCOV base rate estimates provide indirect evidence about the shapes of the curves from which they were calculated, there is still a need to determine the accuracy of structural inferences drawn directly from the shapes of MAXCOV curves that are generated across a realistic range of data conditions, ideally varied in a fully crossed factorial manner.

MAXEIG

Logic of the Procedure

A conceptual and mathematical understanding of the MAXCOV procedure paves the way for an understanding of the multivariate generalization of MAXCOV known as MAXEIG (Waller & Meehl, 1998). The differences between these two procedures, as they have been introduced and traditionally implemented, can be organized into five areas (see Table 6.2). First, instead of calculating the covariance between two output indicators (as in MAXCOV), MAXEIG determines the association between two or more output indicators by calculating the first eigenvalue of a modified covariance matrix, a variance–covariance matrix whose diagonal is replaced by zeros to remove variances and leave only covariances. Thus, all supplied variables contribute to every MAXEIG curve. The calculation of conditional eigenvalues rather than conditional covariances is the core procedural difference between MAXEIG and MAXCOV.

The second distinction is that, rather than using all available variables in every possible input–output indicator triplet, MAXEIG uses each variable once as an input indicator, with all remaining variables simultaneously serving as output indicators. This yields fewer MAXEIG curves

TABLE 6.2
Comparison of MAXCOV and MAXEIG
as Traditionally Described and Implemented

Methodological Issue	*MAXCOV*	*MAXEIG*
Calculation of association among output indicators	Covariance between two output indicators	First (largest) eigenvalue of modified covariance matrix for two or more output indicators
Allocation of variables to roles as input–output indicators	Variables serve in all possible input–output indicator triplets	One variable serves as input indicator, with all remaining variables serving as output indicators
Division of cases into subsamples along the input indicator	Nonoverlapping intervals of variable sample size (e.g., based on *SD* units or intact scale values) or fixed sample size (e.g., deciles)	Overlapping windows of fixed sample size (e.g., specified amount of overlap and number of windows determines sample size per window)
Estimation of taxon base rate	Proportion of taxon members in each interval (estimated by solving quadratic formula based on General Covariance Mixture Theorem) aggregated across intervals	Proportion of cases scoring above hitmax cut on the input indicator
Assignment of cases to latent classes	Using base rate estimate plus the estimated valid and false positive rates achieved by hitmax cut on each indicator, Bayes' Theorem yields probability of taxon membership for each individual	No classification method has been proposed for use with MAXEIG

than MAXCOV curves for a given number of indicators. However, because MAXEIG involves all indicators in the calculation of each curve, it can yield results that more clearly distinguish taxonic and dimensional structure.

Third, whereas MAXCOV has been performed by dividing the sample into nonoverlapping intervals along the input indicator, MAXEIG has been performed by dividing the sample into overlapping windows. The latter approach yields all of the data points that would be obtained using equal-sized intervals, plus many intermediate data points as well. For example, if a sample of 1,000 cases is divided into intervals via deciles, this would yield 10 intervals containing cases 1–100, 101–200, 201–300, . . . , 901–1,000. If, instead, windows of the same size (n = 100) but with 90% overlap were used, subsamples would contain cases 1–100, 11–110, 21–120, . . . , 901–1,000. The original 10 intervals would appear on this list along with 9 windows between each of the 10 intervals, yielding 10 + 9 × 9 = 91 windows. Waller and Meehl (1998) described the general relationship between the number of windows W, sample size N, subsample size n_w, and proportion of overlap O between adjacent windows, which can be expressed as:

$$W = \frac{\frac{N}{n_w} - O}{1 - O} \tag{6.5}$$

For a given subsample size, the greater the number of windows, the more interpretable a graph tends to be because the use of overlapping windows is a smoothing technique. The use of windows can increase the number of data points on a curve without sacrificing the number of cases used to calculate each point because a larger number of windows can retain the subsample size of a smaller number of intervals (e.g., $n_i = n_w = 100$ in the present example). Rearranging the terms in Eq. 6.5 allows one to calculate the sample size within each window as a function of the other relevant parameters (Waller & Meehl, 1998, p. 42):

$$n_w = \frac{N}{W \times (1 - O) + O} \tag{6.6}$$

For example, whereas a MAXCOV analysis of N = 1,000 cases using decile input intervals yields n_i = 100, a MAXEIG analysis with 20 windows that overlap 90% with their neighbors yields n_w = 345. Thus, the MAXEIG curve would not only contain twice as many data points as the MAXCOV curve, but also would involve a greater number of cases in the calculation of each curve (n_w [345] > n_i [100]). The benefit of the larger number of points on a MAXEIG curve may be offset to some extent, or even overridden, if the sampling error of conditional eigenvalues is larger than that of conditional covariances. This possibility, which only applies to analyses with more than two output indicators, was suggested by William Grove and remains an important issue for comparative studies of MAXEIG and MAXCOV.

Estimating the Taxon Base Rate

The first three differences in the traditional implementation process for MAXCOV and MAXEIG reflect procedural differences that affect the nature of the obtained curves. However, the interpretive process still proceeds in the same manner for both procedures: Peaked curves are suggestive of taxonic structure, whereas nonpeaked curves are suggestive of dimensional structure. The fourth difference between the procedures relates to the estimation of latent parameters rather than to curve shapes. Whereas the technique for estimating the taxon base rate from MAXCOV draws on the GCMT and the quadratic formula (see Appendix B for details), the technique introduced for MAXEIG (Waller & Meehl, 1998) estimates the taxon base rate by calculating the proportion of cases falling above the midpoint of the window with the largest eigenvalue (the hitmax window). Henceforth, this technique is referred to as the "hitmax" method for MAXEIG base rate estimation.

It should be noted, however, that the logic of the GCMT applies equally well to MAXEIG. Therefore, one can adapt the MAXCOV base rate estimation approach for use with MAXEIG (J. Ruscio, 2005). This technique is referred to as the "GCMT" method for MAXEIG base rate estimation, and its detailed presentation in Appendix C highlights the parallels to MAXCOV base rate estimation.

Class Assignment and Estimation of Other Latent Parameters

The fifth and final difference between MAXCOV and MAXEIG involves the techniques that are used to classify cases (and, on that basis, to estimate other latent parameters). Whereas Bayes' Theorem can be used for this purpose following MAXCOV analysis (Meehl, 1973; Meehl & Yonce, 1996; Waller & Meehl, 1998), no method has previously been proposed to classify cases on the basis of MAXEIG results. We now introduce two techniques for assigning cases to taxon and complement classes following MAXEIG.

First, having obtained estimates of the proportion of taxon members within each window using the technique shown in Appendix C, one can adapt the technique used for MAXCOV to estimate the *VP* and *FP* rates achieved by the hitmax cut on the input indicator (see Appendix B and simply substitute a subscripted w for each subscripted i), then use this information to assign individuals to the taxon and complement classes using Bayes' Theorem (also in Appendix B).

Unfortunately, Bayes' Theorem cannot be used when MAXEIG (or MAXCOV) is performed with composite input indicators because such analyses will not provide the necessary estimates of the hitmax cut or *VP* and *FP* rates for each *individual* indicator, which are required by Bayes' Theorem. In this situation, a second—and simpler—technique may provide satisfactory results: Sort cases based on their total scores on all available indicators, then assign the highest scoring proportion P to the taxon, where P is the taxon base rate estimated in the taxometric analysis. Although this technique hinges on only a single parameter estimate (the taxon base rate), we have found it to produce surprisingly accurate results across many configurations of data parameters. Later in this chapter, after empirically illustrating the MAXEIG procedure, we revisit this base rate classification technique and examine its efficacy.

Having assigned cases to groups using either Bayes' Theorem or the base rate method, a variety of latent parameters can be computed. For example, one can estimate (a) the score distributions of taxon and complement members on each indicator (and thus the *M*, *SD*, skew, and kurtosis for each class), (b) the indicator correlations within each group, and (c) the validity of each indicator in both unstandardized and standardized units.

Implementation Decisions

The first implementation decision in MAXEIG concerns how variables are assigned to the required input–output indicator roles. Traditionally, MAXEIG is performed by using one variable as the input indicator and all remaining variables as output indicators, yielding k curves for k indicators. Alternatively, one can use indicator triplets, yielding $k(k-1)(k-2)/2$ curves, or composite input indicators, yielding $k(k-1)/2$ curves, as was demonstrated for MAXCOV. Because MAXEIG already incorporates all available data in the generation of each curve, there may be relatively little increase in the clarity of results if composite input indicators are used. Moreover, all else being equal in their implementation, MAXEIG and MAXCOV may yield highly similar results when composite input indicators or triplets are used because the eigenvalues in MAXEIG are virtually identical to the covariances in MAXCOV. The only difference would involve the rare instances in which MAXCOV yields a negative value, which would become positive in MAXEIG because the largest eigenvalue is retained. Nonetheless, using composite input indicators would allow ordered categorical variables to be used in MAXEIG analyses.

A second implementation decision concerns how cases should be divided into overlapping windows along the input indicator. The number of windows, as well as their degree of overlap, may be varied. By convention, windows are set to overlap 90% with their neighbors (e.g., Waller & Meehl, 1998). Although this value could be reduced to create lower n windows that may better detect small taxa (e.g., Waller & Meehl, 1998, p. 45, suggest using 50% overlap), a wiser choice may be to increase the number of windows and leave the overlap value at 90%. Increasing the number of windows allows one to obtain windows of the same size achieved by reducing the degree of overlap while yielding a far larger number of points on the MAXEIG curve that should provide more interpretable results. Thus, the ability to vary the number of windows in a MAXEIG analysis is perhaps more important than the ability to vary the overlap between windows. At the same time, although a larger number of windows will provide more data points for the MAXEIG curve, eigenvalues will be subject to increased sampling error as more windows are used. Thus, researchers might begin by selecting a desired sample size for the windows (i.e., setting n_w = 25, 50, or another preferred value such that sampling error is judged acceptable) and then using Eq. 6.5 to calculate the corresponding number of windows. Alternatively, they could select a desired number of windows and use Eq. 6.6 to check that the sample size within windows is not too small (i.e., that it should not yield an unacceptable amount of sampling error).

Just as with MAXCOV, the number of subsamples in a MAXEIG analysis may exceed the number of distinct scores on the input indicator. When this occurs, it will be inevitable that at least some of the divisions between subsamples will fall between equal-scoring cases. However, even if many individuals across different subsamples have the same input indicator score, the *mean* scores for the subsamples will remain ordered, and the association of the output indicators across these subsamples will be informative. Even in the extreme case in which entire adjacent subsamples consist of equal-scoring cases (i.e., the mean x value for neighboring subsamples is identical), the use of internal replications will minimize the resulting noise by yielding highly similar y values (eigenvalues) for the subsamples (J. Ruscio & Ruscio, 2004b, 2004c). With a sufficiently large number of internal replications, the y values for these subsamples will lie in the same spot on the graph. Elsewhere on the curve, internal replications will produce subtle yet meaningful differences across windows. For example, adjacent windows may average x = 4.20, 4.25, and 4.30 on the input indicator as each sliding cut includes a few more individuals with a score of 5 and excludes a few more individuals with a score of 4. Even differences as small as these can help to reveal the shape of the curve. Thus, analyses using large numbers of windows buttressed by internal replications may yield equal-scoring subsamples in some input regions, subtly and meaningfully differing subsamples in other regions, and more substantially differing subsamples in the remaining regions. The net effect will be a smoother, better-defined curve that should be easier to interpret than a curve resulting from a smaller number of intact scale values, *SD*-unit intervals, or equal-size intervals. As we illustrate shortly, the use of a large number of windows may provide superior results even when there are few distinct scores on the input indicator.

A third implementation decision concerns the scaling of the *y* axis for each MAXEIG plot. The options and issues related to this decision are identical to those presented earlier for MAXCOV.

Empirical Illustrations

A first series of MAXEIG analyses was performed on all eight data sets in a traditional manner by using one variable as the input indicator and all remaining variables as output indicators and dividing cases into 20 windows that overlapped 90% with their neighbors. Because this yielded four curves for each data set, we present a single averaged curve to conserve space; these results appear in Fig. 6.4. Estimates of the taxon base rate generated by both the hitmax and GCMT estimation techniques appear in Table 6.3. TS#1 yielded a characteristically dimensional curve and relatively inconsistent base rate estimates, whereas TS#2 and TS#3 yielded clearly peaked curves and accurate, consistent base rate estimates. DS#1, DS#2, and DS#3 all yielded flat curves, correctly suggesting dimensional structure. Whereas DS#1 and DS#3 produced discrepant base rate estimates, DS#2 yielded highly consistent estimates. Both TS#4 and DS#4 yielded ambiguously rising curves and fairly consistent base rate estimates. Thus, the results of traditional MAXEIG analyses appeared to suggest erroneous structural inferences for TS#1 and DS#4 and somewhat ambiguous conclusions for DS#2 and TS#4.

A second series of MAXEIG analyses was performed using composite input indicators, again employing 20 windows that overlapped 90% with their neighbors. This technique yielded six curves per data set; the averaged curves are shown in Fig. 6.5, and base rate estimates appear in Table 6.3. Relative to results of the previous series of MAXEIG analyses, these results were clearer in nearly every case. In particular, TS#1 yielded a curve that was somewhat more peaked, while the peaks for small taxa in TS#3 and TS#4 were better defined. The results for DS#4, however, remained

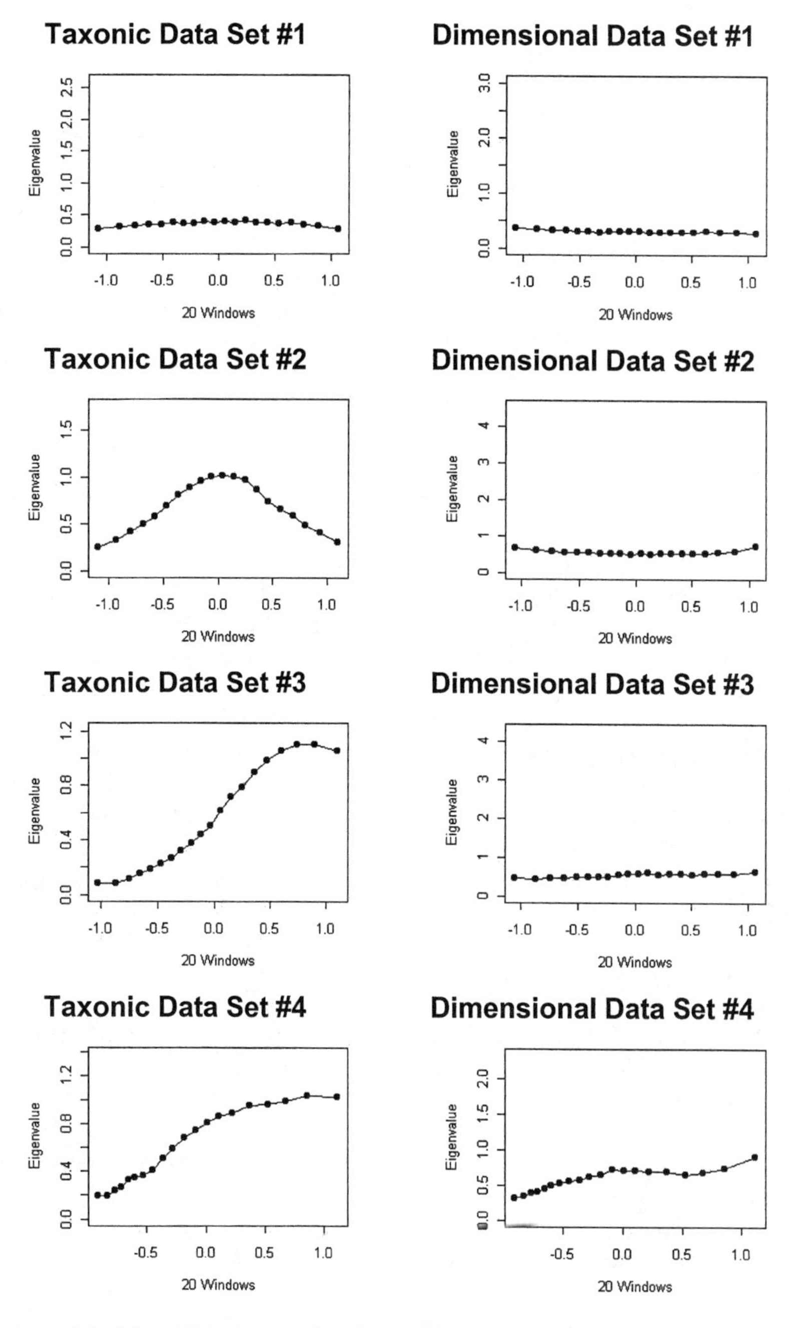

FIG. 6.4. MAXEIG curves for all eight data sets. One variable was assigned to the role of input indicator, and the remaining three variables served as output indicators. Twenty windows that overlapped 90% with their neighbors were formed along the input indicator, and each panel of four curves was averaged for presentation.

TABLE 6.3
Taxon Base Rate Estimates for MAXEIG Analyses

	Estimates for Fig. 6.4			Estimates for Fig. 6.5		
Data Set	*# Estimates*	*Hitmax M (SD)*	*GCMT M (SD)*	*# Estimates*	*Hitmax M (SD)*	*GCMT M (SD)*
TS#1	4	.43 (.12)	.45 (.09)	6	.53 (.13)	.52 (.12)
TS#2	4	.47 (.03)	.48 (.01)	6	.49 (.03)	.49 (.02)
TS#3	4	.22 (.02)	.19 (.01)	6	.26 (.02)	.18 (.01)
TS#4	4	.18 (.02)	.23 (.01)	6	.29 (.04)	.23 (.02)
DS#1	4	.59 (.30)	.59 (.22)	6	.66 (.23)	.67 (.20)
DS#2	4	.17 (.00)	.26 (.02)	6	.28 (.27)	.30 (.22)
DS#3	4	.37 (.15)	.40 (.09)	6	.24 (.08)	.30 (.06)
DS#4	4	.24 (.14)	.22 (.03)	6	.17 (.00)	.18 (.03)

ambiguous; the curve again sloped upward at the right and yielded consistently low base rate estimates. Thus, as in MAXSLOPE, MAMBAC, and MAXCOV analyses, skewed indicators of a latent dimension tended to produce consistent estimates of the taxon base rate. In chapter 7, we discuss the implications of this problem within the broader context of cautious interpretation of consistency of base rate estimates. Nevertheless, aside from lingering ambiguity for DS#4, composite input indicators did appear to provide superior results to the more traditional manner of implementing the MAXEIG procedure. As noted earlier, these MAXEIG curves are virtually identical to those that would have been obtained had MAXCOV analyses been performed with composite input indicators and overlapping windows.

Systematic Study of the Procedure

The MAXEIG procedure has received less attention in Monte Carlo studies than the MAXCOV procedure. As a result, less is known about the conditions under which MAXEIG can distinguish taxonic from dimensional structure. No studies have directly examined the validity of MAXEIG curve interpretations across varying conditions. However, the previously described factorial Monte Carlo study of taxon base rate estimates (J. Ruscio, 2005) suggested broad similarities between the estimates for MAXEIG and MAXCOV. Base rate estimates of both procedures were influenced in similar ways by factors such as indicator validity, indicator skew, within-group correlations, and the actual taxon base rate. Indeed, across the cells of this Monte Carlo study, the mean base rate estimates for MAXCOV and MAXEIG (the latter calculated using the adapted GCMT technique) were correlated $r = .85$. Thus, each of the earlier conclusions about influences on MAXCOV base rate estimates also held for MAXEIG base rate estimates. Because base

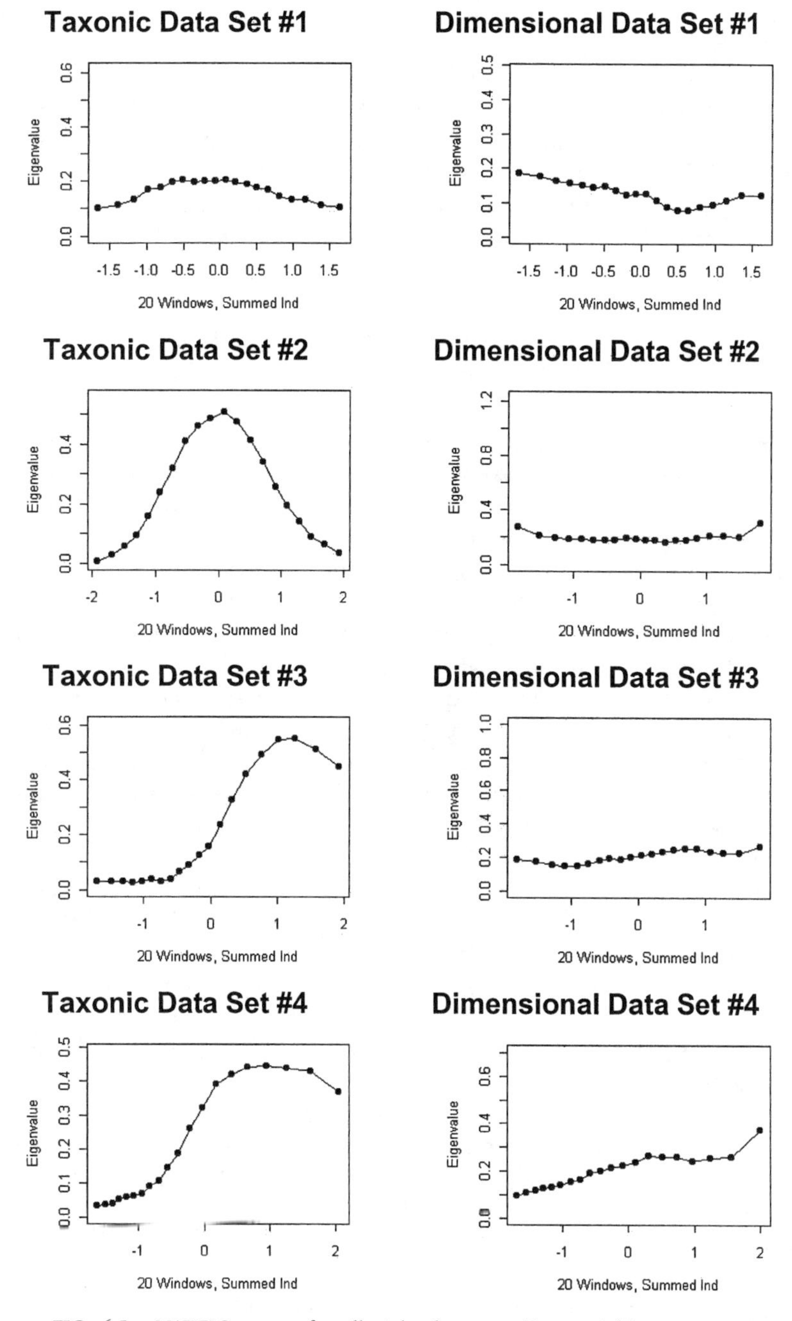

FIG. 6.5. MAXEIG curves for all eight data sets. Two variables were assigned to the role of output indicators, and the input indicator consisted of the sum of the remaining two variables. Twenty windows that overlapped 90% with their neighbors were formed along the input indicator, and each panel of six curves was averaged for presentation.

rate estimates are derived from curve shapes, these results also suggest that MAXEIG curves may be fairly similar to MAXCOV curves under a variety of conditions. However, in light of the potential gains in statistical power afforded by the multivariate nature of MAXEIG as well as its increasing popularity in taxometric studies, there is a great need for more direct and systematic study of MAXEIG curves and parameter estimates.

EVALUATING THE ACCURACY OF THE BASE RATE CLASSIFICATION TECHNIQUE

Having obtained a taxonic result and estimated the base rate of the taxon in the sample, users of MAXCOV and MAXEIG may wish to classify cases into the taxon and complement groups. The simplified classification method introduced earlier requires only an estimate of the taxon base rate to proceed with classification: Assign to the taxon a proportion P of cases who score highest on the sum of all indicators, and assign to the complement the remaining (lower scoring) cases. This base rate classification method shares some of the virtues of the fast and frugal heuristics described by Gigerenzer, Todd, and the ABC Research Group (1999). It is fast in that it takes very little time and computational effort to sum the available indicators and apply a single cutting score. It is frugal in that it requires far less information and estimation than the use of Bayes' Theorem.

Because a fast and frugal method is of little value if it does not yield sufficiently valid results, we performed a small-scale Monte Carlo study to evaluate the classification accuracy of this method. Taxonic data sets were simulated using procedures described by Meehl and Yonce (1994, 1996) and detailed in Appendix A. Each sample contained 1,000 cases and four indicators. Three factors were crossed in this study: taxon base rate (P = .50, .25, .10), indicator validity (d = 2.00, 1.50, 1.00), and within-group correlations (r = .00, .25, .50). For each of the 27 conditions, 100 samples were generated. All cases were classified using Bayes' Theorem. Then cases were classified using the base rate method. To assess the deteriorating effects of incorrect base rate estimates on the base rate method, seven degrees of base rate error were used: −50%, −25%, −10%, 0%, +10%, +25%, and +50%. For example, when the base rate was truly .50, cases were classified using base rate estimates of .25, .38, .45, .50, .55, .63, and .75, respectively.

As can be seen in Fig. 6.6, the base rate method was surprisingly accurate under many conditions. When the base rate estimate contained no error (i.e., when it was identical to the true taxon base rate), this simple method was superior to Bayes' Theorem under all of the conditions tested. The base rate method remained superior under nearly all condi-

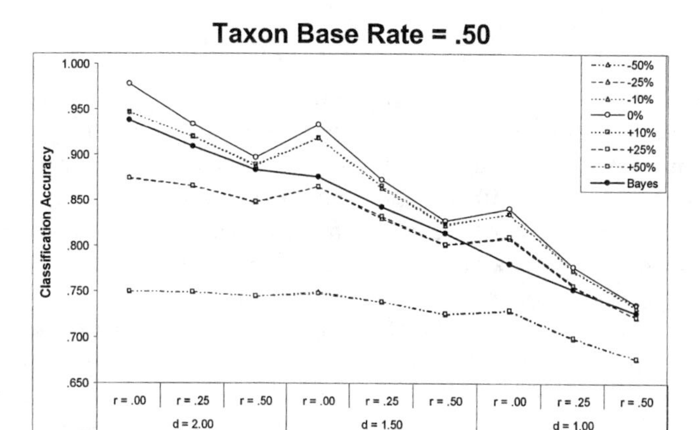

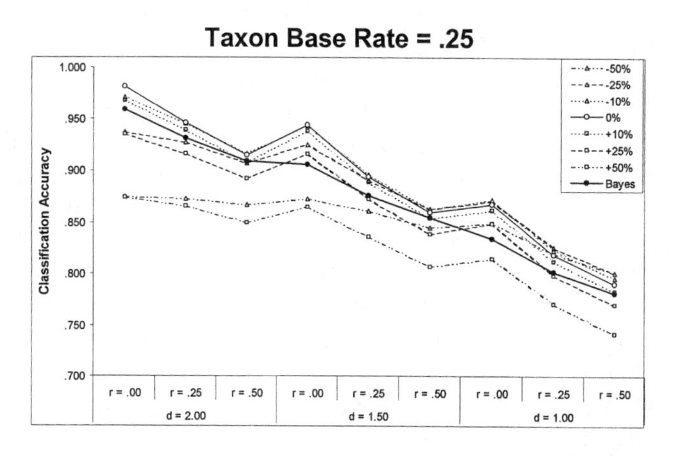

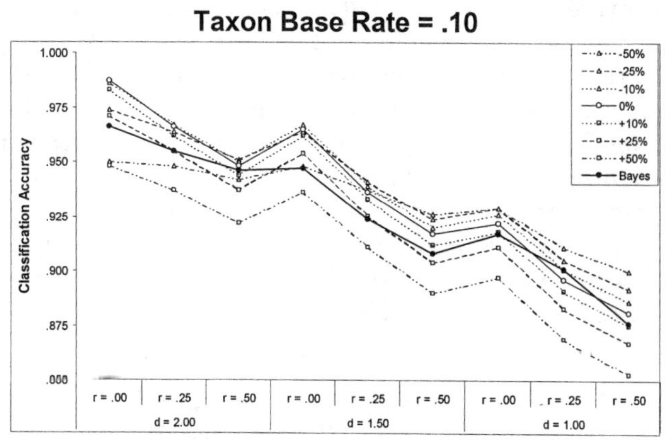

FIG. 6.6. Results for a Monte Carlo study of the accuracy of the base rate classification method. The graphs show results for taxon base rates of P = .50, .25, and .10. Within each graph, results for Bayes' Theorem and the base rate method (with seven degrees of error in the supplied base rate) are displayed for indicator validities of d = 2.00, 1.50, and 1.00 as well as within-group correlations of .00, .25, and .50.

tions after ±10% error was introduced into the base rate estimate. Even with ±25% error in the base rate estimate, accuracy levels were, on the whole, quite comparable to those of Bayes' Theorem. It was not until base rates were estimated with ±50% error that Bayes' Theorem displayed a consistent advantage over the base rate classification method. It should be emphasized that, to make this a stringent test, Bayes' Theorem was always supplied with the correct taxon base rate.

The present results can also be compared to some of those in Beauchaine and Beauchaine's (2002) recent study of MAXCOV classification accuracy. In one instance, their data parameters were identical to a condition in the present simulation study: Their Fig. 4 (Panel c) contains classification data for samples with $N = 1{,}000$, $P = .50$, four indicators, $d = 2.00$, and within-group correlations of $r = .00$. The present results for Bayes' Theorem were in agreement with Beauchaine and Beauchaine's results for Bayes' Theorem following MAXCOV—both analyses classified 94% of cases correctly. Beauchaine and Beauchaine also found that *k*-means cluster analysis classified 95% of cases correctly under these conditions. By contrast, using the correctly specified base rate alone achieved 98% accuracy under the same conditions in the present study. Even when the base rate was specified with ±10% error (i.e., base rates of .45 or .55 were used instead of .50), classification accuracy was 95%.

It is worth noting that the base rate classification method can be used with any taxometric procedure because each procedure provides an estimate of the taxon base rate. When the average taxon base rate estimated across multiple taxometric procedures is suspected to be more accurate than individual estimates, this could be used to classify cases for subsequent analyses. Before concluding, however, we must emphasize that nothing is presently known about the efficacy of the base rate classification method relative to Bayes' Theorem when both are provided with the potentially biased latent parameter estimates of any taxometric procedure. In addition to the lack of data on the robustness of either technique to biased parameter estimates, no studies have compared the classification accuracy achieved by different taxometric procedures using either of these techniques. Thus, more definitive guidelines for the most appropriate way to classify cases await further empirical study.

BLENDING ELEMENTS OF THE TRADITIONAL MAXCOV AND MAXEIG PROCEDURES

As they are traditionally described and implemented, the MAXCOV and MAXEIG procedures each possess strengths and weaknesses. It would be valuable if one could capitalize on the strengths while avoiding the weaknesses of each. Indeed, some of the features conventionally associated with each procedure can be readily applied to the other. For example, there is nothing to prevent MAXCOV from being performed with overlapping windows rather than nonoverlapping intervals. The only fundamental difference between the two procedures is the measure of association that they employ—conditional covariances (MAXCOV) or conditional eigenvalues (MAXEIG). The same three broad decisions must be made when implementing either MAXCOV or MAXEIG: (a) how to allocate variables to the roles of input–output indicators, (b) how to divide cases into subsamples along the input indicator, and (c) how to graph the results. Research traditions prescribe how each of these decisions is usually made for MAXCOV or MAXEIG analyses, but one can mix and match the options that have become associated with each procedure by making these decisions in a more deliberate manner.

Assigning Variables to Input–Output Indicator Roles

The first decision involves the method by which variables in a data set are allocated to the required roles of input and output indicators for analysis. Because there must be one input indicator and at least two output indicators to perform either procedure, there must be $k \geq 3$ variables available in the data. There are at least three ways in which variables can be assigned to input and output roles. First, variables can be placed into each possible input–output indicator triplet, as is typically done in MAXCOV. This yields $k(k-1)(k-2)/2$ curves. Second, each variable can serve once as an input indicator, with all remaining variables serving as output indicators, as is typically done in MAXEIG. This yields k curves. Third, one can remove two variables to serve as output indicators and sum all remaining variables to serve as a composite input indicator. This yields $k(k-1)/2$ curves.

It is easy to see that the three methods outlined earlier are equivalent to one another when $k = 3$ (in which case the composite input indicator is a single indicator). When $k > 3$, however, several considerations must be weighed in deciding how best to allocate variables to input and output roles. First, the chosen method must be able to accommodate the available data. In particular, it is imperative that an input indicator possess sufficient variation in scores to allow a reliable rank ordering of cases for division into subsamples. If individual variables include too few values, they may need to be combined to form composite input indicators.

Second, to estimate certain latent parameters and perform some of the consistency tests described in chapter 7, one must assign cases to the putative taxon and complement groups—and the method by which variables are assigned to the input–output indicator roles can constrain one's choice of a classification technique. Whereas the base rate classification method can be used regardless of how variables are assigned to input–output roles, Bayes' Theorem requires that each individual variable serve as the input indicator for at least one MAXCOV/MAXEIG curve because one cannot estimate the *VP* and *FP* rates achieved by the hitmax cut on each variable when variables are combined into composite input indicators. Thus, if one wishes to classify cases using Bayes' Theorem, composite input indicators cannot be used. Instead, the available variables must be assigned to input–output indicator roles as is traditionally done in either MAXCOV or MAXEIG analyses.

Third, consideration should be given to the likely clarity of results under the three methods. Whereas the indicator triplet method yields the largest number of curves—and thus provides more checks on the consistency of results and the opportunity to determine whether certain combinations of indicators yield more or less interpretable results—the other methods incorporate all available data into the calculation of every curve. These multivariate approaches should increase the power with which taxonic and dimensional structure can be differentiated, *provided* that each indicator was wisely chosen and distinguishes the taxon and complement with sufficient validity. In some cases, this gain in power may offset the reduction in the number of consistency checks and ultimately lead to more accurate structural inferences. However, the relative value of increased power versus increased consistency testing in this context is an empirical question requiring further study.

Fourth, although developers of the taxometric method have strongly favored theory-driven over exploratory taxometric research (e.g., Meehl, 1992, 1995a; Meehl & Golden, 1982; Waller & Meehl, 1998), they have also suggested exploratory strategies that can be used to scan a set of indicators for those appearing suitable for taxometrics (Meehl, 1999). If researchers choose to conduct *initial* taxometric analyses in this exploratory mode, they may prefer to perform MAXCOV/MAXEIG analyses using indicator triplets (and, similarly, MAXSLOPE or MAMBAC analyses using indicator pairs) at the early stage of indicator selection. Once a final set of valid indicators has been chosen, researchers may proceed to test latent structure using a technique that includes all indicators in every curve. The advantage of this approach is that it permits examination of the relative performance of each indicator at the exploratory phase of analysis. When all indicators are involved in the calculation of every curve, it can be difficult to isolate the influence of any particular indicator on taxometric results. By contrast, using indicator triplets (or, in MAXSLOPE or MAMBAC, indicator pairs) enables a researcher to determine whether one or more indicators consistently produces ambiguous results, whereas other indicators consistently produce more interpretable results. By *interpretable* we do not mean that the results must be taxonic; rather, we refer to results that clearly distinguish taxonic from dimensional structure.

When taxometric procedures are performed in a hypothesis-testing mode, it may be desirable to include all available indicators in the calculation of each curve to increase the clarity of results, taking care to establish in advance that each indicator distinguishes a taxon and its complement with sufficient validity to warrant its inclusion in the analyses. It bears repeating that the power of such analyses is likely to suffer when one or more indicators of dubious validity are included in the analysis. However, the decision to exclude an indicator must be weighed against its importance in defining the target construct, leading to a difficult judgment call when omission of that indicator would effectively change the construct under study.

Dividing Cases Into Subsamples Along the Input Indicator

Once it has been decided how variables will be allocated to the roles of input and output indicators, the next decision concerns the method by which cases will be divided into subsamples along the selected input indicator. Although several approaches exist, we believe that one—overlapping windows—is generally preferable to the others. We review the available options before arguing in favor of our preferred approach.

First, subsamples can be constructed on the basis of *SD* units. For example, Meehl and Yonce (1996), in their MAXCOV monograph, created intervals by placing cuts every .25 *SD* units along the input, as did Beauchaine and Beauchaine (2002) in their Monte Carlo study of MAXCOV classification accuracy. Second, subsamples can be constructed using intact scale values. For example, researchers who have summed dichotomous items into composite input indicators have typically used each value of the summed indicator to define a subsample. When eight dichotomous items are available, setting aside two output indicators and summing the remaining six allows input subsamples to be created at scores of 0, 1, 2, 3, 4, 5, and 6. Third, subsamples can be formed using equal-*n* intervals, such as deciles, as has been done with MAXCOV. Fourth, overlapping windows can be used, as is traditionally done with MAXEIG.

We believe that constructing subsamples using nonoverlapping intervals—whether these intervals are based on *SD* units, intact scale values, or an equal-*n* partitioning of cases—is unnecessarily limiting in a MAXCOV/MAXEIG analysis. Earlier we explained why equal-*n* intervals are inferior to overlapping windows: They yield only a subset of the data points that are generated by identically sized windows, with equal sampling error in the calculation of each point. Therefore, the following discussion focuses on the use of intervals based on intact scale values or *SD* units, which also have several important limitations.

First, both of these interval construction techniques yield relatively few points on the curve—a potentially significant drawback in many taxometric investigations. Curve shapes can be difficult to interpret, and latent parameters can be difficult to accurately estimate when one proceeds on the basis of curves containing a small number of data points. This may be especially problematic in the investigation of small taxa, where expected taxonic peaks at the upper end of MAXCOV/MAXEIG curves can be difficult to distinguish from the rising curves of dimensional data with positively skewed indicators (J. Ruscio, Ruscio, & Keane, 2004). Given the frequency with which taxometric investigations have targeted small taxa, the use of intervals rather than windows could pose a substantial limitation.

Second, because indicators may be differentially sensitive to a taxon, curve peaks may emerge in different intervals (corresponding to different scale values or *SD* units) across different graphs. This may not be a problem when the full panel of curves is inspected and presented, but it may become a problem when these curves are averaged. The relative benefits and drawbacks of averaging curves are discussed later; for now it should be noted that there are times when averaging can be useful in a taxometric investigation. At such times, it is advantageous to use a subsample construction technique that facilitates the meaningful averaging of curves, and the overlapping windows technique serves this purpose well. Rather than dividing cases based on their actual scores on the input indicator—whether these be raw or standardized scores, in scale value or *SD* units—windows are formed on the basis of equal-sized successions of cases, an approach that lends itself nicely to subsequent averaging.

Finally, perhaps one of the most significant disadvantages of using intervals is that they would make it impractical, if not impossible, to perform the inchworm consistency test. To perform this test, one conducts MAXCOV/MAXEIG with a systematically increasing number of subsamples to observe whether an apparent peak at either end of the curve becomes more clearly defined (suggesting the presence of a taxon) or does not (suggesting a dimensional conclusion). Because intervals do not overlap, any increase in the number of intervals will rapidly reduce the number of cases within each, leading to a potentially problematic increase in sampling error. In contrast, the overlap between windows allows the number of points on the curve to be increased with a far less dramatic decline in the number of cases per window. This is essential for successful implementation of the inchworm consistency test.

One potential argument in favor of using intervals based on *SD* units or intact scale values, rather than overlapping windows, is that this may reduce the "tilting" of MAXCOV/MAXEIG curves caused by indicator skew. The interpretive challenges posed by skew have been noted several times already and are discussed more extensively in chapter 8, but may be summarized by noting that skew creates a differential restriction of range on the output indicators within subsamples formed along the input indicator. Because windows are equal in size, each output indicator possesses less variance within lower scoring subsamples than within higher scoring subsamples in the presence of positive indicator skew (the reverse is true for negative skew). This means that, regardless of the structure of the underlying construct, a rising curve would be expected due to the increasing range of scores within higher scoring subsamples. It would seem that using intervals that are allowed to contain unequal numbers of cases (e.g., those defined by *SD* units or intact scale values) may alleviate this problem by aggregating larger numbers of low-scoring cases and smaller numbers of high-scoring cases into subsamples. Although the output indicators may still have greater variance in higher scoring intervals, the resultant MAXCOV or MAXEIG curve may be less tilted than when equal-*n* subsamples (intervals or windows) are used.

To evaluate this possibility, we examined the MAXCOV and MAXEIG curves for TS#4 and DS#4 (the data sets with positively skewed indicators), comparing analyses performed using *SD*-unit intervals to those performed using other methods of subscale construction. This comparison revealed that analyses performed with *SD*-unit intervals yielded curves (see Fig. 6.1 for the averaged curve) that were at least as tilted, if not more so, than those produced by overlapping windows (see Figs. 6.4 and 6.5). Of course this comparison of a small number of curves for analyses of just two data sets does not conclusively resolve this issue. At the same time, it provides little support for choosing *SD*-unit intervals over windows to avoid the tilted MAXCOV/MAXEIG curves produced by skewed indicators. As is elaborated in chapter 7, we believe that the best way to distinguish a small taxon from a latent dimension with positively skewed indicators is to use the inchworm consistency test. Because the inchworm consistency test is most appropriately performed with overlapping windows, we again recommend the use of overlapping windows when performing MAXCOV/MAXEIG analyses.

Dividing cases into windows can be a somewhat arbitrary process that creates noise in the resultant curves. Particularly when large numbers of windows are used, divisions are often drawn between individuals who have identical scores on the input indicator. To the extent that many cases are thus affected by the division process, simply resorting the tied cases will redistribute them across windows and alter the curve to some extent. Likewise, the more divisions are made, the more likely it is that entire subsamples will consist of individuals with identical scores—identical not only to one another, but to many or all members of adjacent windows. Thus, increasing the number of windows not only results in greater sampling error within each, but also exacerbates the problem of the arbitrary assignment of equal-scoring cases to windows. Fortunately, the use of internal replications can minimize the latter problem. With a sufficient number of internal replications, the distorting effect of such arbitrary case assignments can be reduced to the extent desired; the only practical constraint is the speed with which one's computer will carry out a large number of replications. In practice, we have found that using just 5 or 10 replications often goes a long way toward eliminating erratic covariances or eigenvalues from MAXCOV/MAXEIG curves.

We recommend using overlapping windows even if the number of windows far exceeds the number of distinct scores on the input indicator. To illustrate the feasibility of this approach, we performed MAXCOV analyses of the same eight items from MMPI Scale 5 (Masculinity–Femininity) that were analyzed with MAMBAC in chapter 5. These were MAXCOV (rather than MAXEIG) analyses because covariances (rather than eigenvalues) were used as the measure of association. Analyses were performed using composite input indicators because individual dichotomous items cannot be used as input indicators. Figure 6.7 shows the averaged MAXCOV curves for several series of analysis, with each curve based on 28 analyses. In the first series, the input was subdivided using intact scale values (top panel). This curve was peaked, although there was a worrisome spike toward the right end of the curve that made it difficult to rule out sampling error as the overarching cause of covariance fluctuations across subsamples. The taxon base rate of .59 was not estimated well in these 28 analyses (M = .43). In the second series, the input was subdivided using 50 windows that overlapped 90% with one another, the approach traditionally associated with MAXEIG (middle panel). This averaged curve was smoother and more clearly peaked, and the taxon base rate was estimated more accurately in these 28 analyses (M = .61). In the final series, the input was again subdivided using 50 windows, and five internal replications were performed to reduce the obfuscating influence of assigning equal-scoring cases to different windows (bottom panel). This averaged curve was smoother still, and estimates of the taxon base rate were even more accurate (M = .59) than those generated by the previous analyses. Thus, the use of many more subsamples than there are distinct input scores, particularly when accompanied by the use of internal replications, can yield more interpretable MAXCOV curves and more accurate estimates of latent parameters such as the taxon base rate than when MAXCOV is implemented in the traditional fashion.

Studies have not yet established an optimal number of data points for maximally informative taxometric curves. Indeed, an optimal number is unlikely to be found, given the probable influence of sample size, taxon size, indicator validity, indicator skew, and a host of other factors on curve interpretability. Although we cannot offer specific advice regarding the

Intact Scale Values

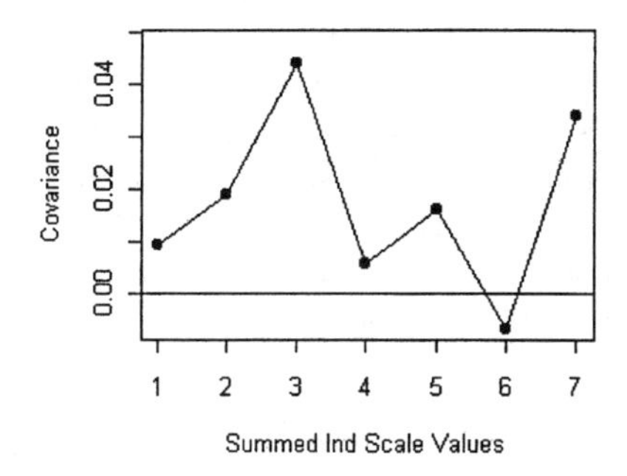

50 Windows

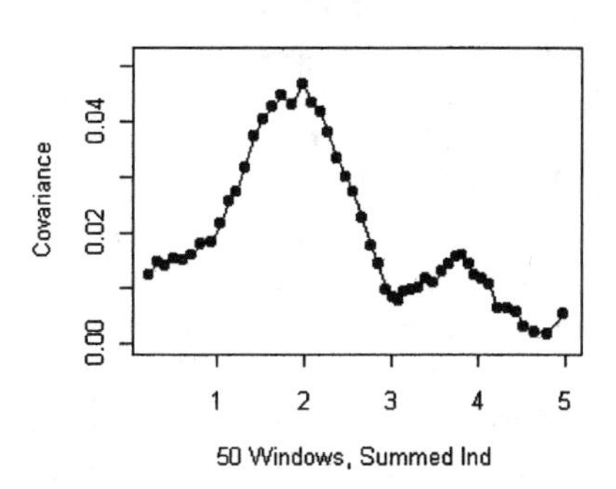

50 Windows with 5 Internal Replications

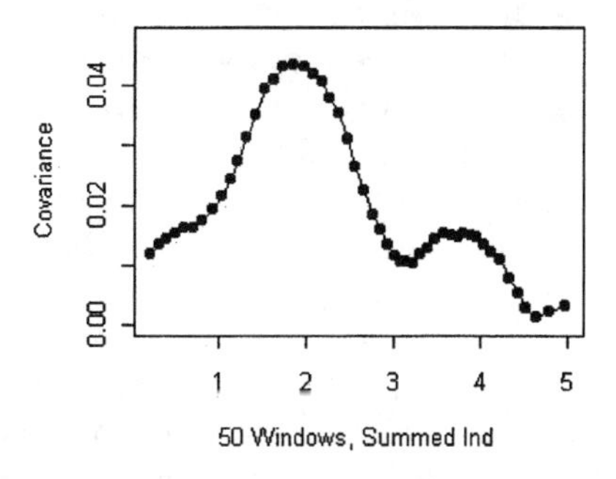

FIG. 6.7. MAXCOV analyses of eight MMPI items used to distinguish 593 women from 407 men. Analyses varied in the division of the input indicator into subsamples. In the top panel, subsamples consisted of intact scale values, whereas in the middle panel subsamples consisted of 50 windows that overlapped 90% with their neighbors. The bottom panel displays the results for 50 windows with five internal replications used to reduce the obfuscating effects of assigning equal-scoring cases to different windows.

number of windows that will be appropriate for any particular investigation, we strongly recommend the use of overlapping windows in MAXEIG or MAXCOV analyses. Moreover, we recommend the adjunctive use of simulated taxonic and dimensional comparison data to generate empirical sampling distributions and thereby determine whether the number and overlap of windows selected for analysis is likely to pose an informative test between taxonic and dimensional structure.

Graphing the Results

A final implementation decision concerns the way in which results are graphed. Several graphing-related factors may influence the interpretation of MAXCOV/MAXEIG curves. First, as noted earlier, inappropriate scaling of the y axis can pose a threat to the accurate interpretation of taxometric curves. A second, controversial issue concerns whether MAXCOV/MAXEIG curves should be smoothed. Many researchers have smoothed such curves using running medians (Tukey, 1977) or LOWESS methods (Cleveland, 1979). Smoothing can help to reveal a curve's shape when it might otherwise be obscured by sampling error; however, smoothing also carries the risk of leveling genuine peaks. None of the MAXCOV/MAXEIG curves presented in this book has been smoothed. That is, curve smoothing was not performed subsequent to the calculation of covariances or eigenvalues within subsamples along the input indicator; as noted earlier, the use of overlapping windows is a smoothing technique. We chose instead to enhance the clarity and interpretability of graphs by plotting a sufficient number of data points and, in some cases, by taking advantage of internal replications.

In each taxometric investigation, the decision of whether to smooth should be based on a careful consideration of pros and cons, perhaps informed by consulting empirical sampling distributions generated for smoothed and unsmoothed curves. In addition, taxometric reports should always inform readers of any smoothing that was performed as well as the technique used for this purpose. Ideally, both raw and smoothed curves would be presented to reveal the effects of smoothing on the shapes of the curves. Finally, researchers may need to examine the program code that they use to perform taxometric analyses (or its documentation file) to determine whether smoothing has been performed. For example, whereas both running medians and LOWESS smoothing are available as user-specified options in Waller and Meehl's (1998) MAXCOV program, LOWESS smoothing is automatically applied in their MAXEIG program even though the use of overlapping windows already constitutes one smoothing technique. Both of these programs plot a single series of data points—smoothed values for MAXEIG and either raw *or* smoothed values for MAXCOV. In our own program for performing MAXCOV or MAXEIG analyses, smoothing is an option, but not the default. In the event that users request running medians or LOWESS smoothing, the program plots both raw and smoothed points on each graph in a full panel of curves.

A final graphing concern involves the relative merits of averaging curves versus evaluating and presenting the full panel of individual curves. Because each curve suffers from some degree of sampling error, averaging across curves can sometimes yield a more interpretable final graph. In addition, researchers who possess a large number of indicators or who perform a large number of analyses may obtain far too many curves to present in a standard journal article. However, because graphical results are so central to a taxometric investigation, it is important that they be displayed for independent evaluation by readers and reviewers. In such cases, the averaged curves can be presented to conserve space, with full panels of curves made available on request (other options are discussed and illustrated in chap. 8).

However, averaging does have the potential to obscure results in several ways. First, some combinations of indicators may yield ambiguous results, whereas others yield interpretable results; if the curves produced by these different indicator combinations are averaged, potentially important information may be diluted by less useful results. Second, a small number of extremely aberrant curves can exert a disproportionate influence on the averaged curve. Third, depending on how cases are divided into subsamples along the input, it is possible that peaks may emerge in different subsamples for different input indicators due to differing score distributions. Such peaks may be artificially flattened in the averaging process. Fourth, some types of input indicators do not lend themselves well to averaging. For example, if intact scale values are used to construct subsamples and each indicator varies across a different range of scores, the resulting curves cannot be averaged in any obvious way. At the very least, one might wish to standardize all indicators if curves are to be averaged, although differing amounts of skew may still result in different ranges of *SD*-unit intervals across indicators.

As elaborated in chapter 7, the taxometric method is built on consistency checks. It may be advisable to interpret a full panel of curves that can serve as consistency tests for one another even if they are later averaged to conserve space in a report, as we have done in this book. Chapter 8 illustrates another approach: plot all curves as relatively light lines on a single graph, with an averaged curve superimposed as a darker line. This allows readers to simultaneously consider the consistency of results, the appropriateness of averaging the curves, and the general trend evident in the averaged curve.

CONCLUSIONS

Chapters 5 and 6 reviewed the logic of five taxometric procedures, described how each procedure is performed and how it can be used to estimate a taxon base rate and other latent parameters, discussed the decisions that must be made when each procedure is implemented and its results are graphed, illustrated each procedure through analyses of multiple data sets of varying structure and data characteristics, and discussed what is presently known about the performance of each procedure from systematic studies. We described traditional features of each procedure and demonstrated the ways in which it is typically implemented. We also pointed out several areas where simple adherence to conventional practice may not provide the most informative or most appropriate taxometric analysis. For these areas, we attempted to list the choices that are available to researchers, to highlight the strengths and limitations of each choice, and, where possible, to illustrate the consequences of different choices through empirical demonstrations. It must be emphasized that the number and scope of these choices—and their specificity to particular procedures—prohibit a more comprehensive summary at this time, particularly given the few empirically driven guidelines afforded by existing Monte Carlo research. At the same time, we would like to highlight several techniques presented in these chapters, which, although requiring more systematic study, have the potential to enhance the clarity and utility of taxometric results.

First, we demonstrated the potential advantages of using a composite input indicator—one constructed by summing across all variables not serving as output indicator(s)—instead of using individual variables in the input role. Although this approach has been applied in prior studies (e.g., those employing dichotomous indicators), it has been largely restricted for use with MAXCOV in the manner introduced by Gangestad and Snyder (1985). Our illustrations suggested that composite input indicators not only accommodate data with ordered categorical response scales, but may increase the clarity of MAXSLOPE, MAMBAC, and MAXCOV/MAXEIG results for a variety of indicator distributions.

Second, we introduced the internal replication technique as a potentially useful adjunct when a procedure such as MAMBAC or MAXCOV/MAXEIG is implemented in a way that can lead to the partitioning of equal-scoring cases along the input indicator. To dispel concerns that creating more cuts than there are unique input values may obfuscate taxometric curves, empirical illustrations revealed that even a far larger number of cuts than input scores can yield smoother, more fine-grained curves that are easier to interpret than curves in which the number of points is limited to the number of distinct input scores. This was especially true when internal replications were performed in which cases with equivalent scores on the input indicator were shuffled and the analysis was repeated. With increasing numbers of replications, the distorting effects of cutting between equal-scoring cases diminished, resulting in even clearer curves and more accurate estimates of latent parameters such as the taxon base rate. Thus, the internal replication technique may have substantial utility in many applications.

Third, we introduced a new technique that uses the taxon base rate estimate provided by any taxometric procedure (or an estimate averaged across procedures) to classify cases into the taxon and complement. This base rate classification technique orders cases according to their total (summed) score on all of the indicators, places a cut at the point along this distribution that corresponds to the taxon base rate, and assigns cases falling above the cut to the taxon and cases falling below the cut to the complement. Our initial Monte Carlo study of this simple technique found it easy to classify cases with accuracy rates comparable to—and, in many cases, greater than—Bayes' Theorem under a variety of data conditions, even when the base rate submitted to the technique was over- or underestimated by up to 25%. Although the base rate classification technique may be more useful in some situations than in others, it presents a flexible new tool for taxometric research that allows cases to be classified even when other considerations prohibit the use of Bayes' Theorem (e.g., when composite input indicators are used). In addition, it allows the use of a single base rate estimate calculated as the average across taxometric procedures, rather than a potentially less reliable estimate from one procedure, to classify cases. Although it awaits more extensive empirical evaluation, this technique appears promising as a simple and flexible tool for taxometric investigations.

Two other issues concern the relationship between MAXCOV and MAXEIG and the manner in which these procedures are implemented. Although MAXEIG was developed as a multivariate generalization of MAXCOV and the two procedures are often viewed as related, each is traditionally implemented in its own distinct way, and both are sometimes included in the same taxometric investigation. In this chapter, we framed MAXCOV and MAXEIG as variants of a single overarching approach, rather than as distinct procedures that provide truly independent sources of structural evidence. We further suggested specific ways in which the most valuable elements of each procedure (e.g., overlapping windows in MAXEIG, GCMT-derived base rate estimation in MAXCOV) could be blended. In doing so, we hope we have encouraged future users of taxometrics to consider carefully the decisions involved in conducting a MAXCOV/MAXEIG analysis, and to make these decisions thoughtfully and deliberately on the basis of their data and research questions, rather than convention or research tradition.

Unfortunately, Monte Carlo studies have not evaluated the effects of specific implementation decisions on taxometric results. As a result, the taxometric researcher—who has no choice but to make these decisions—often has little empirical guidance for doing so. Although the analyses presented in these two chapters may provide a few general suggestions, their limited scope means that they should be regarded as illustrations rather than guidelines, and more extensive simulation research is clearly needed for broadly applicable guidelines to be identified. However, even in the absence of such research, investigators need not conduct taxometric analyses blindly. As discussed in chapter 4, by generating empirical sampling distributions through analyses of bootstrap samples of comparison data, investigators can empirically evaluate whether their procedural implementation can validly distinguish taxonic and dimensional latent structure (J. Ruscio, Ruscio, & Meron, 2005). If the planned analyses do not yield distinct sampling distributions of results for taxonic and dimensional comparison data, investigators can experiment with other implementation approaches and test whether they are more appropriate for the data at hand. This technique, known in the bootstrapping literature as *adaptive calibration* (Efron & Tibshirani, 1993), can help researchers to construct and perform the most informative taxometric investigations that their data allow. Such investigations will usefully include not only the taxometric procedures reviewed thus far, but some of the supplementary consistency tests to which we turn in chapter 7.

CHAPTER SEVEN

Consistency Tests

A cornerstone of the taxometric method is the evaluation of agreement among results yielded by as many nonredundant analyses as possible. Confidence in an inference of taxonic or dimensional structure accumulates as these results cohere. The term *consistency test* has been used to refer to a variety of techniques within the taxometric method, and we use it to broadly refer to any approach that allows a researcher to examine the agreement of results from different taxometric analyses. For example, Meehl (personal communication, February 15, 1999) originally developed the MAXCOV procedure as a consistency test to complement the use of other procedures, but it later became one of the primary procedures employed in taxometric studies—with its own array of associated consistency tests. Meehl suggested that the distinction between a taxometric procedure and a taxometric consistency test is highly arbitrary, and our treatment reflects that perspective.

Although the core of any taxometric investigation is the performance of multiple taxometric procedures, other tests have been developed to further evaluate the consistency of taxometric results. To date there has been relatively little systematic study of the utility of most consistency tests, and virtually no research has examined the incremental validity added by each test to inferences drawn from the taxometric procedures. Moreover, whereas some consistency tests appear to offer useful checks of structural conclusions, others have been found to discriminate poorly between taxonic and dimensional structure and may therefore yield uninformative or even misleading results. In this chapter, we consider the relative merits of many consistency tests and discuss conditions under which each may provide relevant, nonredundant, and discriminating information that can enhance a taxometric investigation.

Before proceeding, it is important to stress that none of the consistency tests reviewed here is necessarily useful or appropriate for each and every taxometric investigation. Moreover, some of our recommendations involve cautionary notes about particular consistency tests that may be poorly supported for widespread use. In chapter 4, we introduced a technique that researchers can use to determine, in the context of the unique characteristics of their data, which taxometric procedures and consistency tests afford a genuine test between taxonic and dimensional structure. In the present chapter, we demonstrate and encourage the generation of empirical sampling distributions through analyses of bootstrap samples of taxonic and dimensional comparison data to help researchers determine which consistency checks are most appropriate for the study at hand.

PERFORMING TAXOMETRIC PROCEDURES MULTIPLE TIMES IN MULTIPLE WAYS

Taxometric Procedures as Checks for One Another

Taxometric procedures that are derived in mathematically distinct ways yield nonredundant results that may be checked against one another for consistency.

MAMBAC and L-Mode use very different approaches to evaluate latent structure. Whereas MAMBAC searches for an optimal cutting score along one indicator that maximizes the difference in scores on another indicator, L-Mode searches for bimodality along the distribution of factor scores estimated from three or more indicators. Thus, MAMBAC and L-Mode are quite distinct from one another (and from MAXSLOPE, MAXCOV, and MAXEIG) as sources of structural information. MAMBAC can be performed with as few as two indicators and should probably be considered for use in most taxometric investigations. L-Mode can be performed with only three indicators, but is more powerful as the number of valid, nonredundant indicators increases (Waller & Meehl, 1998). Provided that a sufficient number of indicators is available, each varying across a sufficient range of values, L-Mode can also be considered for inclusion in most taxometric investigations.

Given the conceptual relatedness of MAXCOV and MAXEIG, it might be most appropriate to use only one of these procedures in any given taxometric study. MAXSLOPE has been recommended as a surrogate for

MAXCOV or MAXEIG when only two indicators are available for analysis (Meehl, 1999).

The choice between MAXCOV and MAXEIG may be more complex. As we noted in chapter 6, these procedures function in highly similar ways and may be usefully conceptualized as variants of a single overarching approach to the evaluation of latent structure. The primary decision distinguishing these procedures is whether covariances or eigenvalues will be calculated as the measure of association between indicators. When only three indicators are available—and hence only two output indicators for any given curve—it will make little, if any, difference whether one opts for MAXCOV or MAXEIG. However, it is still important to make reasoned decisions about how cases are divided into subsamples along the input indicator and how the ensuing results are graphed (e.g., how to scale the *y* axes, whether to smooth or average curves).

The choice between MAXCOV and MAXEIG may be more difficult to make when more than three indicators are available. Because MAXEIG usually includes all available indicators in each analysis, the clarity of its results may degrade if a subset of the indicators are not sufficiently valid or contribute to problematically large within-group correlations. Thus, MAXCOV—which can be performed using just one input and two output indicators—may be preferable to MAXEIG when few valid indicators are available or when one wishes to isolate and closely examine the influence of each indicator on the results. If the shape of the MAXCOV curves or the corresponding estimates of latent parameters vary systematically across analyses employing different indicators, it may be necessary to more closely inspect all of the indicators and to consider refining, discarding, or replacing any that are highly problematic. As we note in chapter 6, one approach may be to perform MAXCOV at the earliest stages of indicator selection and construction for the purpose of scrutinizing individual indicators. One might then perform MAXEIG at the final stages of analysis (i.e., to test latent structure) using the set of indicators shown to validly and nonredundantly distinguish the putative taxon and complement.

Repeated Application of Each Taxometric Procedure

Each taxometric procedure can be performed multiple times by using the available indicators in different configurations. The advantage of doing so is that this yields a panel of curves that can be compared for consistency and interpreted as a whole. Curves that consistently suggest either taxonic or dimensional structure enhance confidence in that structural solution, whereas curve shapes that do not consistently suggest either structure reduce confidence in any structural inference. Although some researchers regard inconsistent results as evidence of nontaxonic structure, we prefer to withhold judgment and consider possible reasons for the inconsistency. If individual indicators are insufficiently valid or if combinations of indicators are overly redundant, one could return to the phase of indicator selection and construction, rather than reach what may be a premature conclusion of dimensional latent structure based on inappropriate data. A more carefully constructed final set of indicators may more consistently suggest either taxonic or dimensional structure.

MAXSLOPE or MAMBAC can be conducted multiple times by using different pairs of variables as input and output indicators and by swapping these variables across input–output roles. L-Mode can be performed using as few as three indicators per analysis, so when more than three indicators are available, one can conduct a series of L-Mode analyses. MAXCOV or MAXEIG can be performed using indicators in a variety of configurations. An alternative is to use composites of the variables in analyses—for example, by assigning a single variable to the role of output indicator and sum the remaining variables to serve as the input indicator. Provided that each available variable is sufficiently valid for analysis, the use of composites may enhance the distinction between taxonic and dimensional structure relative to the more traditional use of individual variables as indicators, although this does yield fewer curves. Because composites—like multivariate procedures—may mask the influence of individual indicators on results, researchers might begin by analyzing individual indicators to help select their final indicator set and then conduct structural analyses using composite input indicators.

Although taxometric procedures can be implemented in a variety of ways (e.g., there are many ways to locate cut scores in MAMBAC and to create subsamples in MAXCOV/MAXEIG), we believe that a procedure usually should be implemented in only one way in an investigation—ideally the most powerful and informative way that can be identified through an examination of empirical sampling distributions of results for taxonic and dimensional comparison data (see chap. 4). Executing a given taxometric procedure in multiple ways would provide only the weakest consistency test by supplementing the most powerful form of the analysis with additional forms that are less powerful and largely redundant. Thus, we recommend that researchers generate empirical sampling distributions of results for each procedure to identify the implementation approach that most powerfully distinguishes taxonic from dimensional structure and perform the procedure using this approach. An important exception to this recommendation involves the inchworm consistency test described later, which requires researchers to systematically vary the number of subsamples in a MAXCOV/MAXEIG analysis.

Constructing and Analyzing Multiple Sets of Indicators

In addition to performing multiple taxometric procedures using available indicators in different configurations, each procedure can be performed on multiple sets of indicators. As described in chapter 4, existing variables can be combined to create one or more indicator sets based on theoretical considerations (e.g., representing the target construct by means of one or more accepted theoretical conceptualizations), empirical criteria (e.g., modeling the empirical keying technique of scale construction using a criterion measure), or a blend of theoretical and empirical considerations. Each of the resulting indicator sets can be submitted to the array of taxometric procedures and consistency tests that seem most capable of distinguishing taxonic from dimensional structure given the properties of that indicator set. To the extent that taxometric results cohere across multiple indicator sets, greater confidence can be held in the resulting structural conclusion. Poor agreement suggests either that different conceptualizations of the target construct might represent different constructs, or that further work is needed to define and assess the target construct before its structure can be inferred with confidence.

Splitting a Sample of Data Into Subsamples for Replication

One other method of evaluating consistency that hinges on the repeated application of taxometric procedures is to divide a large sample of data into two or more subsamples and then perform the taxometric analyses in each. Although the goal of replication is an admirable one, there are two important and related limitations to this consistency testing approach. First, because these subsamples include the same set of indicators drawn from the same measures and, even more important, do not come from different populations, this technique only examines the influence of sampling error on results. Given the large sample size required for taxometric analysis, it may be a smarter choice to retain as large a sample as possible for all analyses, rather than increasing sampling error in all analyses by forming subsamples. More informative consistency testing may be achieved through nonredundant analytic techniques, multiple indicator sets constructed in different ways or selected from different measures, multiple samples of data drawn from different types of populations, and so forth.

Second, any sacrifice in sample size can be especially problematic when there are few suspected taxon members in the full sample. Because the number of putative taxon members can critically influence the ability of an analysis to distinguish taxonic from dimensional structure (see chap. 4), spreading a small number of taxon members across subsamples can reduce the odds of their successful detection in any subsample, let alone their consistent detection across all subsamples. Thus, we recommend against splitting a sample into subsamples unless the total sample size is unusually large and the absolute size of the a taxon (not just its base rate) is believed to be substantial.

Inchworm Consistency Test

Waller and Meehl (1998) introduced a technique that can help to identify a small taxon in MAXCOV/MAXEIG analyses. This method, known as the *inchworm consistency test*, is based on predictable changes in curve shape that are observed across a series of otherwise identical analyses with gradually increasing numbers of overlapping windows. To understand why increasing the number of windows can be useful, consider the analysis of a data set in which 100 taxon members are mixed with 900 complement members. Suppose that a researcher begins by performing a MAXEIG analysis using 20 windows that overlap 90% with their neighbors. Equation 6.6 shows that this analysis would include 345 cases per window. Thus, even if all 100 taxon members were included in the uppermost window—which is unlikely given that taxon and complement indicator distributions usually overlap considerably—they would still be outnumbered in that window (100 taxon members vs. 245 complement members). In lower scoring windows, complement members would outnumber taxon members to an even greater extent. The result would be a rising MAXEIG curve that slopes upward at its rightmost point rather than reaching a well-defined peak. Defining the peak would require an equal mixture of taxon and complement members in some window (yielding the maximum eigenvalue) followed by windows in which taxon members begin to outnumber complement members (causing eigenvalues to decline). Thus, an analysis using too few windows—and hence too large a subsample within each—would not allow the researcher to observe the leveling off and subsequent decline in eigenvalues that most clearly defines a taxonic peak.

Suppose that the analysis using 20 windows is followed by another using 100 windows, each containing 92 cases. If taxon members are sufficiently concentrated in the uppermost windows, they may begin to equal and then exceed the number of complement members, yielding a peak rather than a cusp on the MAXEIG curve. Of course increasing the number of windows decreases the sample size within each, and the consequent rise in sampling error can begin to obfuscate curve shape. Thus, the limit to which one can increase the number of windows is determined by one's tolerance for sampling error within each window. Equation 6.5 can be used to determine the largest allowable number of windows for a given subsample size. For example, if one wished to require a minimum of 50 cases per window to hold sampling error to an acceptable level, the final MAXEIG curve in a series implementing the inchworm consistency test in a sample with N = 1,000 could have as many as 191 windows.

Across a series of MAXCOV/MAXEIG analyses conducted with increasing numbers of windows, taxonic curves that are initially ambiguous, merely rising across the graph, should eventually become peaked as more windows are used. The emergence of a well-defined peak depends on several factors, including the full sample size, the number of taxon members, the degree of indicator validity and within-group correlations, and whether a sufficiently large number of windows can be used before sampling error degrades the shape of the curves. Similarly, there exists no algorithm for determining the optimum number of windows for a MAXCOV/MAXEIG analysis nor the intermediate values that should be used; this also depends on a variety of data characteristics. Empirical sampling distributions can be generated to identify a sequence of progressively larger numbers of windows that clearly distinguishes taxonic from dimensional structure in analyses of simulated comparison data. The final number of windows included in an analysis may far exceed the number of distinct scores on the input indicator; as demonstrated in chapter 6, this does not pose a problem for the production or interpretation of MAXCOV/MAXEIG curves.

In contrast to the increasingly well-defined peak that is expected to emerge when taxonic data are analyzed using the inchworm consistency test, dimensional structure is expected to continue producing nonpeaked curves regardless of the number of windows employed. Thus, the change (or lack of change) in a rising curve as the inchworm test is performed should help to resolve its structural ambiguity. To illustrate this point, a series of MAXEIG analyses was conducted using the inchworm consistency test. Only the two taxonic data sets containing the smallest taxa (TS#3 and TS#4) and their corresponding dimensional data sets (DS#3 and DS#4) were submitted to these analyses. Each variable served once as the input indicator, with the remaining variables serving as output indicators; all windows overlapped 90% with their neighbors. The number of windows was increased from 20 to 40 to 60, resulting in a panel of three curves for each indicator configuration. Because these analyses yielded 12 curves for each data set (4 indicator configurations × 3 numbers of windows), the curves were averaged across indicator configurations for presentation in Fig. 7.1. The results for TS#3 were unambiguous: As expected, the taxonic peak became increasingly well defined with a larger number of windows. Likewise, the results for DS#3 were clear: The obtained curves remained relatively nonpeaked regardless of the number of windows.

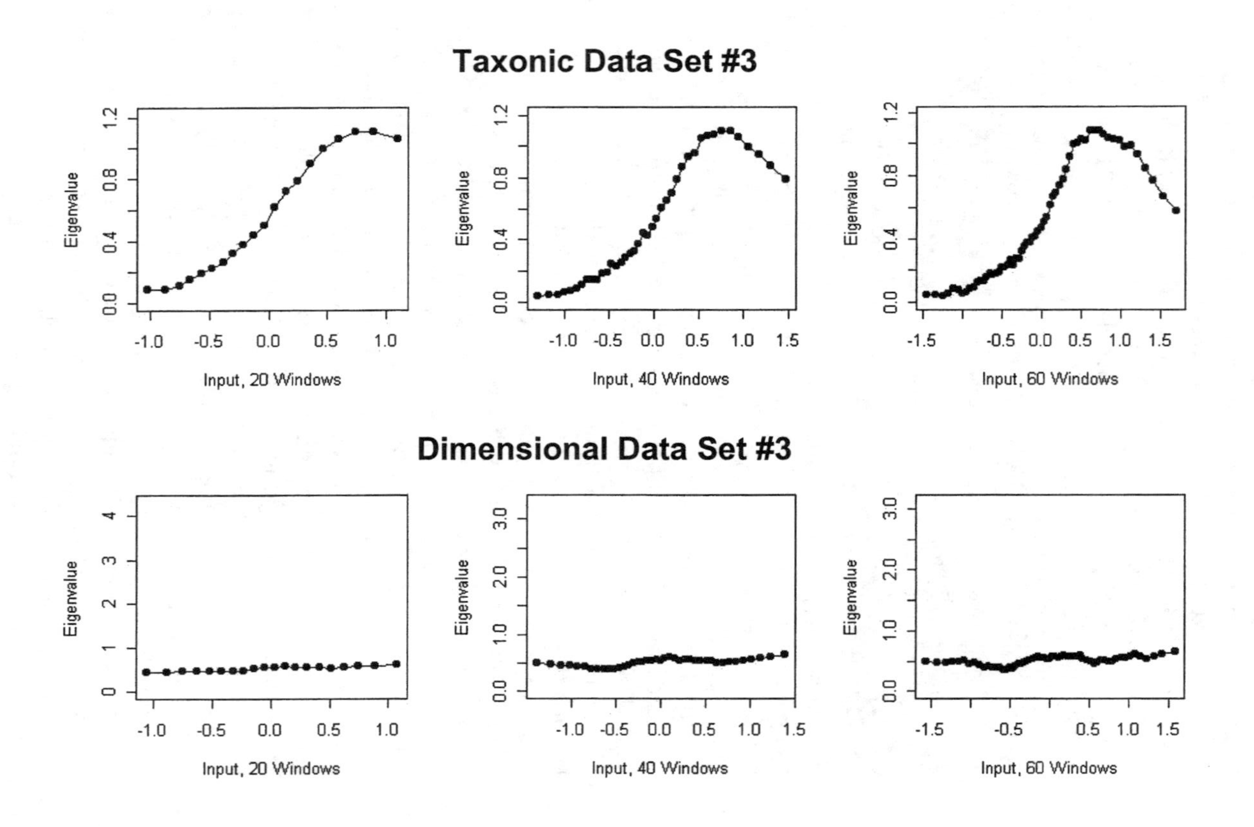
Taxonic Data Set #3
Eigenvalue
Input, 20 Windows
Input, 40 Windows
Input, 60 Windows
Dimensional Data Set #3
Eigenvalue
Input, 20 Windows
Input, 40 Windows
Input, 60 Windows

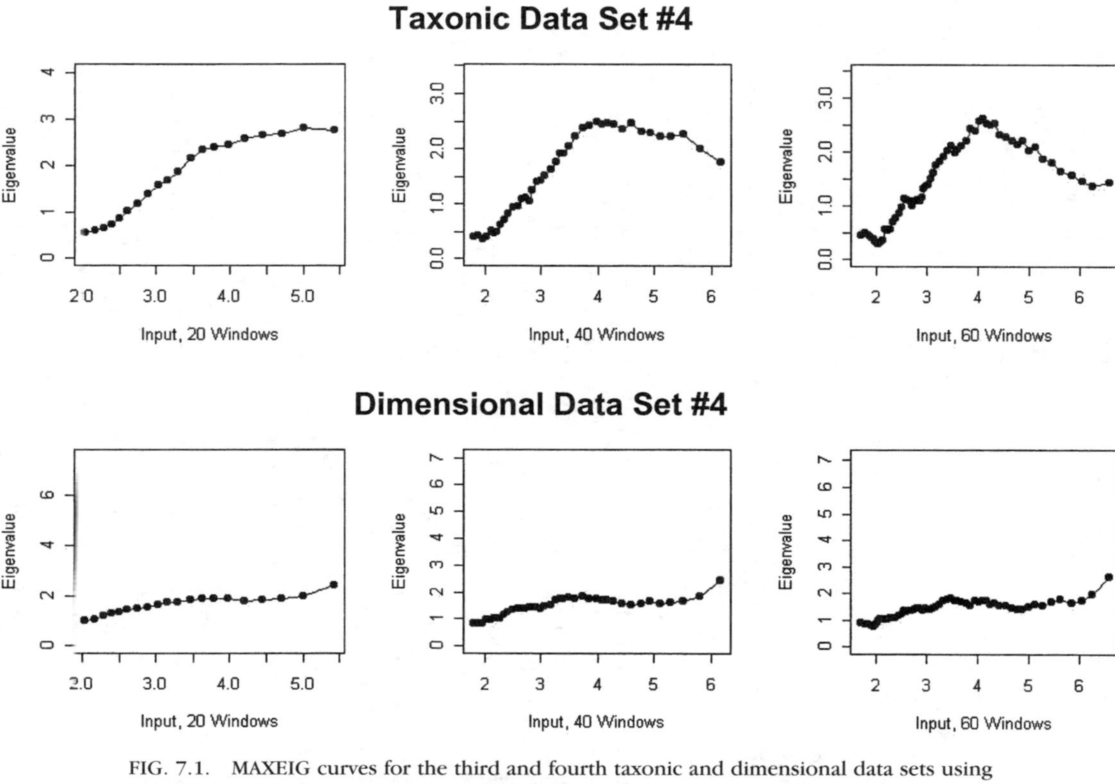

FIG. 7.1. MAXEIG curves for the third and fourth taxonic and dimensional data sets using the inchworm consistency test. Each variable served as the input indicator, with the remaining three variables serving as output indicators; 20, 40, and 60 windows that overlapped 90% with their neighbors were formed along the input indicator; and the four curves generated for each indicator configuration were averaged for presentation.

The results for TS#4 and DS#4 highlight that when indicators of a latent dimension are positively skewed, dimensional data can also produce rising curves with a right-end cusp (A. M. Ruscio & Ruscio, 2002; J. Ruscio, Ruscio, & Keane, 2004). This influence of skew stems from its differential restriction of range on the output indicators across windows. More specifically, the presence of positive skew means there will be considerably less output indicator variance within low-scoring windows than within high-scoring windows. A consequence of this differential restriction of range is a steady increase in the association (whether covariances or eigenvalues) among indicators across windows. The resultant rising curve may cause a dimensional construct assessed by skewed indicators to be mistaken for a small taxon.

Fortunately, the inchworm consistency test can be used to differentiate these structural possibilities. Whereas a small taxon should yield an increasingly well-defined peak with increasing numbers of windows, skewed indicators of dimensional construct should continue to produce merely rising curves. This expectation was borne out in analyses of TS#4, in which a peak did emerge, and DS#4, in which rising curves persisted. Analyses of TS#4 and DS#4 performed in chapters 5 and 6 underscored the challenge of detecting the signal of a small taxon amid the noise of indicator skew—and the consequent difficulty of drawing accurate structural inferences from these taxometric results. Given the ubiquity of indicator skew in research data (Micceri, 1989) and the frequency with which researchers use taxometrics to study rare phenomena, the inchworm consistency test may constitute one of the most powerful tools in the taxometric method. It can be particularly helpful when attempting to distinguish a small taxon from a dimensional construct with positively skewed indicators (J. Ruscio, Ruscio, & Keane, 2004), a situation that is often faced in taxometric investigations of psychopathology.

Waller and Meehl (1998, p. 45) proposed that when the inchworm consistency test is used to test for a small taxon, the overlap between adjacent windows—conventionally set to 90%—may be more appropriately set to 50%. However, as noted in chapter 6, we recommend against reducing the amount of overlap in this way because it sacrifices the number of points on the curve with no discernable payoff. The crucial benefit of the inchworm consistency test lies in creating increasingly smaller subsamples in which the number of taxon members can equal, and then surpass, the number of complement members. Reducing the overlap between windows is one way to achieve this goal, but another way is to simply increase the number of windows. Moreover, for a given sample size within each window, reducing the overlap yields fewer points on the curve than does increasing the number of windows. Consider, for example, a total sample of 1,000 cases. If one wished to create windows that overlap 50% with their neighbors and contain at least 50 cases apiece, one could construct as many as 39 windows. If the windows were allowed to overlap 90%, one could construct 191 windows. Both analyses would afford the same opportunity for taxon members to outnumber complement members in the uppermost windows, and both would allow the same amount of sampling error in the calculation of each point on the resultant curves. However, the latter analysis would provide nearly five times as many points for the MAXCOV/MAXEIG graph, which should form a smoother, more interpretable curve. Thus, although we do not argue that 90% overlap (or any other specific overlap value) is optimal for all analyses, we do recommend against reducing the degree of overlap between windows when performing the inchworm consistency test.

EXAMINING LATENT PARAMETERS AND CLASSIFIED CASES

In addition to generating curves whose shapes can be examined for consistency, each taxometric procedure also yields estimates of latent parameters that can be evaluated for convergence. One particular parameter, the taxon base rate, has been used most frequently for this purpose. Haslam and Kim (2002) observed that the coherence of base rate estimates was used as a consistency test in nearly half of all taxometric investigations that they reviewed, and our review of published studies (see chap. 10) puts the figure at 72%. The popularity of base rate comparisons may be attributed to the fact that, unlike some other parameters of a taxonic model, knowing the size of a taxon in a given population is important for many theoretical and empirical purposes other than arriving at a structural inference. Therefore, it should come as no surprise that estimation of the taxon base rate has been a focal point in the development and presentation of taxometric procedures. Mirroring this emphasis, we focus on the use of taxon base rate estimates as consistency checks. We then identify some additional parameters that might be used for this purpose.

Estimating and Comparing Base Rates of Taxon Membership

Large numbers of base rate estimates can be generated, summarized, and evaluated for consistency in a taxometric investigation. Each taxometric curve can be used to calculate one or more estimates of the proportion of the sample belonging to the taxon: MAXSLOPE, MAMBAC, MAXCOV, and MAXEIG each yield one base rate estimate per curve, whereas L-Mode yields four base rate estimates (see chap. 5 for their calculation and guidelines on selecting the most appropriate estimate to use that were provided by Waller & Meehl, 1998). Table 7.1 summarizes the base rate estimates yielded by MAXSLOPE, MAMBAC, MAXCOV, MAXEIG, and L-Mode analyses of the eight sample data sets in prior empirical illustrations of the procedures; all of these estimates were gathered from tables in chapters 5 and 6.

The standard method of interpreting a set of base rate estimates is to subjectively evaluate their degree of coherence (Meehl, 1973, 1995a; Meehl & Golden, 1982; Meehl & Yonce, 1994, 1996; Waller & Meehl, 1998). It has often been argued that if these estimates agree reasonably well within and across procedures, this supports the existence of a latent taxon whose size is being consistently estimated by each analysis. However, the standard argument goes, if the base rate estimates are discrepant within and between procedures, this suggests that no single entity is consistently being detected—that is, the obtained results are better explained by dimensional than taxonic structure.

Unfortunately, this argument may have several significant limitations. First, it may be overly simplistic. There are a number of circumstances in which dimensional latent structure can give rise to consistent base rate estimates. For example, rising MAXCOV/MAXEIG curves generated in analyses of positively skewed indicators tend to yield consistently low esti-

TABLE 7.1
Taxon Base Rate Estimates for Each Taxometric Procedure

Data Set	*MAXSLOPE* *M (SD)*[a]	*MAMBAC* *M (SD)*[b]	*MAXCOV* *M (SD)*[c]	*MAXEIG* *M (SD)*[d]	*L-Mode* *M*[e]	*Overall* *M (SD)*[f]
TS#1 (P = .50)	.43 (.23)	.47 (.19)	.59 (.25)	.45 (.09)	.48	.48 (.06)
TS#2 (P = .50)	.50 (.03)	.49 (.05)	.48 (.03)	.48 (.01)	.52	.49 (.02)
TS#3 (P = .25)	.25 (.06)	.26 (.05)	.25 (.05)	.19 (.01)	.32	.25 (.05)
TS#4 (P = .30)	.43 (.23)	.33 (.07)	.25 (.10)	.23 (.01)	.46	.34 (.10)
DS#1	.52 (.35)	.52 (.14)	.72 (.30)	.59 (.22)	.51	.57 (.09)
DS#2	.39 (.41)	.49 (.03)	.19 (.19)	.26 (.02)	.50	.37 (.14)
DS#3	.39 (.18)	.45 (.06)	.32 (.24)	.40 (.09)	.51	.41 (.07)
DS#4	.43 (.14)	.40 (.04)	.18 (.09)	.22 (.03)	.68	.38 (.20)

[a]Estimated using pairwise input–output indicator scatterplots for TS#1–TS#3 and DS#1–DS#3, summed input indicator scatterplots for TS#4 and DS#4.

[b]Estimated using pairwise input–output indicator roles and .25 *SD* cuts.

[c]Estimated using indicator triplets and .25 *SD* intervals.

[d]Estimated using the GCMT technique, one input indicator and three output indicators per curve, and 20 windows that overlapped 90% with their neighbors.

[e]Values were chosen following the guidelines of Waller and Meehl (1998) and presented in chapter 5; when none of their criteria was met (e.g., estimates for the two modes differed substantially) and were neither both above nor both below .50, the average estimate for the two modes was used.

[f]Calculated across the estimates for the five procedures listed in this table.

mates. Second, this perspective focuses on the probability that coherent (or incoherent) base rate estimates will be produced by a given structure; it does not address what researchers usually wish to know in practice, which is the probability that a given structure gave rise to the observed coherence (or incoherence) of base rate estimates. These inverse probabilities cannot be equated, and the agreement of multiple base rate estimates may constitute only weak evidence for a taxonic model under many conditions. Third, the notion of reasonable agreement among base rate estimates has seldom been quantified. Schmidt, Kotov, and Joiner (2004) suggested that a threshold of .10 be used to infer latent structure, with *SD*s below this cutoff indicative of taxonic structure, but the most rigorous evaluation of this threshold calls it into question (J. Ruscio, 2005).

Because many investigators have drawn taxonic conclusions from the consistency of base rate estimates, we believe that it is especially important to consider some cautionary notes about their interpretation. First, consider how the taxon base rate is estimated from a MAMBAC curve. As shown in Eq. 5.1, the two endpoints of the curve are the only factors contributing to this calculation. Thus, anything that influences one or both of these points will influence the base rate estimate. For example, skewed indicators will tilt MAMBAC curves, with positive skew producing rising curves and negative skew producing falling curves (A. M. Ruscio & Ruscio, 2002). To the extent that the base rate estimates for a panel of MAMBAC curves are all biased in the same direction, this will artificially constrain their variability. Hence, as was the case for DS#4, it is possible to obtain very consistent estimates of the taxon base rate when highly skewed indicators of a latent dimension are analyzed using MAMBAC (J. Ruscio, Ruscio, & Keane, 2004). Although Meehl and Yonce (1994) suggested that MAMBAC base rate estimates lower than .40 (or, by virtue of the symmetry of taxometric curves for small and large taxa, greater than .60) support a taxonic conclusion, this suggestion may have been premature, resulting from their exclusive analysis of normally distributed data. When skewed data are analyzed, far more extreme base rate estimates are possible even for dimensional data, and these estimates can be quite consistent with one another.

Even when indicator skew is not a problem, MAMBAC tends to yield similar shapes for each curve in a panel, thus providing fairly consistent base rate estimates for dimensional as well as taxonic data. As is apparent in Table 7.1, the base rate estimates for each dimensional data set that did not contain skewed indicators (DS#1 through DS#3) were about as consistent as those for its corresponding taxonic data set (TS#1 through TS#3). The estimates for TS#1 and DS#1 were the least consistent because their indicators were the least valid, leading to greater variability in curve shapes.

It should also be noted that within-group correlations can bias the size of base rate estimates. For example, indicator correlations within the taxon will cause the right end of a MAMBAC curve to rise more sharply, thereby deflating base rate estimates; correlations within the complement will have the opposite, inflating effect; and correlations in both the taxon and complement will elevate both ends of the curve, thereby biasing estimates toward .50 (see chap. 5). Thus, differential within-group correlations among the indicators that generate each MAMBAC curve can decrease the consistency of base rate estimates. In summary, unless one can establish that both indicator skew and within-group correlations are negligible, one should interpret the size and consistency of MAMBAC base rate estimates with caution.

As noted in our discussion of the inchworm consistency test, skewed indicators can also cause a MAXCOV/MAXEIG curve to rise (in the case of positive skew) or fall (in the case of negative skew; A. M. Ruscio & Ruscio, 2002; J. Ruscio, Ruscio, & Keane, 2004). Just as in MAMBAC, this can yield highly consistent estimates of the taxon base rate even for dimensional data, as was the case once again for DS#4. Therefore, not only is consistency of base rate estimates *within* a taxometric procedure potentially misleading, but consistency *across* procedures can also be misleading when multiple procedures are influenced in similar ways by the same data characteristics.

Another parallel to MAMBAC is that within-group correlations can bias the size of base rate estimates obtained from MAXCOV/MAXEIG. To the extent that there are comparable indicator correlations in both the taxon and complement, all points on the curve will be elevated by a similar amount (Meehl, 1995b). This does not bias the location of the hitmax cut (Meehl, 1995b), but it does reduce the difference in height between the maximum covariance/eigenvalue and other covariances/eigenvalues on the curve. As shown in Appendix B, the proportion of taxon members in a particular subsample is determined by the extent to which the observed covariance (or eigenvalue, see Appendix C) is less than the maximum covariance. For subsamples below the hitmax, lower covariances generate lower p estimates; for subsamples above the hitmax, lower covariances generate higher p estimates. Thus, if all covariances on the curve are elevated due to within-group correlations, none of the p estimates differs as much from .50 as they would have in the absence of such correlations, and the overall estimate of P is biased toward .50. To the extent that indicators are correlated more strongly within the taxon than the complement (or vice versa), taxon base rate estimates may be biased upward or downward rather than toward .50.

Each of these hypothesized influences on the accuracy and consistency of MAMBAC base rate estimates were borne out in a recent Monte Carlo study (J. Ruscio, 2005). In fact, all else being equal, the *SD*s of MAMBAC base rate estimates usually were lower for dimensional than taxonic data, suggesting that base rate consistency should be interpreted with great caution. Results for MAXCOV and MAXEIG base rate estimates were less problematic than those for MAMBAC, in the sense that taxonic data often resulted in lower *SD*s than dimensional data. However, the difference between *SD*s for taxonic and dimensional data sets possessing otherwise comparable properties (e.g., indicator correlations and skew) was rather modest at times and was considerably reduced as within-group correlations increased.

Two broad conclusions were well supported by the results of this study. First, for a given taxometric procedure, there was no single threshold *SD* below which one could safely infer taxonic structure and above which one could safely infer dimensional structure. The coherence of taxon base rate estimates depended on the complex interplay of several data conditions, hence the threshold used for interpretation would need to vary with a number of data parameters. Indeed, because the differences between the *SD*s for taxonic and dimensional data were highly variable—and often very small—across the conditions studied, even a scheme that allows for variable thresholds that take into account the relevant data properties may perform poorly. Second, across taxometric procedures, there was again no single threshold that distinguished the two structures. In fact, substantial discrepancies between MAMBAC and MAXCOV/MAXEIG estimates were the rule, not the exception, even for taxonic data sets. This finding not only underscores the challenge of applying either fixed or variable thresholds to cross-procedure *SD*s, but suggests that, under certain circumstances, agreement between MAMBAC and MAXCOV/MAXEIG base rate estimates may actually constitute evidence *against* taxonic structure.

As a case in point, consider the base rate estimates obtained through analyses of 100 taxonic data sets in which a relatively small taxon (P = .10, N = 1,000, hence n_{taxon} = 100) was identified by four highly valid (d = 2.00) indicators that were normally distributed within groups and correlated within groups at r = .20 (a nontrivial yet small value). On the whole, these were highly favorable, even enviable, data conditions for taxometric analysis. Averaged across full panels of curves and all replication samples, base rate estimates ranged from M = .07 for MAXEIG analyses (using the hitmax estimation method) to M = .13 for MAXCOV analyses to M = .26 for MAMBAC analyses. As a point of comparison, consider the extent of cross-procedure agreement for analyses of 100 dimensional data sets with N = 1,000, in which four nonskewed indicators correlated at a comparable level to those in the taxonic data sets (r = .41). Base rate estimates yielded by these dimensional data sets ranged from M = .48 for MAXEIG analyses to M

= .50 for both MAXCOV and MAMBAC analyses. Thus, even under favorable conditions, taxonic data can lead to highly discrepant base rate estimates, whereas dimensional data can lead to highly coherent estimates across procedures. This example—not at all atypical of the results of this Monte Carlo study—was presented to reinforce the need for caution when drawing structural inferences from base rate estimates. Although the expected low *SD* of estimates for taxonic data were observed in many cases, large *SD*s were also observed in a surprising number of cases, and the expected difference in *SD*s for taxonic and dimensional data was neither consistently observed nor particularly strong under many conditions.

Although this fully crossed, factorial Monte Carlo study yielded a wealth of information about the accuracy and consistency of MAMBAC, MAXCOV, and MAXEIG base rate estimates, the influence of factors such as indicator validity, within-group correlations, and indicator skew on base rate estimation techniques for MAXSLOPE or L-Mode is less understood and requires further investigation. We return to this controversial consistency test in a Monte Carlo study employing a representative sampling design toward the end of this chapter.

Estimating and Comparing Other Latent Parameters

In addition to the taxon base rate, other latent parameters can be estimated from the results of a taxometric analysis. In principle, any of these parameters can be compared for consistency, although this has seldom been done in taxometric investigations.

As noted in chapter 6 and procedurally detailed in Appendix B, after a MAXCOV or MAXEIG analysis is performed, one can estimate the valid positive (*VP*) and false positive (*FP*) rates achieved by the hitmax cut on the input indicator. If each indicator serves as the input more than once (e.g., when four or more indicators are used in all possible input–output indicator triplets in MAXCOV), multiple estimates of the *VP* and *FP* rates will be available for each indicator. These estimates can be examined for consistency and then averaged for subsequent use.

Next, using an estimate of the taxon base rate and the *VP* and *FP* rates achieved by the hitmax cut on each indicator, the probability of taxon membership can be calculated for each case. Cases can be assigned to the taxon and complement groups using Bayes' Theorem, the base rate classification method, or, for L-Mode, classification based on the minimal absolute distance of each case's estimated factor score to the scores beneath the two modes in the distribution of factor scores. Once each case has been assigned to the more probable latent class, one can estimate: (a) the distribution of taxon and complement members' scores on each indicator; (b) summary statistics such as the *M*, *SD*, skew, and kurtosis of the taxon and complement classes on each indicator; (c) the indicator correlations within each group; and (d) the validity of each indicator, in terms of both raw mean differences (e.g., $\bar{x}_t - \bar{x}_c$) and standardized mean differences (expressed in the more familiar metric of Cohen's *d* as calculated in Eq. 4.1). Any of these estimates can be compared across analyses to assess the consistency of results. Meehl and Yonce (1994, 1996) presented extensive findings on the accuracy with which latent parameters such as indicator validity can be estimated from MAMBAC and MAXCOV results for taxonic data sets. Unfortunately, as is the case for taxon base rate estimates, little empirical guidance is available for deciding when parameter estimates are consistent enough to warrant a taxonic interpretation or divergent enough to warrant a dimensional interpretation.

Notably, besides the Monte Carlo study of taxon base rate estimates described earlier, no research has yet examined whether parameter estimates really are more consistent in the presence of taxonic than dimensional structure. To begin a preliminary examination of this issue, we computed estimates of indicator validity (in standardized units) for our eight data sets; these appear in Table 7.2. For the taxonic data sets, MAXCOV, MAXEIG, and L-Mode analyses yielded reasonably accurate and consistent estimates of indicator validity; this was particularly true for TS#2 through TS#4, both of which possessed greater indicator validity than TS#1. However, contrary to expectations, the *SD*s of cross-procedure validity estimates for the dimensional data were about as low as those for the taxonic data. These results suggest that the consistency of indicator validity estimates yielded by different taxometric procedures may not provide a very useful clue to latent structure. However, further research is needed to investigate this possibility using estimates of indicator validity or other latent parameters to determine which, if any, may afford informative consistency tests. Like all investigations of consistency tests, such studies should demonstrate not only that taxonic and dimensional data yield sufficiently different results across a range of data conditions typical of actual research data, but also that the results are not simply redundant with those yielded either by the taxometric procedures or by other consistency tests. The incremental validity of results provided by a particular taxometric procedure or consistency test, over and above any others that were previously performed, has received virtually no empirical attention in the taxometrics literature.

Case Removal Consistency Test

Earlier, we cautioned researchers against randomly dividing a sample into subsamples to examine consistency through replication of results. However, one technique performed in subsamples of data does appear to have

TABLE 7.2
Indicator Validity Estimates for MAXCOV, MAXEIG, and L-Mode

Data Set and Procedure	*Indicators in Taxonic Data Sets*				*Indicators in Dimensional Data Sets*			
	1	*2*	*3*	*4*	*1*	*2*	*3*	*4*
#1 Actual	.98	.91	1.09	.95	—	—	—	—
MAXCOV	1.28	1.12	1.24	1.21	1.23	1.19	1.13	1.14
MAXEIG	.83	.65	.94	1.58	1.37	.38	.86	.76
L-Mode	1.26	1.09	1.32	1.16	1.24	1.27	1.07	.96
SD[a]	.25	.26	.20	.23	.08	.49	.14	.19
#2 Actual	2.06	2.12	1.89	1.99	—	—	—	—
MAXCOV	2.13	2.06	1.92	2.05	1.73	1.58	1.64	1.73
MAXEIG	2.21	1.82	1.60	1.96	1.60	1.50	1.54	1.66
L-Mode	2.07	2.11	1.93	2.03	—[b]	—[b]	—[b]	—[b]
SD[a]	.07	.16	.19	.05	.09	.06	.07	.05
#3 Actual	2.04	2.03	1.96	1.83	—	—	—	—
MAXCOV	1.99	2.12	1.94	1.98	1.54	1.66	1.55	1.56
MAXEIG	1.94	2.01	1.92	1.77	1.19	1.53	1.56	1.32
L-Mode	2.00	2.05	1.92	1.91	1.49	1.62	1.59	1.48
SD[a]	.03	.06	.01	.10	.19	.07	.02	.12
#4 Actual	2.10	2.36	2.01	2.21	—	—	—	—
MAXCOV	2.15	2.35	2.15	2.47	1.90	2.01	2.10	2.20
MAXEIG	1.94	2.18	2.07	1.94	1.78	1.94	2.04	1.86
L-Mode	1.67	2.19	2.16	2.36	1.35	1.59	1.79	1.96
SD[a]	.24	.10	.05	.28	.29	.23	.16	.17

Note. Indicator validity is expressed as Cohen's *d*. MAXCOV classifications were based on analyses with indicator triplets and .25 *SD* intervals. MAXEIG classifications were based on analyses with one input indicator and three output indicators using 20 windows that overlapped 90%.

[a]These values were computed across the three procedures to assess consistency.

[b]Indicator validities could not be estimated for this analysis because the L-Mode procedure classified all cases into the taxon.

some promise as a consistency test. This technique, the *case removal consistency test*, removes a *targeted* subsample of cases from the data to see how taxometric results—particularly estimates of the taxon base rate—change (Meehl & Yonce, 1994; J. Ruscio, 2000). For example, if taxometric analyses in the full sample of data suggest the presence of a small taxon, one might remove some of the lowest scoring cases (probable complement members) to determine whether the curve shape and taxon base rate estimates change in ways predicted by taxonic structure. Following such targeted removal of cases, the location of taxonic peaks in MAMBAC or MAXCOV/MAXEIG graphs may be expected to change, whereas the estimated taxon base rate would be expected to increase by an amount commensurate with the proportion of cases removed.

To illustrate this technique, we performed MAXEIG analyses of TS#3, DS#3, TS#4, and DS#4 in the full sample (N = 1,000) and then computed total scores across all four indicators and re-ran MAXEIG on the highest scoring 800 and 600 cases, respectively. The results appear in Fig. 7.2. For both TS#3 and TS#4, the taxonic peak that initially emerged toward the right end of the curve appeared to shift toward the left with the removal of low-scoring cases. Close inspection of the x axis of each graph, however, revealed that the hitmax cut remained in about the same absolute location, with the apparent leftward migration of the taxonic peak caused by the elimination of lower scoring values from the input indicator following case removal. Examination of the taxon base rate estimates found them to decrease in an orderly and predictable manner with the targeted removal of probable complement members. In the full sample of TS#3, the estimated base rate was .23 (close to the correct value of .25). If, in fact, there were 230 taxon members in the full sample of 1,000 cases, the removal of 200 complement members would be expected to yield a base rate estimate of (.23 × 1000)/800 = .29. However, because the lowest scoring 200 cases might include some taxon members, the base rate would be unlikely to increase all the way to .29. Consistent with these expectations, the observed base rate estimate for the MAXEIG analysis performed in the subsample of 800 highest scoring cases was .26. Using the estimate of .26 to predict the likely outcome of removing the next 200 lowest scoring cases—most of whom presumably belong to the complement—yielded an upper limit base rate estimate of (.26 × 800)/600 = .35. The observed value for the MAXEIG analysis performed in the subsample of 600 highest scoring cases was .32, again in reasonable agreement with expectations.

In contrast, DS#3 yielded very different results. The MAXEIG curves continued rising toward the right despite the removal of low-scoring cases. Moreover, estimates of the taxon base rate *decreased* across analyses as presumed complement members were removed, yielding values of .34, .18, and .17, respectively. It could be argued that the change in results from the full sample (N = 1,000) to the highest scoring 800 cases is consistent with the existence of a small taxon, undetected in the full sample due to its small base rate, but accurately detected in the reduced sample, yielding an apparent right-end peak and low estimated taxon base rate. However, follow-up results for the highest scoring 600 cases contradicted this interpretation. If, in fact, there was a small taxon among the highest scoring 800 cases, one would expect its base rate to increase when MAXEIG was performed among the highest scoring 600 cases, approaching (.18 × 800)/600 = .24. Instead, the observed value decreased to .17. What seemed to occur here was that the removal of low-scoring cases accentuated the influence of positive indicator skew among the re-

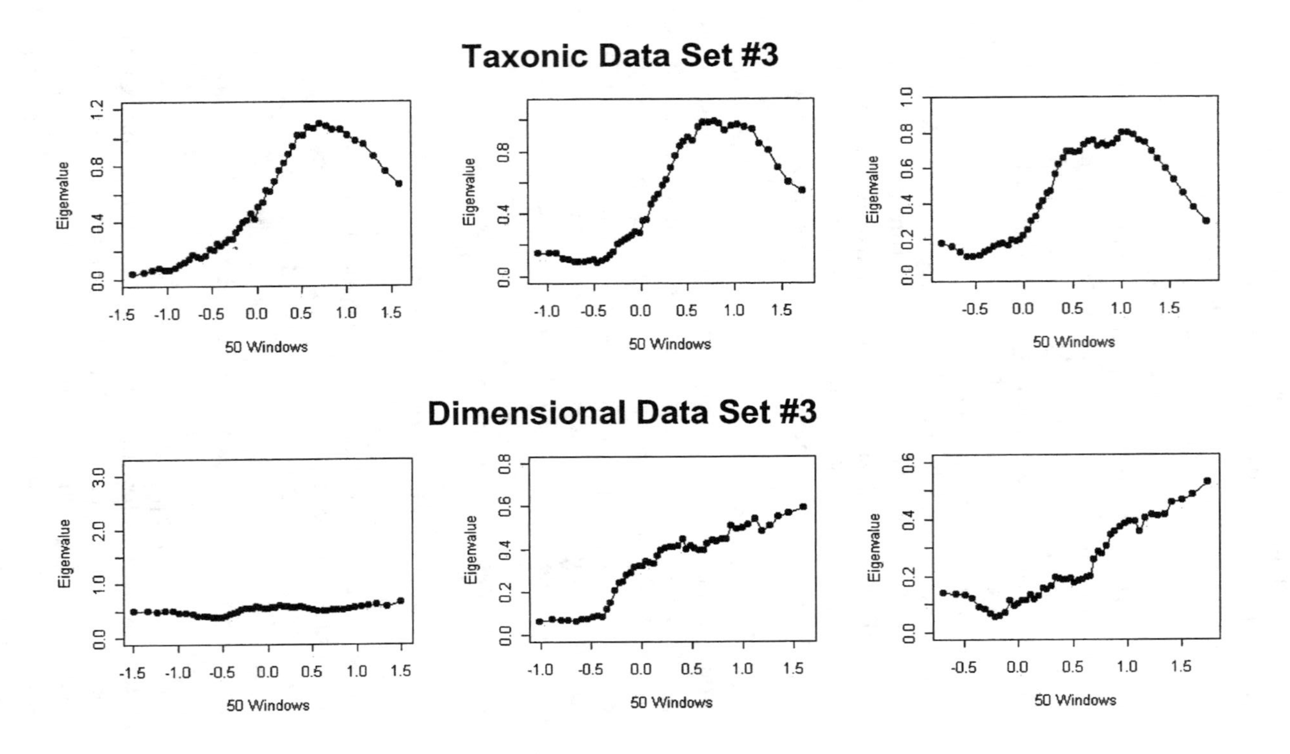
Taxonomic Data Set #3
Dimensional Data Set #3
Eigenvalue
50 Windows

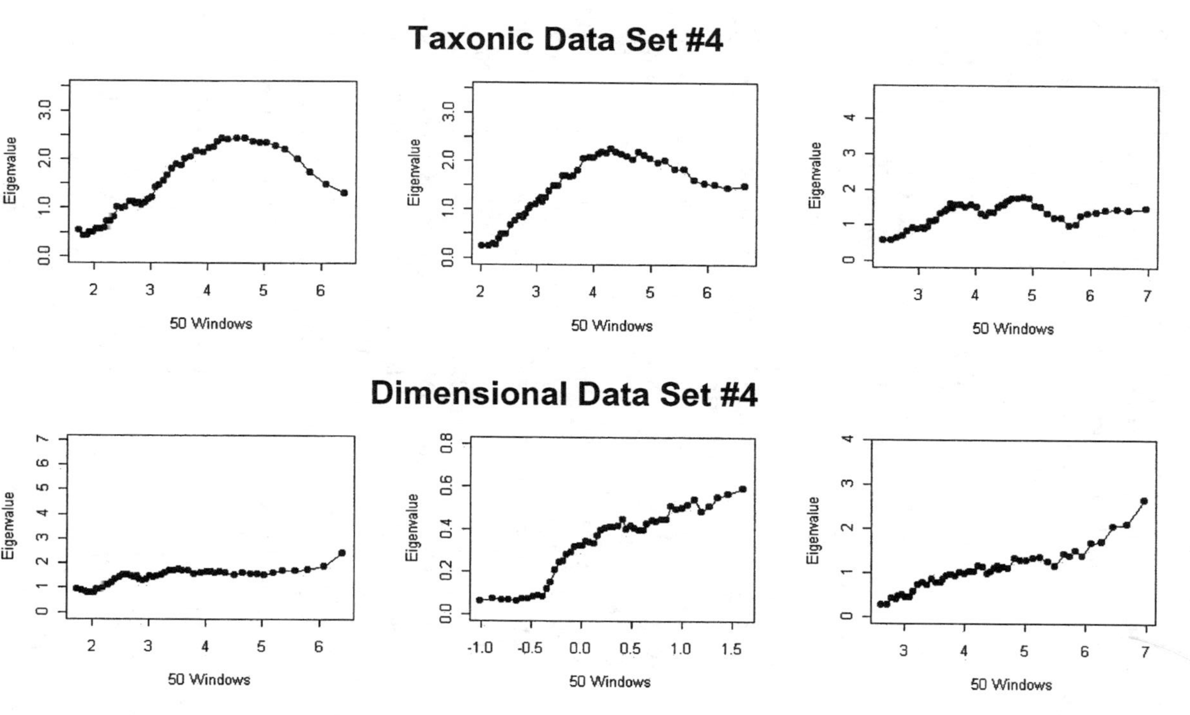

FIG. 7.2. MAXEIG analyses performed on the full sample (N = 1,000) followed by subsamples of the highest scoring 800 and 600 cases, respectively. Each variable served as the input indicator, with the remaining three variables serving as output indicators, 50 windows that overlapped 90% with their neighbors were formed along the input indicator, and the four curves generated for each series of analyses were averaged for presentation.

maining cases, with the rising MAXEIG curves yielding lower estimates of the taxon base rate.

Similar results emerged in analyses of TS#4 and DS#4. A taxonic peak was present in each MAXEIG analysis of TS#4 and appeared to migrate leftward with the removal of low-scoring cases. The estimated taxon base rates for analyses in the full sample (N = 1,000), the highest scoring 800 cases, and the highest scoring 600 cases were .29 (near the correct value of .30), .34 (close to the upper limit expected value of [.29 × 1000]/800 = .36), and .39 (rising toward the upper limit expected value of [.34 × 800]/600 = .45). In contrast, a rising curve shape was accentuated in MAXEIG analyses of DS#4 as low-scoring cases were removed; estimates of the taxon base rate decreased from .23 in the full sample to .15 among the highest scoring 800 cases and .12 among the highest scoring 600 cases. Again, the progressively steeper rising curves across the three panels suggested that indicator skew was driving curve shape, with the decreasing taxon base rate estimates further contradicting a taxonic conclusion.

Reanalysis following the targeted removal of cases appears to be a promising way to evaluate the consistency of results indicative of taxonic structure. However, although the visual inspection of curves following case removal may be informative, researchers should exercise caution when examining curves obtained from subsamples of data because many parameters other than the taxon base rate are affected by sample restriction. For example, the indicator means and variances of one or both latent classes will likely be altered; changes in the indicator distributions can influence the validities and within-group correlations of the indicators and may yield rising curves. Thus, it may be more valuable to focus on changes in taxon base rate estimates, rather than changes in curve shapes, as analyses are performed with increasingly small subsamples of cases. Whereas a rising MAXCOV/MAXEIG curve obtained in a full sample of data may be suggestive of a small taxon, failure of estimates of the taxon base rate to increase with the removal of low-scoring cases suggests that the curve shape may stem from indicator skew rather than taxonic structure.

In addition to cautiously interpreting changes in curve shape, investigators should also bear in mind that removing cases to perform follow-up analyses will increase sampling error, which can degrade the clarity of obtained results. For example, when analyses were performed using only 60% of the cases in TS#4, the taxonic peak emerged with considerably less clarity than in analyses of the full sample or the highest scoring 80% of cases. It may be difficult to determine whether the disappearance of an apparent taxonic peak across analyses of increasingly small subsamples constitutes evidence of dimensional latent structure or merely an artifact of increased sampling error and decreased indicator validity.

Distribution of Bayesian Probabilities of Taxon Membership

One frequently used consistency test based on the classification of cases involves examining the distribution of Bayesian probabilities of taxon membership. To perform this test, one calculates the probability of taxon membership for each case in the data set using Bayes' Theorem and then inspects the frequency distribution of these probabilities. If cases pile up near probabilities of 0 or 1, with few cases scoring at intermediate values, this is taken to support a taxonic interpretation (Waller & Meehl, 1998, pp. 29–30). In other words, if cases can be classified into either the taxon or the complement with high probability, this seems to suggest that the two groups actually exist. Provided that indicators are sufficiently valid, taxonic structure should produce such a U-shaped distribution of Bayesian probabilities.

Unfortunately, a U-shaped distribution can occur even when latent structure is dimensional. When indicators of a latent dimension are analyzed using MAXCOV/MAXEIG, a maximum covariance/eigenvalue can be identified for each curve despite that a hitmax cut does not denote a meaningful parameter in a dimensional model. Because the indicators are correlated with one another, many or most of the cases who score above the hitmax cut on one indicator will also score above this cut on the other indicators. Thus, probabilities estimated using Bayes' Theorem can approach 0 or 1 in the presence of a large number of correlated indicators. Consequently, there are a priori reasons to suspect that this consistency test may not distinguish taxonic and dimensional structure particularly effectively.

In an unpublished manuscript, Beauchaine and Beauchaine (2003) quantified the extent to which a distribution of Bayesian probabilities generated from MAXCOV results is U-shaped by calculating the proportion of cases whose estimated probability of taxon membership was less than .10 or greater than .90. Using a threshold of .80 to infer taxonic structure, Beauchaine and Beauchaine found that this consistency test poorly discriminated taxonic from dimensional structure across a wide range of data parameters (e.g., taxon base rate, indicator validity, within-group correlations). Whereas 98% of taxonic data sets surpassed the threshold, 75% to 80% of dimensional data sets also did so. On the basis of these results, these authors recommended against its use in taxometric investigations.

As Beauchaine and Beauchaine (2003) have shown, the Bayesian distribution consistency test clearly has the potential to yield ambiguous or misleading results. Unless Monte Carlo studies performed under realistic data conditions reveal a way to generate and evaluate such distributions with adequate discriminating power, we recommend that researchers avoid drawing inferences of taxonic structure on the basis of U-shaped curves. However, it is possible that obtaining non-U-shaped (e.g., fairly flat or normal-shaped) distributions when sufficiently valid indicators are analyzed may provide reliable evidence *against* a taxonic interpretation. However, this potential asymmetry in the utility of the test has not yet been examined and requires systematic investigation. This consistency test is also explored in the Monte Carlo study reported toward the end of this chapter.

T Tests Between Cases Classified Into the Taxon and Complement

A handful of taxometric reports have employed another approach based on the classification of cases. This involves assigning individuals in the sample to taxon and complement classes, then performing statistical comparisons (e.g., *t* tests) between the two classes on variables of interest—including the indicators or external measures—as a further test of latent structure. Although this practice may provide useful descriptive information about a taxon following its consistent detection in taxometric analyses, we contend that significant *t* tests cannot, in and of themselves, corroborate or refute a taxonic inference. This is not only because of the circularity of using the same variables to form groups and to perform structural tests, but also because mean differences should be observed on any valid indicator or correlate of the target construct, whether these reflect differences between taxa or differences between high and low scorers on one or more latent dimensions. In fact the absence of mean differences on the indicators would call into question the validity (and hence the appropriateness) of these variables for taxometric analysis. Thus, although significant differences between groups can highlight potentially important associations between the target construct and specific variables of interest, such differences should not be regarded as validating any particular structural conclusion. Two anonymous reviewers of this book expressed astonishment that anybody would have performed *t* tests as a consistency test in a taxometric investigation. We share their surprise. However, because it is not our goal in this book to critique individuals studies, we do not cite examples here.

To illustrate the futility of this technique as a consistency test, we performed *t* tests on all indicators in the four-dimensional data sets (see Table 7.3). Using the taxon base rate estimates shown in Table 7.1, we assigned a proportion P of the highest scoring cases to the taxon and the remaining cases to the complement. As is apparent from these results, even dimensional data produce substantial group differences (all ts $\geq$ 16.93) and statistically significant p values (all ps $< .001$) on variables re-

TABLE 7.3
T Tests for Dimensional Data Sets

Data Set	*Indicator*	*Taxon M (SD)*	*Complement M (SD)*	*t*	*d*
DS#1	1	.57 (.88)	−.43 (.86)	18.01	1.15
(*P* = .57)	2	.59 (.87)	−.44 (.85)	18.81	1.20
	3	.56 (.87)	−.42 (.88)	17.47	1.12
	4	.54 (.84)	−.41 (.91)	16.93	1.08
DS#2	1	.47 (.80)	−.80 (.78)	24.63	1.61
(*P* = .37)	2	.48 (.81)	−.82 (.72)	25.73	1.68
	3	.47 (.81)	−.80 (.76)	24.40	1.60
	4	.45 (.83)	−.77 (.76)	23.09	1.51
DS#3	1	.49 (.85)	−.71 (.75)	23.11	1.49
(*P* = .41)	2	.51 (.82)	−.73 (.75)	24.22	1.56
	3	.51 (.79)	−.73 (.79)	24.50	1.58
	4	.49 (.83)	−.71 (.78)	23.08	1.48
DS#4	1	3.06 (1.00)	2.08 (.31)	18.73	1.22
(*P* = .38)	2	3.92 (1.41)	2.24 (.59)	22.14	1.44
	3	4.77 (1.76)	2.36 (.92)	24.68	1.61
	4	5.74 (2.00)	2.34 (1.00)	30.79	2.01

Note. For each data set, cases were sorted by their total scores on all indicators. Using the overall taxon base rate estimate for that data set (shown in parentheses; see Table 7.1), a proportion *P* of the highest scoring cases served as the taxon and a proportion 1 − *P* of the lowest scoring cases served as the complement.

lated to the latent construct. Therefore, such differences should not be taken as evidence of taxonic structure. This may be the only consistency test reviewed here that we believe should never be used.

ASSESSING MODEL FIT

Goodness of Fit Index (GFI)

Waller and Meehl (1998) introduced a consistency test that has been employed in a number of taxometric investigations. The test consists of comparing the fit of an indicator variance–covariance matrix predicted by the taxonic structural model (S_{pred}) to the observed variance–covariance matrix (S_{obs}). To generate S_{pred}, one begins by estimating the taxon base rate (P) and the complement base rate ($Q = 1 - P$). Then one assigns each case to the taxon or complement class to estimate each indicator's variance within the taxon (s^2_{xt}) and complement (s^2_{xc}) as well as the validity with which each indicator distinguishes these two groups ($D_x = \bar{x}_t - \bar{x}_c$). Assuming that nuisance covariance is zero (that indicators are uncorrelated within groups), Eq. 6.2 expresses the covariance between two indicators in the following way:

$$\text{cov}(xy) = PQD_xD_y$$

Waller and Meehl (1998, p. 12) presented a similar equation for each indicator's variance:

$$s_x^2 = Ps_{xt}^2 + Qs_{xc}^2 + PQD_x^2 \tag{7.1}$$

Using these two equations, one constructs $\boldsymbol{S}_{pred}$ as the predicted values for each indicator's variance and covariance in the mixed-group sample. For example, with three indicators x, y, and z, $\boldsymbol{S}_{pred}$ is:

$$S_{pred} = \begin{bmatrix} Ps_{xt}^2 + Qs_{xc}^2 + PQD_x^2 & PQD_xD_y & PQD_xD_z \\ PQD_xD_y & Ps_{yt}^2 + Qs_{yc}^2 + PQD_y^2 & PQD_yD_z \\ PQD_xD_z & PQD_yD_z & Ps_{zt}^2 + Qs_{zc}^2 + PQD_z^2 \end{bmatrix} \tag{7.2}$$

To quantify the fit between the observed and predicted variance–covariance matrices, Joreskog and Sorbom's (2001) GFI is calculated. This index is well known in the literature on structural equation modeling (SEM). Whereas SEM researchers often supplement the GFI with other indexes that take into account the number of free parameters in the model being evaluated, to date none of these alternative indexes has been introduced into taxometric research.

On the basis of preliminary simulations, Waller and Meehl (1998, p. 27) reported that in a small-scale Monte Carlo study, taxonic data usually yielded GFI values above .90, whereas dimensional data rarely did so. However, subsequent research has failed to support this conclusion. For example, Cleland, Rothschild, and Haslam (2000) and Haslam and Cleland (2002) performed Monte Carlo studies that included evaluations of the performance of GFI values. Their results afford two tentative observations. First, GFI values were found to poorly discriminate taxonic from dimensional structure. Second, no particular cutoff was found to be universally optimal for differentiating the two structures: For some data conditions, both taxonic and dimensional data tended to yield GFI values below .90, whereas for other conditions, both types of data tended to yield values above—sometimes well above—.90. In Beauchaine and Beauchaine's (2003) unpublished study of MAXCOV consistency tests, an even more extensive evaluation of GFI values yielded similarly negative conclusions.

For illustrative purposes, we computed GFI values from MAXCOV analyses of our eight data sets (using indicators in all possible triplets, with 50 overlapping windows to accommodate the ordered categorical values in TS#4 and DS#4). Although GFI values were quite high for the taxonic data sets (.99, 1.00, .99, and .97 for TS#1 to TS#4, respectively), they were also high for the dimensional data sets (1.00, .84, .94, and .84 for DS#1 to DS#4, respectively), surpassing the suggested threshold in two cases. This underscores the cautions raised by prior research about the potential for drawing mistaken structural inferences from GFI values. A more rigorous test of the GFI than is possible using only our eight illustrative data sets is presented toward the end of this chapter.

Using Empirical Sampling Distributions to Examine Model Fit

We have been exploring alternatives to the GFI that may more usefully evaluate the fit between results obtained for the research data and results expected under the competing structural models. To understand the logic of our approach, it is important to consider two drawbacks inherent to the approach taken by the GFI. First, although the unique correlational and distributional characteristics of particular research data sets have been shown to exert an influence on the GFI (Beauchaine & Beauchaine, 2003; Cleland, Rothschild, & Haslam, 2000; Haslam & Cleland, 2002; J. Ruscio, Ruscio, & Meron, 2005), the threshold value that was suggested for the GFI (.90) does not take these influences into account. Second, the GFI provides a single measure of fit to a taxonic model, but no measure of fit to a dimensional model. This prohibits direct comparisons of fit to the two models.

Our approach addresses these drawbacks by quantifying the extent to which taxometric curves for the research data match those in the empirical sampling distributions generated through analyses of simulated taxonic and dimensional comparison data that reproduce the distributional and correlational characteristics of the research data (J. Ruscio, 2004; J. Ruscio & Ruscio, 2004a, 2004b, 2004c; J. Ruscio, Ruscio, & Meron, 2005). The approach begins by using the technique described in chapter 4 to bootstrap multiple samples of taxonic and dimensional comparison data. Next, these data sets are submitted to the same analysis that is applied to the research data. For example, after performing a series of MAXEIG analyses to obtain an averaged curve for the research data, two additional series of identical analyses are performed to obtain averaged curves for each set of simulated taxonic and dimensional data. Finally, in addition to visually assessing whether the research curves are more consistent with those produced by analyses of simulated taxonic or dimensional data, this comparison is quantified. We have explored the performance of various indexes of curve fit (e.g., J. Ruscio, 2004; J. Ruscio & Ruscio, 2004b), all of which work in similar ways, and we present the one that has been examined most thoroughly next (J. Ruscio, Ruscio, & Meron, 2005).

To calculate this comparison curve fit index (CCFI), one begins by calculating the root mean square residual (RMSR) of the y (ordinate) values on the averaged curves of the research data and the simulated data. This is done once to evaluate the fit of the taxonic comparison data and once to evaluate the fit of the dimensional comparison data:

$$Fit_{RMSR} = \frac{\sqrt{\sum (y_{res.data} - y_{sim.data})^2}}{N}, \qquad (7.3)$$

where $y_{res.data}$ refers to a data point on the curve for the research data, $y_{sim.data}$ refers to the corresponding data point on the curve for taxonic or dimensional comparison data, and N is the number of points on each curve. *Lower* values of Fit_{RMSR} reflect better fit, with perfect fit represented by a value of 0. The two fit values, $Fit_{RMSR\text{-}tax}$ and $Fit_{RMSR\text{-}dim}$, are then integrated into a single index as follows:

$$\text{CCFI} = Fit_{RMSR\text{-}dim} \,/\, (Fit_{RMSR\text{-}dim} + Fit_{RMSR\text{-}tax}) \qquad (7.4)$$

CCFI values can range from 0 to 1, with lower values suggesting better fit for dimensional structure and higher values suggesting better fit for taxonic structure. The index is symmetric about .50 in that this middle value represents equivalent fit for both structures. It is important to note that the CCFI indexes the relative fit of taxonic and dimensional structural models, not the absolute goodness of fit of either model. CCFI values close to .50 are ambiguous in that the fit may, in an absolute sense, be good *or* poor for both structures. For example, two groups poorly separated by each indicator may yield ambiguous CCFI values because there is little difference between a taxonic model with highly overlapping groups and a purely dimensional model. Likewise, data with a more complex structure (e.g., four groups) may yield ambiguous CCFI values because neither the (two-group) taxonic model nor the dimensional model fits well. To the extent that the CCFI deviates from .50 in either direction, this suggests that one structural model fits better than the other. When the CCFI closely approaches either 0 or 1, this provides suggestive evidence that the absolute fit of one model may be good. Again, we underscore the point that the CCFI—like the taxometric method more broadly—aims to determine the relative fit of two specified structural models, not the absolute fit of either.

It may be useful to revisit a point raised in chapter 3: Some fit indexes penalize researchers for working with large samples. Specifically, very high statistical power can sometimes lead to the rejection of an otherwise acceptable structural model when deviations from perfect fit are assessed or when model fit is confounded with the extent to which assumptions are satisfied (Meehl, 1967). It is important to note that the CCFI is not susceptible to such problems. This index does not lead to acceptance or rejection decisions based on the closeness of fit to a structural model. Instead, as sample size increases, both the research curve and the simulated comparison curves become more stable, which should increase the validity of the fit index. All else being equal, the reduction of sampling error with larger samples should afford more valid visual inspections of curves and CCFI values that more accurately discriminate between taxonic and dimensional latent structure.

Illustrating the CCFI

To illustrate the use of this fit index, we performed MAMBAC and MAXEIG analyses on four of our target data sets as well as taxonic and dimensional comparison data simulated to reproduce the distributional and correlational characteristics of each target data set. First, we generated taxonic and dimensional comparison data for TS#3 and DS#3, data sets that previously yielded clearly distinguishable results for each taxometric procedure. Next, we applied this technique to TS#4 and DS#4, data sets that previously yielded difficult-to-interpret results for most taxometric procedures and consistency tests. For each analysis, we generated $B = 10$ bootstrap samples each of taxonic and dimensional comparison data. To simulate taxonic comparison data, fallible criteria denoting membership in the putative taxon were generated by means of the base rate classification method using the estimated taxon base rate from the averaged research curve.

The averaged curves for all analyses are shown in Fig. 7.3. Within each pair of MAMBAC or MAXEIG graphs, the dark curve for the research data is superimposed on the light curves for comparison data sets simulated to have either taxonic (left) or dimensional (right) structure; the two comparison curves in each graph fall ±1 *SD* (which in this context represents one standard error) from the average of the curves for 10 simulated data sets of that structure. In all analyses, the research data produced a curve that more closely resembled that of its known latent structure (i.e., taxonic for TS#3 and TS#4, dimensional for DS#3 and DS#4). Each research curve fell within—or very close to within—the ±1 *SD* bounds for curves produced by simulated comparison data of the correct structure. Thus, a simple visual comparison of the results produced by research data and simulated comparison data can provide a valuable consistency check that facilitates accurate inferences of latent structure, even under data conditions that produced ambiguous or misleading results for many other analytic techniques.

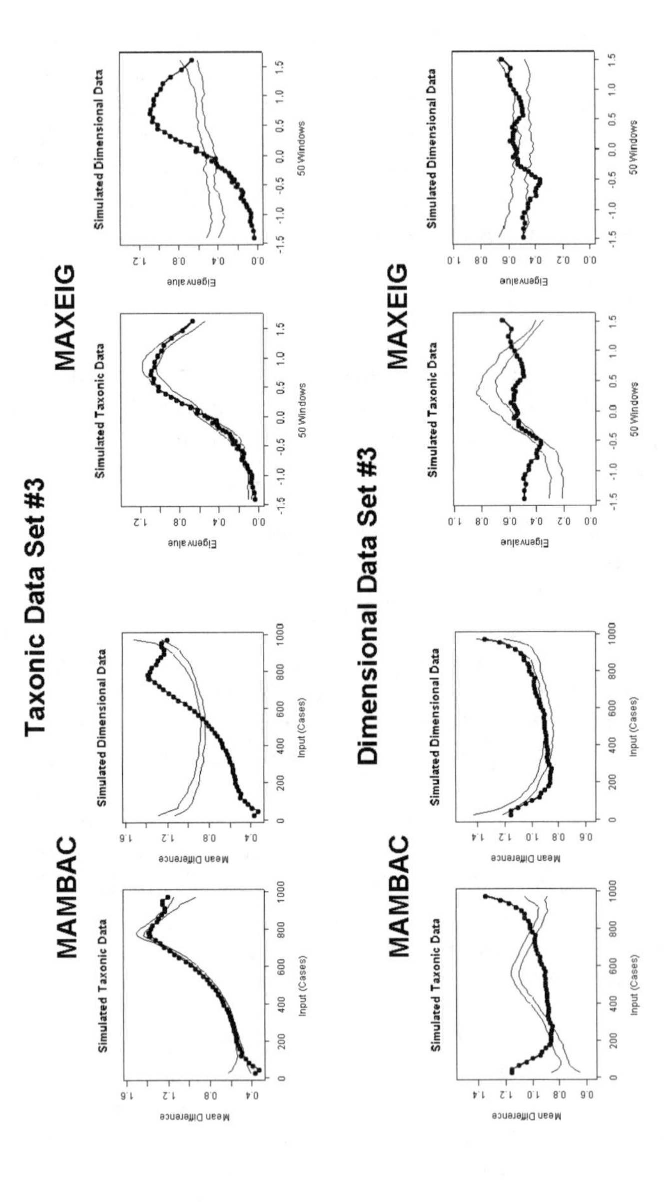

Taxonic Data Set #3
MAMBAC
MAXEIG
Simulated Taxonic Data
Simulated Dimensional Data
Mean Difference
Input (Cases)
Eigenvalue
50 Windows
Dimensional Data Set #3
MAMBAC
MAXEIG
Simulated Taxonic Data
Simulated Dimensional Data
Mean Difference
Input (Cases)
Eigenvalue
50 Windows

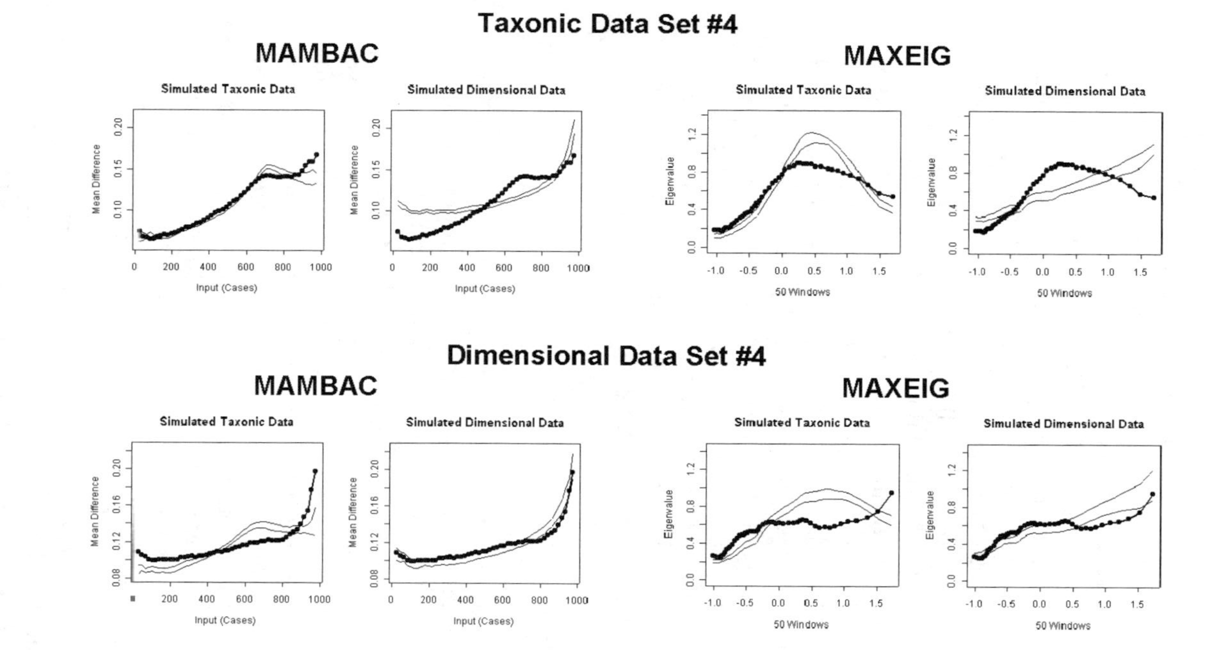

FIG. 7.3. MAMBAC and MAXEIG analyses performed on the research data plus simulated taxonic and dimensional comparison data sets to evaluate the relative fit of the curves for each structure. Ten sets each of taxonic and dimensional comparison data were generated for every series of analyses, and the results were averaged for subsequent graphing and calculation of fit indexes. Within a pair of graphs, each shows the results for the research data (dark line with data points) superimposed on the results for comparison data sets simulated using either taxonic (left) or dimensional (right) structure; lighter lines represent ±1 *SD* from the average of the curves for 10 simulated data sets of that structure. For MAMBAC analyses, composite input indicators were used, 50 evenly spaced cuts were placed along the input indicator (beginning and ending 25 cases from each end), and the four curves generated for each series of analyses were averaged for presentation. For MAXEIG analyses, each variable served as the input indicator, with the remaining three variables serving as output indicators, 50 windows that overlapped 90% with their neighbors were formed along the input indicator, and the four curves generated for each series of analyses were averaged for presentation. Analyses of TS#4 and DS#4 were performed with 10 internal replications to smooth the curves. Within each panel of curves, the scaling of the y axis was held constant.

Using the CCFI to quantify the relative fit of curves yielded by the taxonic and dimensional comparison data affords an objective assessment of latent structure that takes the unique parameters of the research data into consideration. The cost of such objectivity is that potentially useful information is inevitably lost when something as nuanced as the similarity of taxometric curves is summarized by a single number. Despite the risk of information loss, the CCFI performed well in these analyses, correctly identifying latent structure in each case. The results for TS#4 and DS#4 seemed particularly impressive in light of the challenge of distinguishing a small taxon from a dimensional construct with positively skewed indicators—a challenge that stumped most of the conventional taxometric procedures and consistency tests described thus far. MAMBAC analyses effectively revealed the taxonic structure of TS#4 (CCFI = .742); MAXEIG analyses also suggested the correct structural inference (CCFI = .571), although not as strongly. Likewise, DS#4 was clearly and correctly identified as dimensional by MAMBAC and MAXEIG analyses (CCFI = .290 and .327, respectively). Given the ease with which most techniques correctly identified the taxonic structure of TS#3 and the dimensional structure of DS#3, it came as little surprise that the curve fit index was accurate for both MAMBAC (CCFI = .849 for TS#3 and .269 for DS#3) and MAXEIG (CCFI = .898 for TS#3 and .342 for DS#3) analyses.

More Systematic Study of Curve Fit Indexes

In addition to the illustrative analyses presented here, J. Ruscio (2004) evaluated precursors to the CCFI in two ways. First, the indexes were calculated in MAXEIG and MAMBAC analyses of each sample of taxonic and dimensional data generated by Meehl and Yonce (1994, 1996) for their Monte Carlo studies. Across the four fit indexes evaluated, the two taxometric procedures, and the 14 data configurations, accuracy rates for taxonic structure ranged from 88% to 100%. The mean accuracy was greater than 99%, with the only mistakes occurring when some of the indicators possessed low validity. Accuracy rates for dimensional samples were far more variable, ranging from 48% to 100%. However, average accuracy for these samples was greater than 88%, and when the two data configurations with $N < 300$ were excluded, average accuracy rose to 92%. These accuracy rates compared favorably with those of human judges in the Meehl–Yonce studies (1994, 1996), as discussed later.

Second, in addition to analyses of the Meehl–Yonce samples, the fit indexes were evaluated in a series of Monte Carlo simulations that varied a larger number of data conditions across a broader range of values (J. Ruscio, 2004). Specifically, five properties were varied: sample size (N =

300, 400, 500, . . . , 1000), number of indicators (3, 4, 5, . . . , 10), taxon base rate (P = .05, .10, .15, . . . , .50), indicator validity (d = 0.2, 0.4, 0.6, . . . , 2.0), and within-group correlations (r = .00, .05, .10, . . . , .50); for dimensional data sets, expected indicator correlations were derived from the base rate, validity, and within-group correlation values in the corresponding taxonic data sets (see Eq. A.4). Due to the intense computational demands of generating and analyzing multiple sets of simulated taxonic and dimensional comparison data for each target data set, conditions were not varied factorially; 100 samples of target data were generated and analyzed within each data condition.

When fit indexes derived from MAXEIG and MAMBAC analyses were examined, a number of trends emerged. For taxonic data, fit indexes yielded stronger results—and higher accuracy rates—in the presence of a larger number of indicators, higher indicator validity (gradients were especially steep for this parameter), lower within-group correlations, and low ($P \leq .10$) or moderate ($P \geq .40$) base rates, rather than more intermediate values. For dimensional data, fit indexes yielded stronger results with larger sample sizes, with larger numbers of indicators (although this effect was much smaller than for taxonic data), and when indicator correlations were either low ($r \leq .10$) or high ($r \geq .40$) rather than intermediate in size.

In a more extensive study that varied data parameters simultaneously, rather than individually, J. Ruscio, Ruscio, and Meron (2005) found that the CCFI distinguished taxonic and dimensional structures much better than several traditional consistency tests. These results, which are discussed in greater detail later, provide the strongest evidence to date in support of the utility of curve fit indexes such as the CCFI that take advantage of simulated comparison data to help interpret results for the research data.

Interpreting Curves Using Fit Indexes Versus Traditional Visual Inspection

Although curve fit indexes worked well in the present illustrative analyses as well as in more systematic tests, we believe that well-informed investigators who visually inspect taxometric results yielded by their research data, simulated taxonic data, and simulated dimensional data may arrive at more accurate structural inferences than any quantitative index of curve fit. Preliminary results support this conjecture. Meehl and Yonce (1994) obtained ratings of 150 panels of MAMBAC curves—75 taxonic and 75 dimensional—from 12 naive judges (i.e., no advanced training in psychology) and 5 individuals with training in psychology (4 PhD psychologists and one clinical psychology graduate student). These raters achieved accuracy rates of 98% (naive judges) and 100% (psychologically trained judges) for taxonic structure, and 98% (naive judges) and 99% (psychologically trained judges) for dimensional structure. Within the same configurations of data parameters, J. Ruscio (2004) found that curve fit indexes averaged accuracy rates of 100% for taxonic structure and 97% for dimensional structure.

In a subsequent report, Meehl and Yonce (1996) found three psychologists and two naive judges to achieve 100% accuracy in evaluating 90 panels of MAXCOV curves. Because the authors did not specify the data conditions from which these 90 panels were drawn, the corresponding accuracy rate for curve fit indexes could not be calculated. However, fit indexes based on MAXEIG (a close analogue of MAXCOV; see chap. 6) achieved 100% accuracy for taxonic data and 98% accuracy for dimensional data under the conditions that Meehl and Yonce (1994) employed for their rating task (J. Ruscio, 2004). Thus, for MAMBAC and MAXCOV/MAXEIG analyses, curve fit indexes matched the classification accuracy of Meehl and Yonce's raters for taxonic data and lagged slightly behind for dimensional data.

We suspect that human raters will generally prove better at pattern recognition than quantitative fit indexes. Unfortunately, this hypothesis is difficult to test. All of the data conditions that produced the panels of curves judged by raters in the Meehl and Yonce (1994, 1996) studies were highly favorable for taxometric analysis, which caused ceiling effects for raters and curve fit indexes. In addition, although Meehl and Yonce (1994, 1996) obtained ratings from individuals with training in psychology, no study has yet used raters with substantial training in the taxometric method to determine whether knowledgeable investigators are able to achieve still higher accuracy levels. The relative accuracy of structural inferences yielded by the computation of curve fit indexes versus visual inspection by knowledgeable taxometrics researchers, drawn on the basis of more realistic and challenging data characteristics than have hitherto been used in such investigations, constitutes an important area for future research.

Curve fit indexes may be of particular value in Monte Carlo investigations, where researchers often cannot feasibly inspect the overwhelming number of curves generated across the many cells of the experimental design and the large number of replication samples analyzed within each cell. The information loss intrinsic to the use of quantitative indexes suggests that such Monte Carlo studies may tend to underestimate the validity of the structural inferences that can be reached when knowledgeable researchers visually inspect taxometric curves; however, this sacrifice may be necessary for large-scale Monte Carlo studies to be conducted. The use of indexes such as the CCFI in such studies would enable researchers to evaluate the performance of taxometric procedures under a wide variety of realistic research conditions in a fully automated, objective fashion.

Within the context of a single taxometric investigation, however, the number of curves requiring presentation and interpretation is ordinarily quite manageable, obviating the need to rely solely on curve fit indexes for structural inferences. Meehl and Yonce (1994, 1996) demonstrated that taxometric curves can be interpreted with acceptable levels of interrater agreement and an impressive degree of accuracy. Until a quantitative index is shown to surpass the accuracy of trained taxometric investigators, we recommend that researchers interpret taxometric curve shapes primarily through visual inspection and present these curves even if curve fit indexes are reported. Nevertheless, when they subjectively interpret taxometric curves, we urge researchers to keep two important qualifiers in mind. First, we caution researchers to remember that much of what is often written about the expected shapes of taxometric curves for taxonic and dimensional data stems from a consideration of highly idealized data conditions and from the results of two particular studies (Meehl & Yonce, 1994, 1996) that did not broadly cover a realistic range of conditions. In chapter 8, we discuss several potentially influential factors that can complicate the interpretation of curve shapes in ways that do not appear to be appreciated by the authors of all taxometric reports. Second, we encourage researchers to routinely generate empirical sampling distributions of taxometric curves, not only to assess the adequacy of data for analysis, but to serve as an interpretive aid. Such distributions can provide an informative benchmark for comparison, especially when the research data differ markedly from those of the Meehl–Yonce samples, by taking into account the unique data configuration and procedural implementation involved in a particular taxometric analysis.

A MONTE CARLO STUDY OF MAXCOV CONSISTENCY TESTS

Before concluding our discussion of consistency tests, we sought to more closely examine four consistency tests to inform their use in future investigations. As documented in chapter 10, MAXCOV has been used far more widely than any other taxometric procedure, and four consistency tests based on MAXCOV, results have enjoyed particular attention and use. One of these tests—a "nose count" of peaked MAXCOV curves—was recently suggested by Schmidt et al. (2004). The three other tests—the *SD* of taxon base rate estimates, the GFI, and the proportion of cases with Bayesian probabilities of taxon membership close to 0 or 1—have been employed extensively in taxometric investigations.

Data Sets

Because it was not feasible to perform a crossed factorial Monte Carlo study covering the full range of realistic data conditions, a random sampling technique was used to generate data sets for analysis, with data parameters varying as follows: sample size (N) = 300 to 1,000; taxon base rate (P) = .10 to .50; indicator validity (d) = 1.25 to 2.00; within-group correlation (r) = .00 to .30; indicator skew (S) = 0, 1, or 2; number of indicators (k) = 3 to 8. For each data parameter, a random value was drawn from a uniform distribution; only whole numbers were allowed for N, S, and k. For dimensional data sets, indicators were correlated at the level equivalent to the values of P, d, and r that had been randomly chosen (see Eq. A.4). It is important to note that all data parameters varied across levels that, according to conventional standards in the taxometric literature (e.g., Meehl, 1995a), would be considered to range from minimally acceptable (e.g., N = 300, P = .10, d = 1.25, r = .30) to highly favorable (e.g., N = 1,000, P = .50, d = 2.00, r = .00). In other words, researchers estimating their data parameters to fall within these ranges probably would conclude that their data meet the requirements for informative taxometric analyses. In all, 10,000 samples of data were generated, including 5,000 taxonic and 5,000 dimensional samples.

Data Analysis and Scoring

MAXCOV was performed using .25 *SD* intervals with a minimum subsample size of 25. Because far too many curves were generated to ask expert raters to judge which ones were peaked, a scoring algorithm was developed by following the suggested interpretive guidelines of Schmidt et al. (2004) as closely as possible. A curve was scored as peaked if the maximum covariance (a) did not emerge in the lowest or highest interval (to avoid interpreting ambiguously "cusped" curves as peaked), (b) had a z score of at least 1.50 relative to all covariances for that curve (to avoid overinterpreting small deviations in covariance as peaks), (c) was at least 10% larger than covariances more than 1 interval away, and (d) was at least 25% larger than covariances more than two intervals away. Extensive pilot testing suggested that the scores yielded by this algorithm for which curves were peaked corresponded well with blind ratings made by an experienced rater. The proportion of peaked curves was tallied across all curves for each data set. In addition, the *SD* of the taxon base rate estimates calculated from the curves, the GFI, and the proportion of cases whose Bayesian probabilities fell near the extremes (i.e., $< .10$ or $> .90$) were recorded for each data set. For ease of exposition, these four consistency tests will be abbreviated as PPeak, SD, GFI, and PBayes, respectively, throughout this section.

Threshold values for classifying data sets as taxonic or dimensional were drawn from published recommendations for each of the four tests. Specifically, Schmidt et al. (2004) suggested that a data set be scored as taxonic if there are at least as many peaked as nonpeaked curves in a panel (i.e., a threshold of PPeak $\geq .50$ in the present study), if the *SD* of taxon base rate estimates is $< .10$, or if the GFI is $\geq .90$ (a threshold adopted from Waller & Meehl, 1998, p. 27). There is no published recommendation for the proportion of extreme Bayesian probabilities indicative of taxonicity, but a threshold of .80 was used in one published (Beauchaine & Beauchaine, 2002) and one unpublished (Beauchaine & Beauchaine, 2003) Monte Carlo study, so this threshold was adopted here. In addition, receiver operating characteristic (ROC) curves were used to examine the validity with which each consistency test distinguished taxonic and dimensional data, independent of threshold. A series of ROC analyses examined each test's performance across levels of each data parameter that was varied in the study design.

Results

The middle portion of Table 7.4 shows the accuracy achieved for each consistency test in the full sample of all 10,000 data sets as well as within subsamples of 5,000 taxonic and 5,000 dimensional data sets. Using the recommended thresholds, overall accuracy rates ranged from 58% (PBayes) to 64% (GFI). Two tests (SD and PBayes) were more accurate for taxonic than dimensional data, another (PPeak) was more accurate for dimensional than taxonic data, and the last (GFI) was similarly accurate for both structures.

Because these results were surprisingly poor, 1,000 samples of data (500 taxonic and 500 dimensional) with more favorable parameters were analyzed and scored by the same program that was used to perform this study. Each of these easy samples of data contained 1,000 cases and four indicators with zero skew (in the full sample for dimensional data, within groups for taxonic data). For taxonic data, $P = .50$, $d = 2.00$, and $r = .00$; for dimensional data, indicators were correlated at $r = .50$ (the correlation expected for the taxonic data; see Eq. 4.1). The accuracy rates for these 1,000 samples are shown in the top portion of Table 7.4. These rates are reassuringly better than those obtained for the more realistic data conditions in the random-sampling design, suggesting that the program used to generate and analyze the data functioned properly. The results also emphasize that performance in easy data conditions does not necessarily generalize to more realistic conditions, even when individual data parameters fall within the range of values considered to be acceptable.

TABLE 7.4
Performance of MAXCOV Consistency Tests

Variable	*Proportion of Peaked Curves*	*SD of Base Rate Estimates*	*GFI*	*Proportion of Extreme Bayesian Probabilities*
Recommended threshold for inferring taxonic structure	.50[a]	.10[b]	.90[a]	.80[a]
Percent correct in "easy" condition (1,000 samples):				
All samples	89	100	52	93
Taxonic samples	95	100	100	85
Dimensional samples	83	99	4	100
Percent correct in more realistic conditions (10,000 samples):				
All samples	62	62	64	58
Taxonic samples	43	74	65	66
Dimensional samples	81	50	64	50
ROC analyses:				
Area under curve	.71	.66	.70	.60
Optimal threshold	.38[a]	.15[b]	.90[a]	.70[a]
Percent correct at optimal threshold	66	66	64	61

[a]Values above this threshold are indicative of taxonic structure.
[b]Values below this threshold are indicative of taxonic structure.

Returning to the results for the 10,000 more realistic samples, we next examined the possibility that a test might have discriminated taxonic and dimensional structure well, but that its apparent accuracy was reduced by a poorly chosen threshold. ROC analyses were performed to test accuracy independent of threshold. The bottom portion of Table 7.4 shows the area under the ROC curve for each test, the optimal threshold for differentiating taxonic and dimensional structure in these 10,000 samples, and the percent correct classification at this optimal threshold. Results show that none of the tests would have classified a much larger percentage of data sets correctly had their thresholds been set at the optimal level. PPeak and SD would have increased from 62% to 66% correct, PBayes would have increased from 58% to 61%, and the GFI would have remained at 64%, as its recommended threshold coincided with the optimal level for these data sets. In summary, whether using recommended or optimized thresholds, each test only modestly exceeded chance-level accuracy.

These aggregate results mask a number of moderating effects for individual data parameters. Follow-up analyses revealed conditions under which each test performed substantially better or worse than its average level. Figure 7.4 presents the results of ROC analyses performed within subranges along each data parameter that was varied in the study. All four consistency tests were minimally affected by changes in sample size. Two tests (GFI and PBayes) performed considerably better with larger indicator correlations, whereas the other two tests (PPeak and SD) performed better with smaller indicator correlations. Because the indicator correlations in taxonic data are a function of several parameters, additional analyses explored the influence of taxon base rates, indicator validities, and within-group correlations. Two tests (SD and PBayes) performed at or only slightly above chance levels with large base rates and much better with low base rates; the other two tests (PPeak and GFI) performed best with moderate to large base rates and much worse with low base rates. Not surprisingly, increases in indicator validity improved the performance of all four tests. Increases in within-group correlations reduced the performance of all but PPeak, which was relatively unaffected. Indicator skew weakened the performance of all four tests, especially SD and PBayes. Whereas each test performed at its best when indicator skew was 0, taxonic and dimensional structures were distinguished much more poorly as skew was increased to 2. This finding seems especially important given indications that distributions with skew values of 1, 2, or higher are the norm for social science research data (Micceri, 1989). Finally, larger numbers of indicators yielded modest improvements for all tests except the GFI, whose performance varied little across this factor.

The results presented here are consistent with those of a Monte Carlo study of similar magnitude and design that examined the performance of

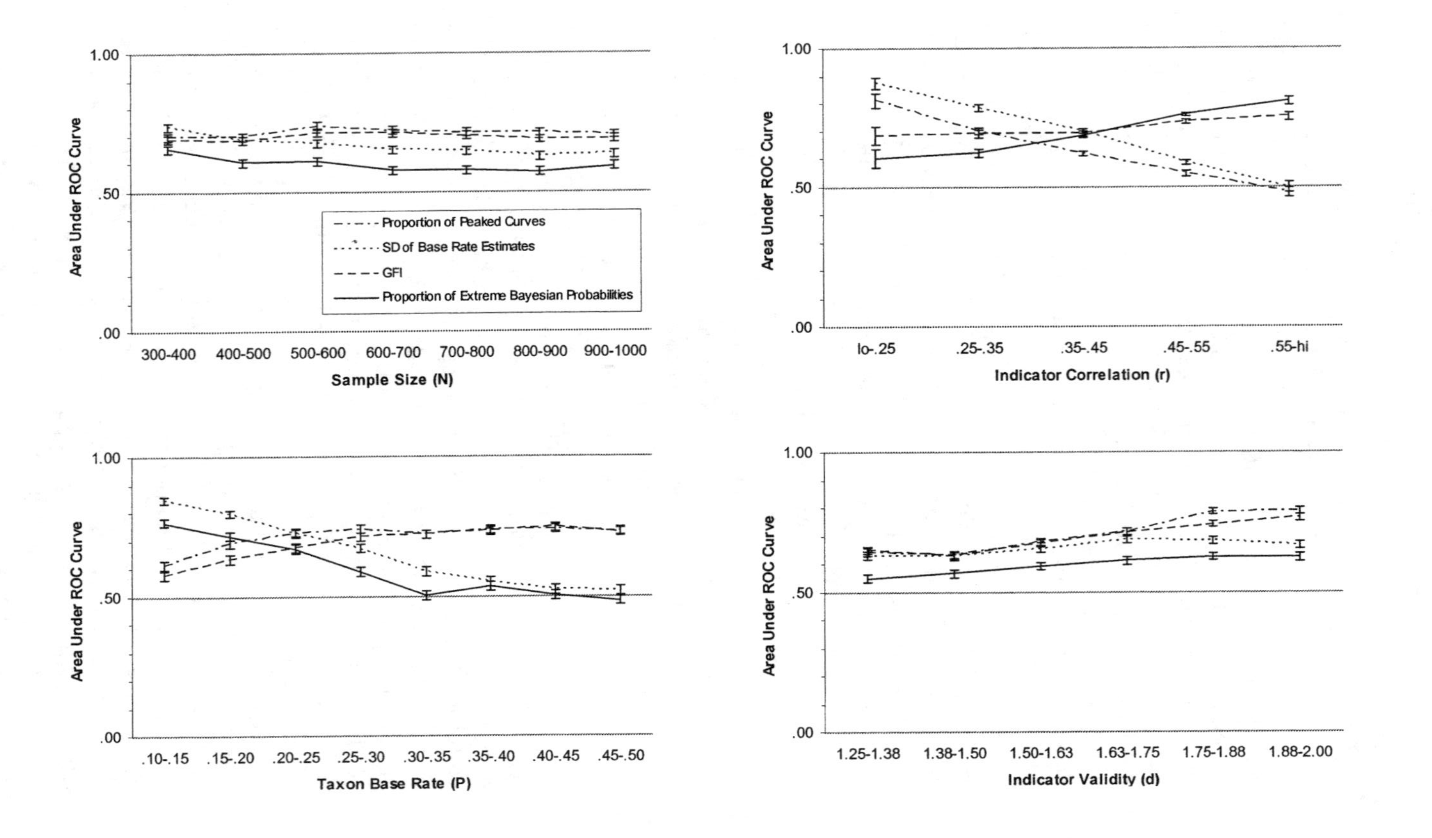

Area Under ROC Curve
1.00
.50
.00
300-400
400-500
500-600
600-700
700-800
800-900
900-1000
Sample Size (N)
Proportion of Peaked Curves
SD of Base Rate Estimates
GFI
Proportion of Extreme Bayesian Probabilities
Area Under ROC Curve
1.00
.50
.00
lo-.25
.25-.35
.35-.45
.45-.55
.55-hi
Indicator Correlation (r)
Area Under ROC Curve
1.00
.50
.00
.10-.15
.15-.20
.20-.25
.25-.30
.30-.35
.35-.40
.40-.45
.45-.50
Taxon Base Rate (P)
Area Under ROC Curve
1.00
.50
.00
1.25-1.38
1.38-1.50
1.50-1.63
1.63-1.75
1.75-1.88
1.88-2.00
Indicator Validity (d)

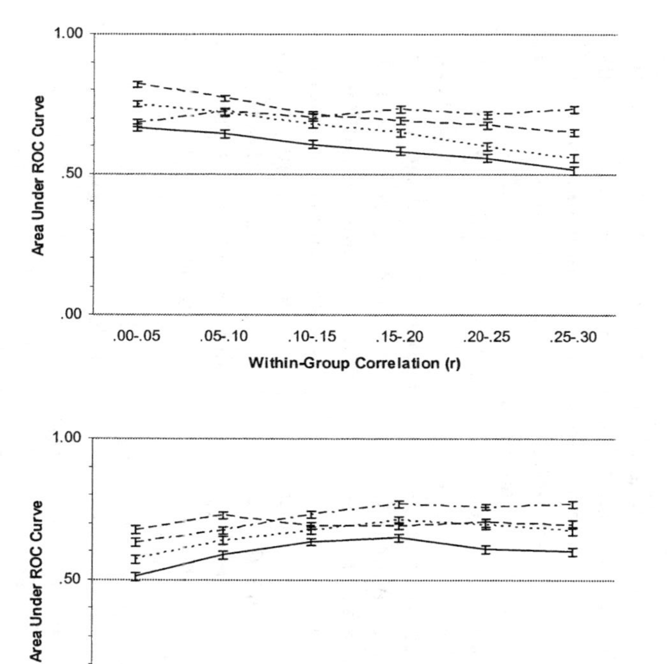

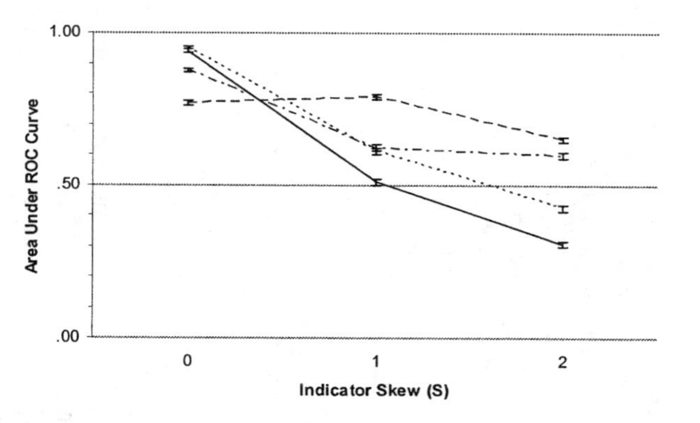

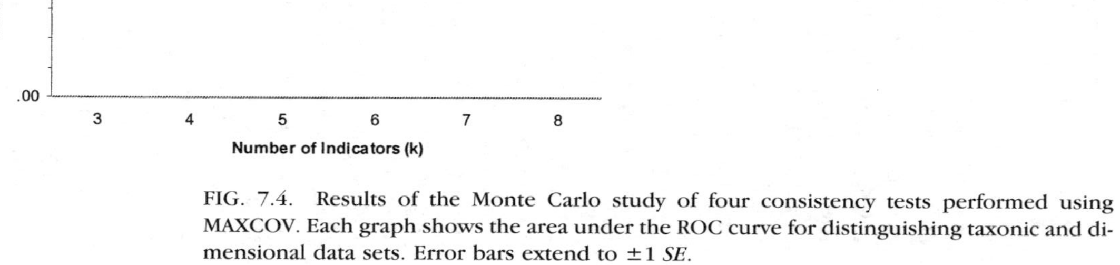

FIG. 7.4. Results of the Monte Carlo study of four consistency tests performed using MAXCOV. Each graph shows the area under the ROC curve for distinguishing taxonic and dimensional data sets. Error bars extend to ± 1 *SE*.

several consistency tests in MAXEIG analyses (J. Ruscio, Ruscio, & Meron, 2005). That study explored the accuracy achieved not only by these consistency tests, but also by the CCFI (the comparison curve fit index introduced earlier in this chapter) using B = 10 taxonic and 10 dimensional comparison data sets per analysis. ROC analyses revealed that the CCFI distinguished taxonic from dimensional structure far more effectively than three of the consistency tests included in the present study, with an area under the ROC curve of .93 for the CCFI as compared with .75 (GFI), .60 (PBayes), and .35 (SD). The latter value is worse than chance (consistent with the findings of J. Ruscio, 2005) and may have been substantially lower than in the present study due to the inclusion of even larger skew conditions in the design. Ruscio et al. examined performance within subranges of each data parameter as in Fig. 7.4, and the previously mentioned rank ordering of discrimination power was observed for most conditions. In no condition did any of the traditional consistency tests reach the accuracy level of the CCFI, and seldom did any even approach it very closely. Thus, there is at least tentative evidence that (a) the conventional consistency tests may not be among the most useful tools available for differentiating taxonic and dimensional structures, and (b) the use of empirical sampling distributions as an interpretive aid can yield superior differentiation of these structures.

We conclude this section with a response to one potential alternative interpretation of the implications of this study. One might argue that these consistency tests are not meant to be interpreted in isolation—that it is the totality of evidence that should inform a structural conclusion, and hence that a poor showing for any individual test does not necessarily recommend against its use. There are two reasons that this alternative interpretation might not rescue the utility of consistency tests that perform poorly in isolation. The first concerns the issue of incremental validity. It seems unlikely that a test with low zero-order validity would improve the validity of structural inferences drawn from other tests of equal or greater validity. It is even possible that structural inferences could become less valid with the addition of a weak consistency test to others that are more valid. Future research should explicitly examine the incremental validity of consistency tests such as the four examined here, especially as a supplement to structural inferences drawn from the curve shapes produced by taxometric procedures. Until positive evidence of incremental validity is demonstrated, we suggest that these tests be interpreted with due caution.

Second, although the validity of structural inferences can be improved by requiring consistently taxonic or dimensional results across multiple tests to reach a conclusion, the substantial drawback to this requirement is that only a small fraction of data sets may attain the degree of consistency necessary for interpretation. For example, among the samples for which all four consistency tests yielded consistently taxonic or dimensional results, results were correct 88% of the time. This represents a great improvement over the 58% to 64% accuracy attained by the individual tests. However, only 19% of all samples met this consistency criterion. Relaxing the criterion to require consistency across at least three of these four tests would permit conclusions to be reached in many more samples (63% rather than 19%), but at a substantial cost in accuracy (76% rather than 88%). Given the poor performance of these individual tests, it is possible that even their joint use will not significantly enhance structural interpretation in taxometric investigations.

CONCLUSIONS

This chapter described the logic underlying a number of techniques that can be used to evaluate the consistency of taxometric results. We loosely grouped these techniques according to three broad approaches to consistency testing: (a) performing multiple taxometric procedures numerous times and in numerous ways, (b) evaluating estimates of latent parameters and classifying cases for consistency checking, and (c) assessing the fit of taxonic and dimensional models to results yielded by the research data. Some of these techniques are more conceptual than analytic, and different methods defend against different sources of inferential error. For example, analyzing multiple indicator sets drawn from multiple measures or samples of data guards against a number of potential measurement and sampling artifacts. Other techniques, such as performing multiple taxometric procedures or reanalyzing data following the targeted removal of cases, guard against the possibility that a particular set of indicators and a particular analytic procedure will combine to suggest an erroneous interpretation. Therefore, we recommend that investigators check the consistency of their results in multiple ways to reduce the likelihood of mistaken inferences and to increase confidence in their structural conclusions.

Some of the consistency tests that we reviewed are better supported than others by logic and by currently available (although admittedly limited) empirical evidence. First, performing multiple taxometric procedures as consistency checks for one another is a cornerstone of the taxometric method, and these consistency checks are further enhanced when each procedure is performed with a variety of samples, indicator sets, data sources, and measures, in as many configurations as are appropriate for the data and procedure. Second, the inchworm consistency test, in which a series of MAXCOV/MAXEIG analyses is performed with expanding numbers of subsamples, can be a powerful tool for clarifying ambiguous taxometric results, especially when a taxon is suspected to be small.

Finally, visual inspection of taxometric curves yielded by the research data, superimposed on empirical sampling distributions of taxometric results yielded by analyses of taxonic and dimensional comparison data, can help researchers evaluate the consistency of their data with each structural model. The CCFI that can be calculated from these results may provide an informative quantitative supplement to visual inspection; this may be particularly useful in Monte Carlo studies that generate too many curves to be inspected by expert raters.

By contrast, several of the consistency tests described in this chapter rest on more tenuous logical grounds, are largely untested, or have been found to poorly differentiate taxonic from dimensional structure in Monte Carlo trials. First, despite its popularity in the taxometric literature, a U-shaped distribution of Bayesian probabilities of taxon membership may provide only weak support for taxonicity. Second, the GFI has also performed poorly in Monte Carlo studies. Both of these indexes yielded only modest accuracy rates in our own Monte Carlo study. Third, the lesser used approach of splitting a research sample into subsamples for purposes of replication may reduce the power of a taxometric analysis without offering a sufficiently strong benefit to offset this cost. Fourth, the use of *t* tests to search for taxonicity is problematic, as significant differences between any ordered groups—either actual or constructed—would be expected under both taxonic and dimensional models. Finally, although the taxon base rate is frequently estimated in taxometric studies and provides the basis for the most popular consistency test of all, the coherence (or incoherence) of such estimates can often be misleading. This, too, is illustrated by the modest performance of the *SD* of base rate estimates in our Monte Carlo study (see also J. Ruscio, 2005). Still less is known about whether the consistency of other latent parameters can successfully distinguish taxonic from dimensional structure. Therefore, we advise caution when drawing structural inferences from any estimates of latent parameters until further research establishes when and how this should be done.

Although we have suggested that certain approaches to consistency testing may be more informative than others, our intention is not to formulate a blueprint for prescribing which consistency tests should be included in any particular taxometric investigation. There are at least two arguments against a formulaic or absolute blueprint of this kind. First, far too little empirical attention has been devoted to evaluating either the zero-order or incremental validity of taxometric consistency tests. That is, relatively little is known about the validity with which various tests can distinguish taxonic from dimensional latent structure, and even less is known about the redundancy of the evidence provided by different taxometric procedures and consistency tests. Consistency checking is only valuable to the extent that the chosen tests increase the likelihood of reaching an accurate structural inference, over and above the tests already performed. At present, close to nothing can be said about this subject on the basis of rigorous empirical investigation.

Second, although we have offered guidelines for the cautious use and interpretation of each consistency test, we believe that final decisions about which techniques are appropriate for a particular taxometric investigation should be made on the basis of empirical sampling distributions. Using the technique described in chapter 4, we recommend that researchers simulate taxonic and dimensional comparison data sets that reproduce indicators' distributions and correlations, then submit these data to the planned taxometric analyses. To the extent that a procedure or consistency test successfully distinguishes taxonic from dimensional structure under the unique conditions of the research data, researchers can feel more confident that it will yield useful information when applied to the research data. Further, the examination of these sampling distributions may suggest an appropriate threshold for distinguishing structures under these conditions.

Conversely, any procedure or consistency test that does not clearly differentiate the simulated taxonic and dimensional comparison data sets is not only less likely to yield informative conclusions about the research data, but may produce misleading results. For example, without realizing that positive indicator skew can produce low, consistent taxon base rate estimates in MAMBAC and MAXCOV/MAXEIG analyses, regardless of the structural model underlying the data, the discovery of such consistent estimates may be taken as support for the existence of a taxon. Thus, we advise researchers to consider and evaluate a large number of taxometric procedures and supplementary consistency tests, but to submit their research data only to those analyses that show themselves capable of differentiating taxonic from dimensional structure, given a unique sample of research data. This analytic approach is further illustrated in each of the following two chapters.

CHAPTER EIGHT

Interpretational Issues

Chapters 5 through 7 described the primary procedures and consistency tests that make up the taxometric method. As was noted throughout these chapters, Monte Carlo evaluation of these analytic techniques is in its infancy, and the few systematic studies that have been published tended to employ relatively idealized data. As a result, the descriptions of results expected for taxonic and dimensional structure—and hence the expectations of researchers who inspect the output of taxometric analyses—are guided largely by the prototypical curve shapes and consistency test results obtained under the limited conditions of prior Monte Carlo studies. The unfortunate consequence is that taxometric findings may be interpreted without a realistic understanding of the influence of critical factors on the appearance and consistency of results.

In this chapter, we draw attention to several understudied factors that can complicate the interpretation of taxometric results. Whereas Chapters 5 through 7 discussed interpretive issues specific to particular procedures or consistency tests, the present chapter explores more general issues involved in presenting and interpreting the results of a taxometric analysis. To this end, we consider alternative ways of graphing taxometric curves for interpretation and more thoroughly describe the ways in which indicator distributions, indicator validity, and within-group correlations can influence taxometric results. Although we offer suggestions based on the available data, the logic of the procedures, and our practical experience in conducting taxometric analyses under numerous data conditions, we emphasize that the dearth of systematic research leaves investigators with many decisions, but only limited guidance in the interpretation process.

Therefore, we recommend three procedural safeguards that can facilitate the accurate interpretation of taxometric results: (a) presenting as many curves as possible in a taxometric report, (b) obtaining systematic judgments of taxometric results from multiple independent raters, and (c) using empirical sampling distributions of taxometric results as an interpretive aid. Until more is known about the influences on taxometric results and more definitive guidelines are afforded by systematic study, we urge researchers to carefully consider factors that may produce ambiguous or misleading taxometric results and to interpret their findings with due caution.

GRAPHING AND PRESENTATION OF TAXOMETRIC RESULTS

In chapters 5 and 6, we discussed several decisions that must be made when graphing the results yielded by each taxometric procedure. However, two graphing options are sufficiently general to warrant broader attention within the context of the interpretational process. Each of these options—curve smoothing and curve averaging—is discussed in turn.

Smoothing Curves

In some investigations, researchers have smoothed the graphs produced by taxometric analyses. Most commonly, this has involved applying one of several variants of Tukey's (1977) running medians technique to MAXCOV curves or applying the LOWESS (Cleveland, 1979) scatterplot smoother to MAMBAC, MAXCOV, or MAXEIG curves. As introduced and traditionally implemented, MAXEIG analyses already incorporate the smoothing technique of overlapping windows. Further smoothing of MAXCOV or MAXEIG curves constructed with overlapping windows (e.g., the MAXEIG program presented in Appendix C of Waller & Meehl, 1998, uses the LOWESS method) may provide little benefit and might increase the risks associated with smoothing that we discuss later. In our experience, the practice of smoothing appears to be somewhat controversial among researchers using the taxometric method and reviewers of such studies. Research has not yet established the conditions under which various smoothing techniques can be beneficial in taxometric studies. Thus, we raise several issues for researchers to contemplate when deciding whether to smooth curves.

Smoothing is often performed to facilitate the interpretation of curve shapes or improve the accuracy with which latent parameters are estimated. Of course smoothing is not guaranteed to achieve these goals and may introduce new difficulties. One salient concern is that inappropriately smoothed curves can obscure genuine taxonic peaks. Consider, for example, a MAXCOV analysis conducted with eight dichotomous items, with the sum of six items serving as the input indicator and the other two items serving as output indicators. Performing MAXCOV with nonoverlapping intervals would result in a 7-point curve, and applying a running medians smoothing technique to a curve with so few data points would carry a substantial risk of leveling a true taxonic peak. By contrast, performing the analysis with overlapping windows would yield a 100-point curve that already incorporates a smoothing process, reducing the need to use an additional smoothing technique that might distort the curve shapes. As discussed in chapter 6, one virtue of employing windows rather than intervals in any MAXCOV/MAXEIG analysis is that this tends to yield a smoother, more interpretable curve by providing all of the data points one would obtain with equal-sized intervals plus a larger number of intermediate points. The benefits of smoothing are achieved without altering the data points that would have been obtained using nonoverlapping intervals, and this approach can reduce the need for an additional smoothing procedure.

Sample size is also a relevant factor in considering the appropriateness of smoothing. When samples are relatively small, the influence of normal sampling error may create rocky or lumpy raw (unsmoothed) curves that may become easier to interpret after smoothing. By contrast, when samples are sufficiently large, there is less sampling error and therefore less justification for smoothing.

Even when smoothing is not needed to increase the interpretability of curves, it may be used to improve the accuracy of latent parameter estimates. As noted in chapter 5, estimates of the taxon base rate are calculated from MAMBAC curves using only the two endpoints of the curve. Unfortunately, these outermost points are subject to a considerable amount of sampling error (i.e., relatively few cases contribute to the means above or below the upper and lower end data points, respectively) and can also be the most heavily influenced by factors such as within-group correlations. Thus, although a smoothing procedure may not be necessary to help discern the shape of a MAMBAC curve, it may be desirable to apply a smoothing technique before estimating the taxon base rate from a MAMBAC curve, particularly if extreme regions of the raw curves exhibit marked instability. Similar reasoning holds true for MAXCOV/MAXEIG analyses: Although latent parameters such as the taxon base rate can be estimated using all data points on these curves (rather than only the two endpoints), sampling error may also reduce the accuracy of these parameter estimates.

Figure 8.1 illustrates the potential benefits of estimating the taxon base rate from smoothed rather than raw taxometric curves. Taxometric analyses were performed on taxonic data sets with N = 300, P = .50, and d =

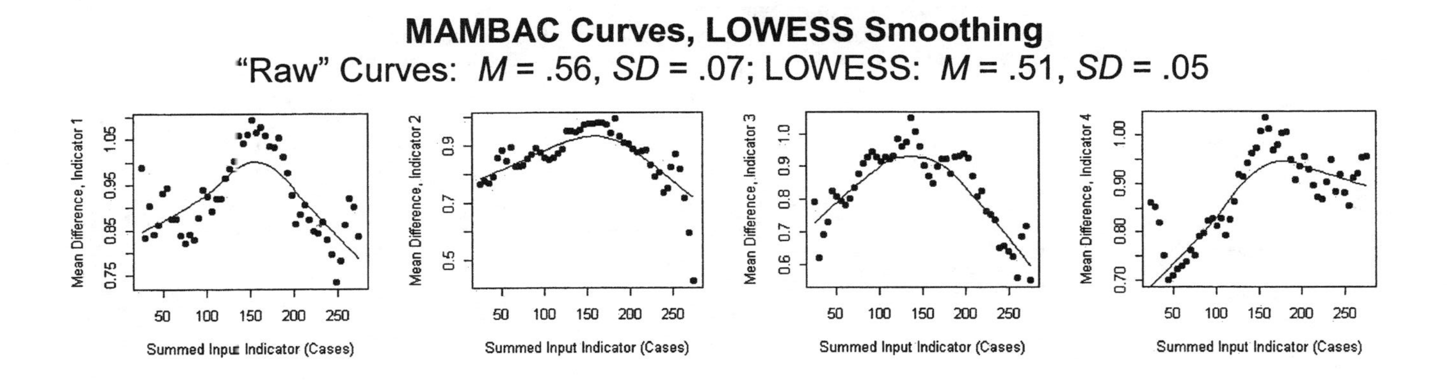

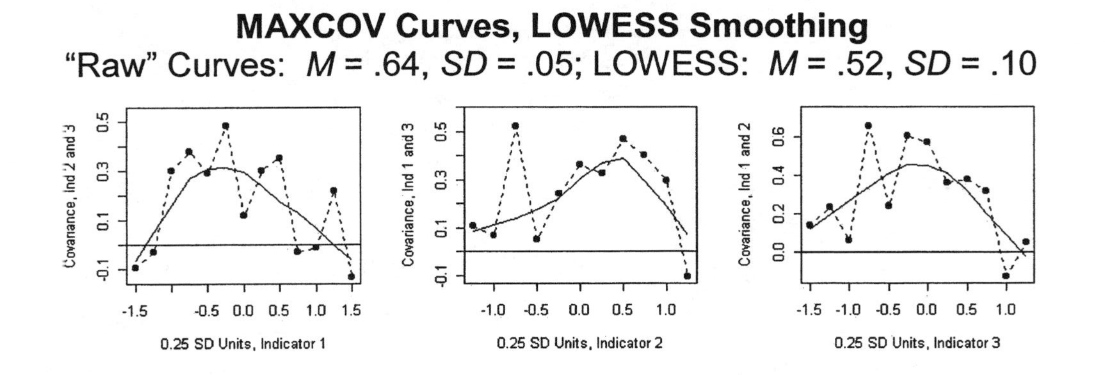

FIG. 8.1. *(Continued)*

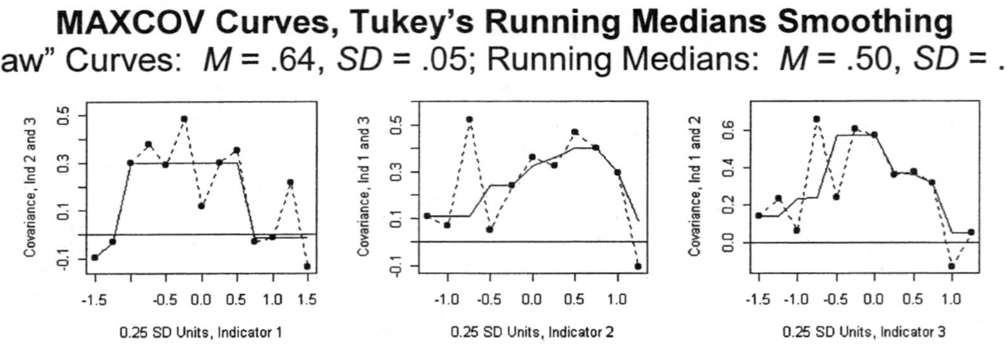

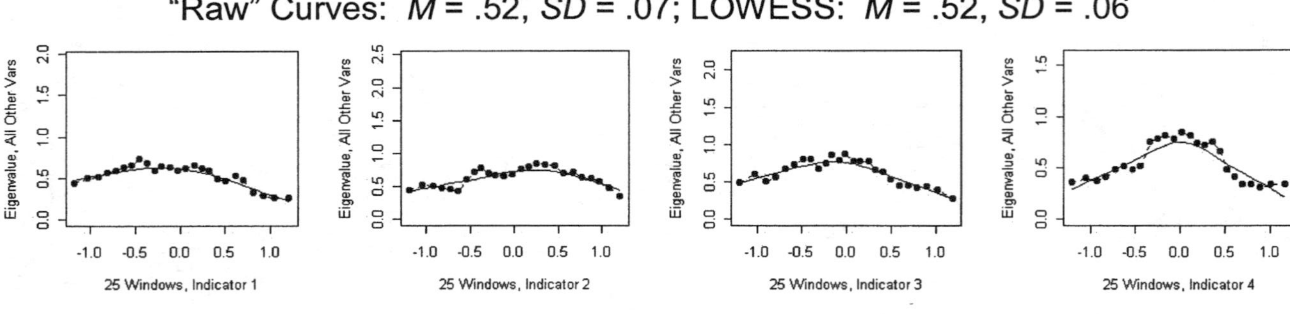

FIG. 8.1. Smoothing MAMBAC, MAXCOV, and MAXEIG curves. Raw curves appear as data points, connected by dashed lines for MAXCOV/MAXEIG curves; smoothed curves appear as solid lines. Above each panel of curves, the *M* and *SD* of taxon base rate estimates—calculated for raw and smoothed curves—is displayed.

1.50. The LOWESS smoothing technique was applied to the MAMBAC, MAXCOV, and MAXEIG curves. In addition, because some researchers have used Tukey's (1977) running medians technique with the MAXCOV procedure, this smoothing technique was also applied to the MAXCOV curves. To allow the presentation of a complete panel of curves, MAMBAC was performed with four indicators using the composite input indicator technique; 50 evenly spaced cuts were located beginning and ending 25 cases from each end of the input indicator. Taxon base rate estimates for the LOWESS-smoothed MAMBAC curves were considerably more accurate and slightly more consistent than those for the raw curves. To again present a full panel of curves, MAXCOV was performed using three of the four indicators, with intervals constructed using .25 *SD* units extending as long as subsamples contained a minimum of 25 cases. Taxon base rate estimates were much more accurate, but less consistent, when calculated from either LOWESS- or Tukey-smoothed curves than from the raw curves. Finally, MAXEIG was performed using each of the four indicators once as the input, with 25 windows that overlapped 90%. Here smoothing had little effect on taxon base rate estimates because the use of overlapping windows had already produced smooth curves.

This illustration suggests that, for some taxometric procedures, smoothing curves can sometimes improve the accuracy with which latent parameters are estimated from these curves. Establishing the conditions under which the benefits of smoothing outweigh the costs is an important task of future investigations. For now we encourage researchers to think carefully about the potential advantages or disadvantages of curve smoothing in any particular taxometric analysis. As illustrated in Fig. 8.1, we recommend that raw curves be presented even when smoothing has been performed so that readers can evaluate the appropriateness of the smoothing technique for themselves.

Averaging Curves

Another unresolved issue in the taxometrics literature involves the averaging of results across a full panel of curves. This approach was first suggested by Gangestad and Snyder (1985), who performed the MAXCOV procedure using composite input indicators, a modification they introduced to accommodate dichotomous indicators. These researchers argued that individual curves were subject to substantial sampling error and that their average should be more meaningful and interpretable. Under some circumstances, this may be true. At other times, however, curve averaging may obscure taxonic results. For example, if only some indicators of a taxonic construct are valid enough to yield taxonic peaks, an averaged curve across all indicators may appear more consistent with dimensional than taxonic structure. Alternatively, depending on how indicators are scaled and analyses are performed, taxonic peaks may emerge in different locations across a panel of curves, and averaging may obscure their consistently peaked shapes. Finally, it has often been suggested that structural conclusions be based on the inspection of a full panel of individual curves in a taxometric analysis, in part, because of the opportunity this affords to assess the consistency of results (see chap. 7). Interpreting averaged curves rather than panels of individual curves forgoes one opportunity for consistency testing.

Thus, as with smoothing, research is needed to determine when taxometric curves can safely and profitably be averaged. For example, from a practical perspective, averaging may be useful when there are simply too many individual curves yielded by the analyses performed to include in a research report. However, there are presentational alternatives to averaging, such as displaying a representative subset of curves or the first few curves from each panel. As another example, the quantitative curve-fit indexes presented in chapter 7 (which compare results for the research data to those for simulated taxonic and dimensional comparison data) have only been applied to averaged curves, and it is unknown how well they would perform when applied to full panels of individual curves. Revisions to our program code are planned that will permit assessment of fit for each individual curve as well as the average fit of all curves. Calculating the CCFI for individual curves would not be difficult, but it is not obvious how these CCFI values might be aggregated across curves.

Curve averaging may confer substantial benefits, but only when it can be performed without sacrificing valuable information or obscuring important results. Therefore, we suggest that researchers consider the appropriateness of averaging for a given study by examining the adequacy with which an averaged curve represents the results obtained for a full panel of curves. Sometimes one can feasibly present both individual and averaged curves in a research report. Figure 8.2 shows the results of MAMBAC and MAXEIG analyses of four of our illustrative data sets; each graph contains the full panel of curves (dotted lines) as well as the average of these curves (solid line). In analyses of TS#3 and DS#3 (top row), the individual curves were sufficiently consistent to make their average an appropriate summary of results. The reason for these impressively consistent curves is that, within each data set, every indicator shared the same metric and validity.

In TS#4 and DS#4, however, this was not the case—each indicator varied along a different metric and possessed a different validity. Analyses of these data sets yielded far less consistent results (middle row). Although the averaged curves reflected the general trend in the data, they did not represent the full panel as well. Fortunately, one can standardize indica-

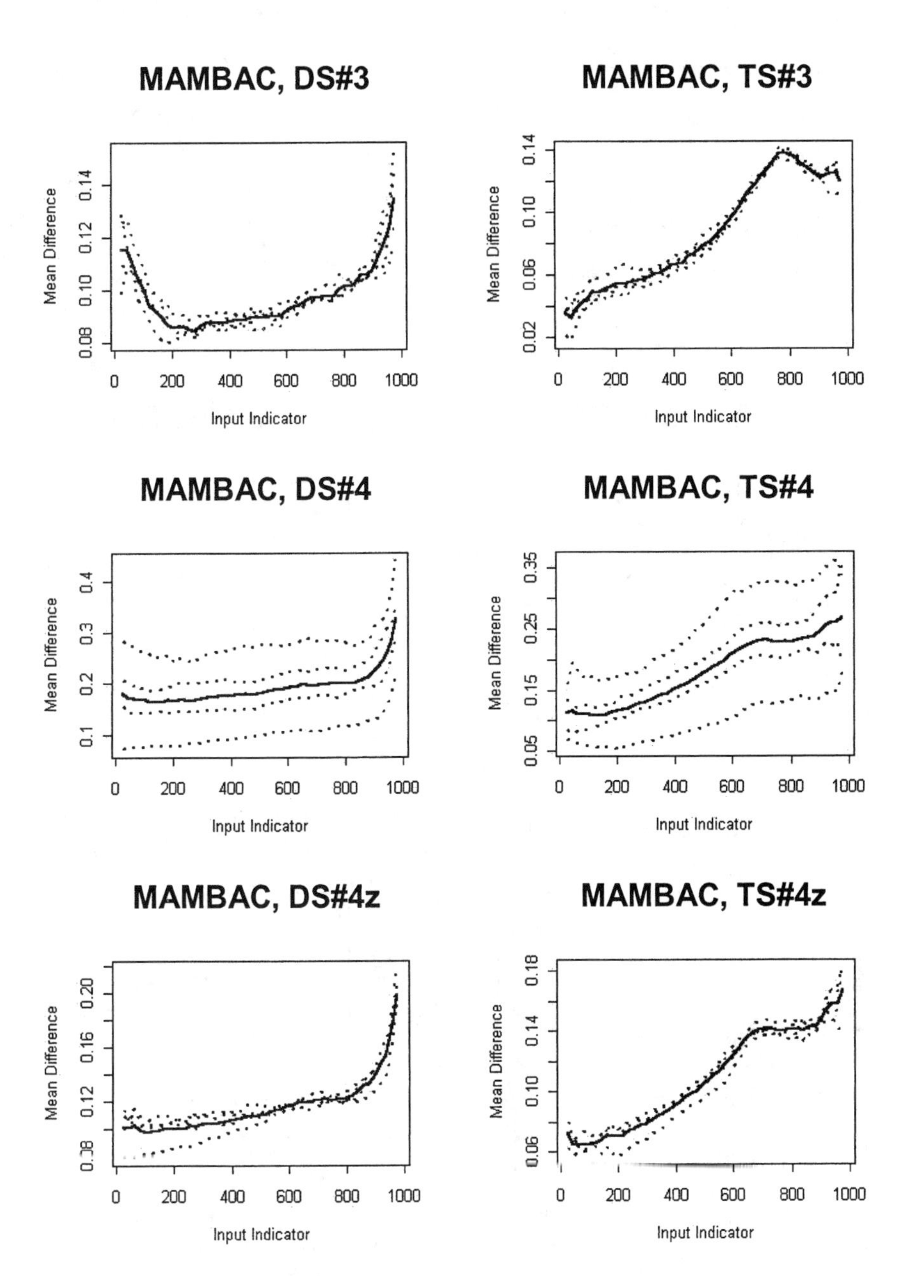

FIG. 8.2. *(Continued)*

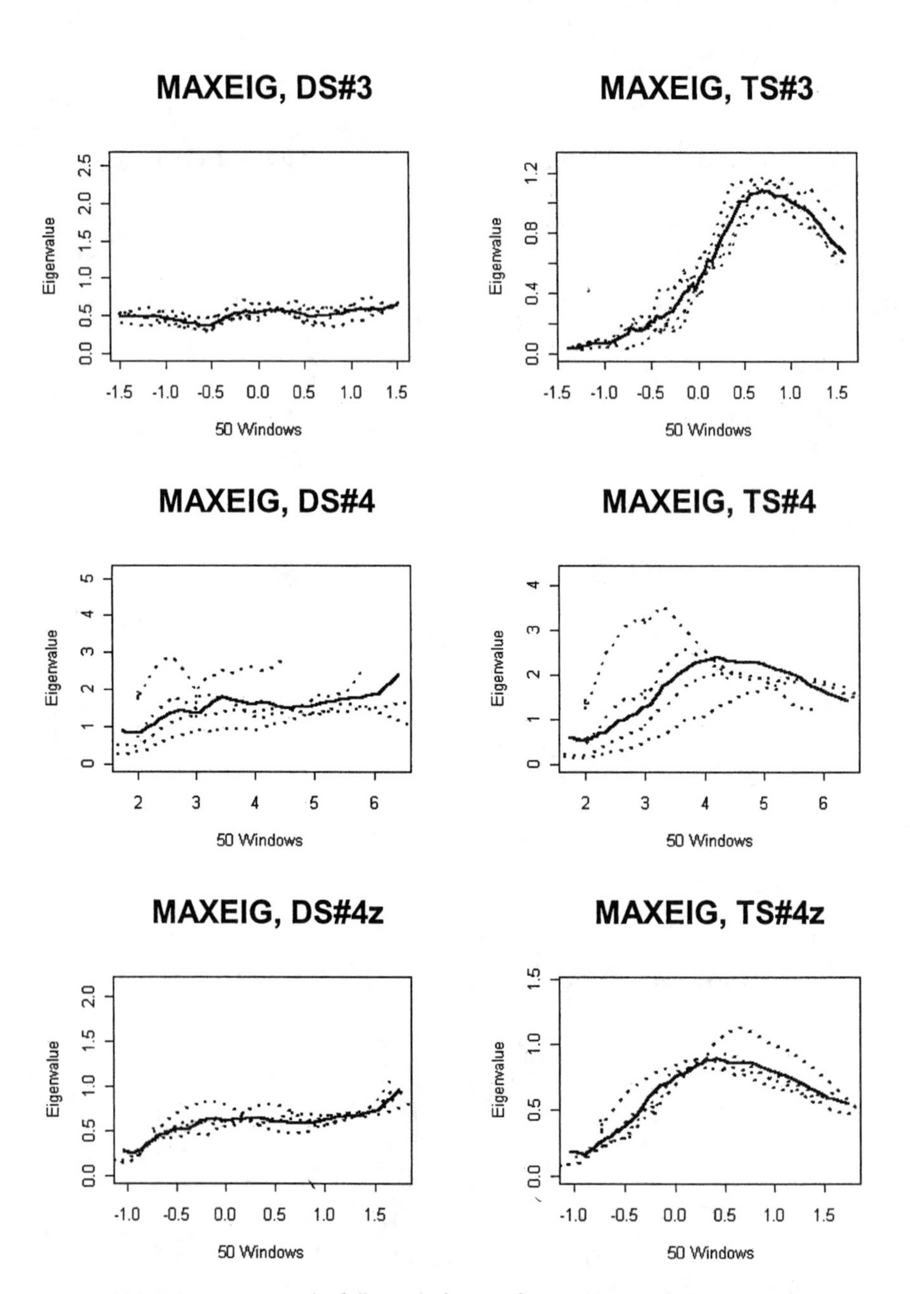

FIG. 8.2. Averaging the full panel of curves for MAMBAC and MAXEIG analyses. Each graph depicts the full panel of individual curves as dotted lines, with the single averaged curve as a solid line. DS#3, DS#4, TS#3, and TS#4 refer to taxonic and dimensional data sets described and analyzed in previous chapters; DS#4z and TS#4z refer to copies of DS#4 and TS#4, in which each indicator was standardized prior to analysis.

tors to equate their metric, often yielding a more consistent panel of curves that is better represented by a single averaged curve. Despite the difference in validities across the indicators of TS#4 and DS#4, analyses performed on the standardized indicators (bottom row) produced more consistent panels of curves that were well summarized by their averages. This presents another argument in favor of performing taxometric analyses with standardized indicators.

Plotting a full panel of curves and an averaged curve on the same graph may not only help researchers judge whether the average is a fair representation of results, but also may be a useful way to present taxometric results in a research report; it conserves space and assures readers that the averaged curve does not obscure important inconsistencies in the full panel of results. To the extent that the individual curves do not cohere well around the averaged curve, the average becomes less useful and may ultimately be inappropriate for use. Of course a graph including both individual and averaged curves will become harder to read as the number of indicators (and hence the number of curves) increases. An additional limitation of such a graph is that it does not specify which indicator configurations produced which individual curves. To the extent that curve shapes systematically differ across indicator configurations, this can be important information. One alternative graphing technique that addresses this concern is to group curves by the input indicator that produced them, with the curves for all output indicator configurations graphed together for each input indicator. Figure 8.3 illustrates this approach for MAMBAC analyses of TS#4 and DS#4, the data sets with varying indicator validities. Indicators were assigned to the required input and output roles in all pairwise combinations, yielding 12 curves for each analysis. Graphs in the top row present the full panel of curves for each analysis. However, given the relatively large number of curves, it is difficult to distinguish the individual curves from one another and, hence, to determine whether some indicators systematically yielded more interpretable results than others. Graphs in the bottom two rows present the same panel of curves organized by input indicator. With fewer curves per graph, it is easier to evaluate the relative interpretability of curves stemming from each input indicator; the tradeoff is that this requires more space for presenting results.

Several additional issues bear on the subject of curve averaging. First, different measures of central tendency can be used to average curves (e.g., mean vs. median; means are used in our programs). Second, averaging can be performed in multiple ways. For example, one might average the mean differences (y values) across MAMBAC curves for each case number (x value) at which a cutting score was located. Alternatively, one might average the mean differences as well as the input scores at each of a fixed number of cutting scores to obtain a set of averaged (x, y) pairs. We prefer

MAMBAC Analyses of TS#4

Full Panel of Curves

Curves for Each Input Indicator

MAMBAC Analyses of DS#4

Full Panel of Curves

Curves for Each Input Indicator

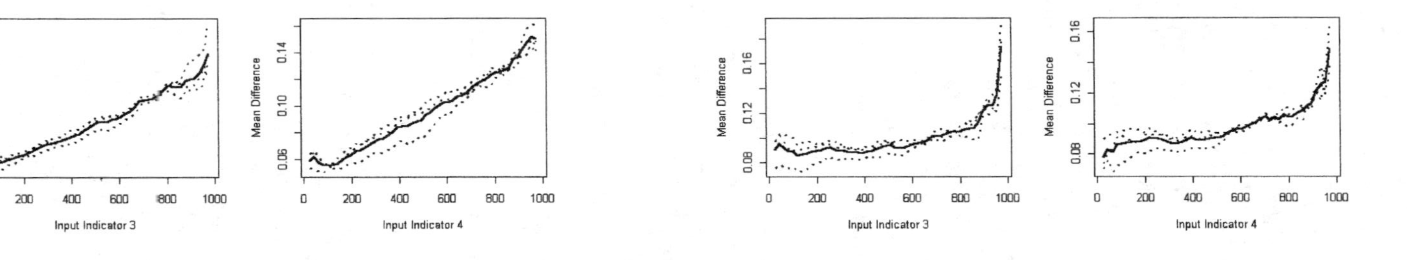

FIG. 8.3. Averaging panels of curves by each input indicator. MAMBAC analyses of TS#4 and DS#4 were performed using indicators in all pairwise configurations, which yielded 12 curves for each analysis. The top row presents the full panels of 12 curves for each analysis (dotted lines) along with the overall averaged curves (solid lines). The second and third rows present the same panels of curves organized by each input indicator, with three curves per graph (dotted lines) along with their average (solid lines).

the former technique because it is consistent with the methods of locating cutting scores and constructing the x axis that we have recommended elsewhere and may help to align taxonic peaks across curves even when indicators differ in their metrics and validities. For instance, if a taxon comprises 25% of a sample of 1,000 cases, then MAMBAC peaks should emerge fairly consistently when cuts are placed near the 750th case (counting from the low end of the input indicator). When averaging MAXCOV/MAXEIG curves, we prefer to average both the y values (conditional covariances or eigenvalues) and subsample means (x values) again because this is consistent with our recommended techniques for forming subsamples (i.e., many overlapping windows) and constructing the x axis (i.e., actual indicator scores, rather than ordinal numbers). Provided that indicators are standardized prior to analysis, taxonic peaks should emerge near the same region of the input indicator scale across curves, thus alleviating the concern that averaging will obscure otherwise consistently peaked curves.

Having stated our preferences, however, we acknowledge that the optimal approach to averaging remains another fertile area for future research on taxometric methodology. Therefore, we offer these options as tentative suggestions and await data on the most appropriate conditions and most effective techniques for averaging curves.

THE INFLUENCE OF INDICATOR SKEW

Some influences on the shape of taxometric curves have been widely recognized. For example, it is generally known that the location of a taxonic peak is related to the taxon base rate, with smaller taxa yielding peaks toward the right end of the curves. However, there are other influences on curve shape that may be less appreciated, and these are the ones on which we focus here. In this section, we discuss the influence of indicator skew; in the next, we discuss the influence of indicator validity and within-group correlations.

In the primary publications that introduced or evaluated the core taxometric procedures (Grove & Meehl, 1993—MAXSLOPE; Meehl & Yonce, 1994—MAMBAC; Meehl & Yonce, 1996—MAXCOV; Waller & Meehl, 1998—MAXEIG and L-Mode), all analyses were performed on data that were normally distributed along continuous scales. Taxonic data sets were generated such that within-group indicator distributions were normal; dimensional data sets were generated such that full-sample indicator distributions were normal. Deviations from normality and continuity have not been systematically investigated. However, it has been shown that when indicators are skewed in the full sample (for dimensional data) or within groups (for taxonic data), this can exert a substantial and predictable influence on curves produced by the most widely used taxometric procedures (A. M. Ruscio & Ruscio, 2002; J. Ruscio, Ruscio, & Keane, 2004).

In chapters 5 and 6, we noted that MAMBAC, MAXCOV, and MAXEIG curves can be expected to rise with positively skewed indicators, and that—given sufficient indicator skew—such rising curves can be mistaken for evidence of a small taxon. Following a parallel process, curves are expected to fall in the presence of negatively skewed indicators, producing curves that may be interpreted as reflecting a large taxon. To more systematically illustrate the influence of positive indicator skew on these three procedures, we performed each procedure using three sets of dimensional data. In the first data set, indicators were normally distributed; in the second, indicators were slightly skewed; and in the third, indicators were more substantially skewed (see the caption for Fig. 8.4 for details on how skew was imparted). Each data set contained four indicators that, prior to skewing, were correlated at an average of r = .50 with one another in a sample of 1,000 cases. MAMBAC was performed using all possible pairwise input–output indicator configurations, with 50 equally spaced cutting scores placed along the input for each of the 12 curves. MAXCOV was performed using all possible input–output indicator triplets, with subsamples formed along the input using decile intervals for each of the 12 curves. MAXEIG was performed using each indicator once as input and the other three variables serving as outputs, with 50 overlapping windows formed for each of these four curves. Figure 8.4 displays a representative frequency distribution for the first indicator in each data set, as well as the full panel of individual curves (dotted lines) and averaged curves (solid lines) for each series of analyses.

As can be seen in the figure, MAMBAC curves took on the concave shape expected of dimensional data when nonskewed indicators were analyzed, but became increasingly J-shaped with larger amounts of indicator skew—a shape that could be easily mistaken for evidence of a small taxon. MAXCOV curves took on the characteristically flat shape of dimensional data with nonskewed indicators, but became right cusped as positive skew was introduced, again producing curves consistent with expectations for a small taxon. Finally, increased skew caused MAXEIG curves to change from a flat line to rising curves that could be misinterpreted as evidence of a small taxon. Thus, if one failed to consider the influence of indicator skew, these consistent curves across multiple taxometric procedures could lead to an erroneous inference of taxonic structure.

Indicator skew affected not only the shapes of the taxometric curves, but also the taxon base rate estimates, which are included in Fig. 8.4. Generally speaking, base rate estimates decreased with larger amounts of skew. In their MAMBAC monograph, Meehl and Yonce (1994) suggested

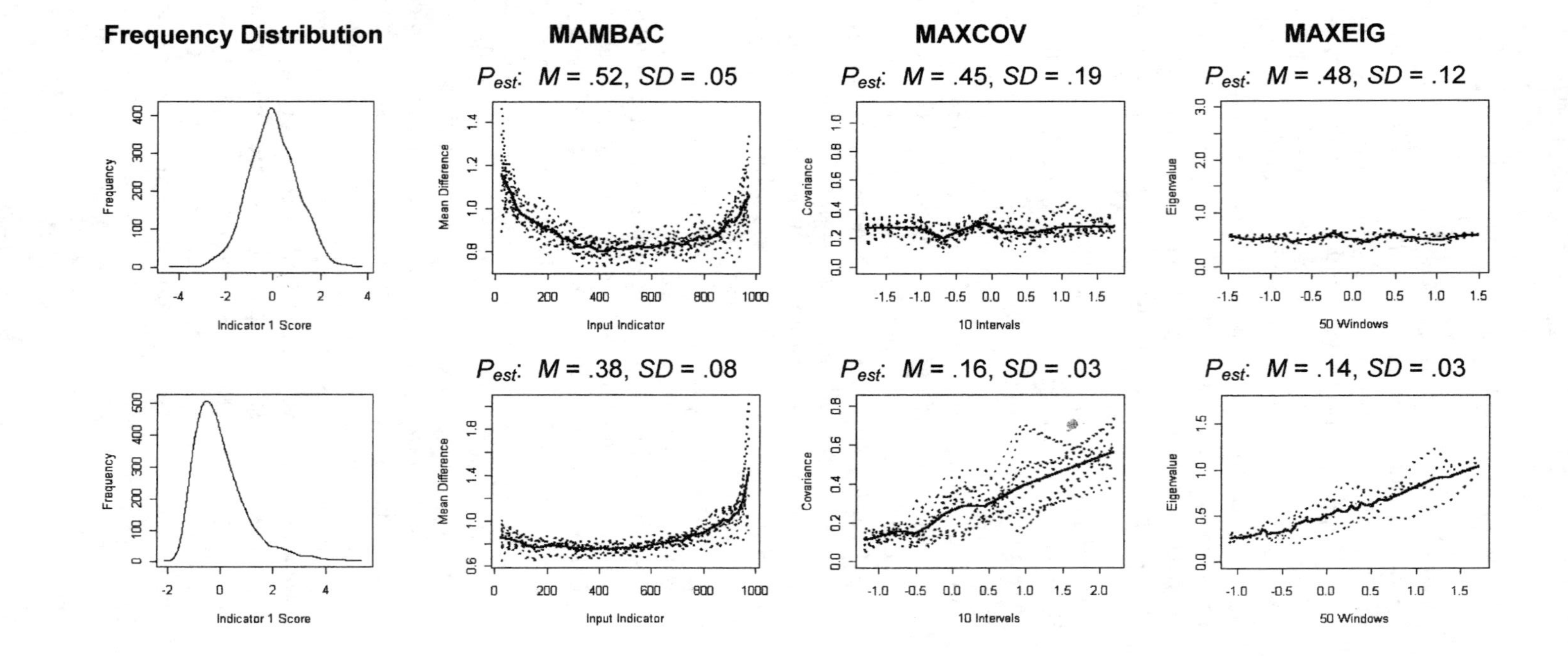
Frequency Distribution
MAMBAC
MAXCOV
MAXEIG
Pest: M = .52, SD = .05
Pest: M = .45, SD = .19
Pest: M = .48, SD = .12
Frequency
Indicator 1 Score
Mean Difference
Input Indicator
Covariance
10 Intervals
Eigenvalue
50 Windows
Pest: M = .38, SD = .08
Pest: M = .16, SD = .03
Pest: M = .14, SD = .03
Frequency
Indicator 1 Score
Mean Difference
Input Indicator
Covariance
10 Intervals
Eigenvalue
50 Windows

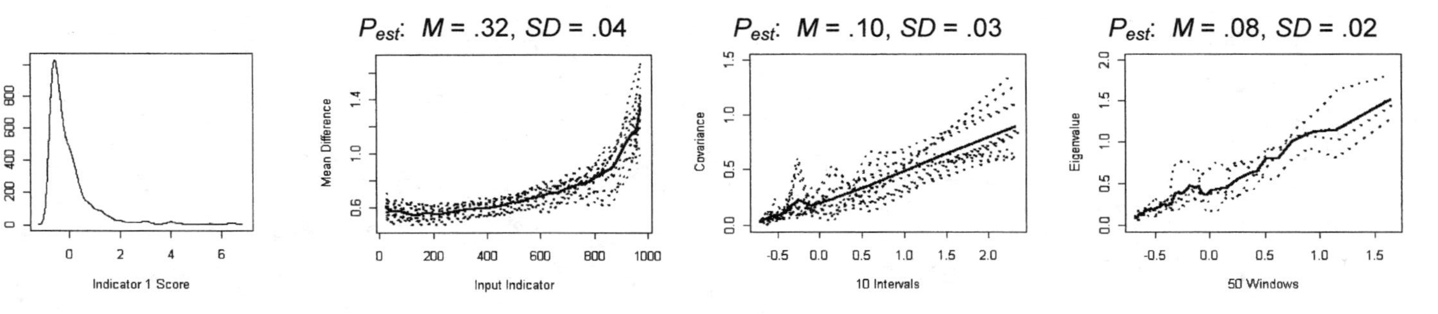

FIG. 8.4. Influence of increasing amounts of indicator skew on MAMBAC, MAXCOV, and MAXEIG results for dimensional data. Each dimensional sample was created by generating four normally distributed indicators that were correlated at an average of $r = .50$ in a sample of $N = 1{,}000$. After standardizing these indicators, low skew was induced by transforming the initially normally distributed scores using the function skew(x) = exp(x/2) and then restandardizing each indicator; high skew was induced using the function skew(x) = exp(x). The first row shows the results for indicators with no skew (M skew = .01, SD = .03), the middle row for low skew (M = 1.67, SD = .24), and the bottom row for high skew (M = 4.87, SD = 1.71). The first column shows a representative frequency distribution from each dimensional sample of data that was analyzed, and the following three columns show the results for MAMBAC, MAXCOV, and MAXEIG analyses. The M and SD of the taxon base rate estimates for full panels of curves are displayed above the averaged curve for each analysis. Each taxometric graph shows the full panel of curves (dotted lines) plus the averaged curve (solid line).

that taxon base rate estimates falling below .40 were evidence of taxonic structure. Several researchers have since cited this statement as support for a taxonic conclusion after obtaining base rate estimates below this threshold in MAMBAC or other procedures. Unfortunately, even the small sampling of illustrative analyses included here underscores the danger of reaching structural conclusions in this way. Given sufficient levels of positive skew, taxon base rate estimates can be quite low for dimensional data. This observation held for each of the three procedures, with MAXCOV and MAXEIG base rate estimates dropping even lower than those for MAMBAC. At least as important is the consistency with which base rate estimates calculated from the full panel of curves cohered around a single value. MAMBAC analyses yielded highly consistent estimates for all three dimensional data sets, whereas MAXCOV and MAXEIG analyses of the nonskewed data were more discrepant, yielding fairly large *SD*s. However, all three procedures produced very low *SD*s in analyses of skewed data. Thus, as was argued in chapter 7, the conventional wisdom that coherent estimates of the taxon base rate support an inference of taxonic structure may not apply to analyses of skewed data.

Given the ubiquity of indicator skew in psychological research (Micceri, 1989), it is important that investigators consider its potential influence on taxometric curves and parameter estimates. The expectation that MAMBAC, MAXCOV, and MAXEIG analyses will yield nonpeaked curves for dimensional data is clearly mistaken to the extent that (a) indicator distributions are skewed, and (b) a right-end cusp is interpreted as a peak. Although the influence of skew on curve shapes has not been systematically studied, its potentially significant impact has been sufficiently documented to recommend against relying too heavily on prototypical taxometric curve shapes when interpreting the results of analyses performed on real data. Investigators should keep in mind that the curves published in the seminal articles cited earlier represent what one can expect under fairly idealized conditions, including nonskewed data.

In many research contexts, the obfuscating influence of skew can be reduced by applying an appropriate nonlinear transformation (e.g., the square root, logarithm, or reciprocal of a set of values) to achieve a distribution that more closely approximates normality. Unfortunately, this is unlikely to be very helpful in a study involving MAMBAC, MAXCOV, or MAXEIG analyses. The reason is that, after sorting cases by their scores along an input indicator, these procedures use the rank ordering of cases—not actual scale values—as the basis for locating cutting scores (in MAMBAC analyses) or for dividing cases into subsamples (in MAXCOV/MAXEIG analyses). It is possible that using *SD* units or intact scale values to divide the input in these analyses—instead of using case numbers in MAMBAC or equal-*N* subsamples in MAXCOV/MAXEIG—would allow a nonlinear transformation to alleviate the influence of indicator skew, a possibility that awaits further study. However, as discussed in chapters 5 and 6, using evenly spaced cutting scores in MAMBAC analyses and equal-*N* overlapping windows in MAXCOV/MAXEIG analyses is desirable for a variety of reasons, so any benefit of transformed indicator scores may be offset by the alternative implementation procedures that they would require.

THE INFLUENCES OF INDICATOR VALIDITY AND WITHIN-GROUP CORRELATIONS

Whereas the influence of indicator skew on taxometric results has only recently come to the attention of taxometric researchers, the influences of indicator validity and within-group correlations have been discussed since the inception of the taxometric method (e.g., Meehl, 1973, 1995a, 1995b; Meehl & Golden, 1982) and are usually described each time a new taxometric procedure is introduced (e.g., Grove & Meehl, 1993; Meehl & Yonce, 1994, 1996; Waller & Meehl, 1998). What these and other studies have shown is that low levels of indicator validity can make it difficult to distinguish taxonic and dimensional structure. For example, if the taxon and complement are not adequately separated on the indicators submitted to analysis, the local regression curve will not take on the expected S shape in MAXSLOPE, only one mode will be apparent in L-Mode, and peaks will not emerge in MAMBAC or MAXCOV/MAXEIG.

The degrading effects of decreasing levels of indicator validity on these procedures are illustrated in Fig. 8.5. Analyses were performed using taxonic data sets with N = 1,000, P = .50, and four indicators that were uncorrelated and normally distributed within groups. Each taxometric procedure yielded clearly taxonic results when indicator validity was high (e.g., d = 2.00) and generated increasingly ambiguous results as indicator validity declined. As this figure shows, when indicators are insufficiently valid, results for taxonic data cannot be distinguished from those for dimensional data. The point at which indicator validity can be considered large enough varies across taxometric procedures and probably depends on the joint influence of a number of additional characteristics that bear on the quality of the data (e.g., sample size, taxon base rate, within-group correlations, indicator skew).

Whether due to shared method variance or substantive content overlap, within-group correlations can also make it more difficult to reach an accurate structural inference. Although most investigators seem aware of the importance of indicator validity, too little attention is often paid to within-group correlations, and to the importance of employing indicators that independently assess relevant aspects of the target construct. The correlation

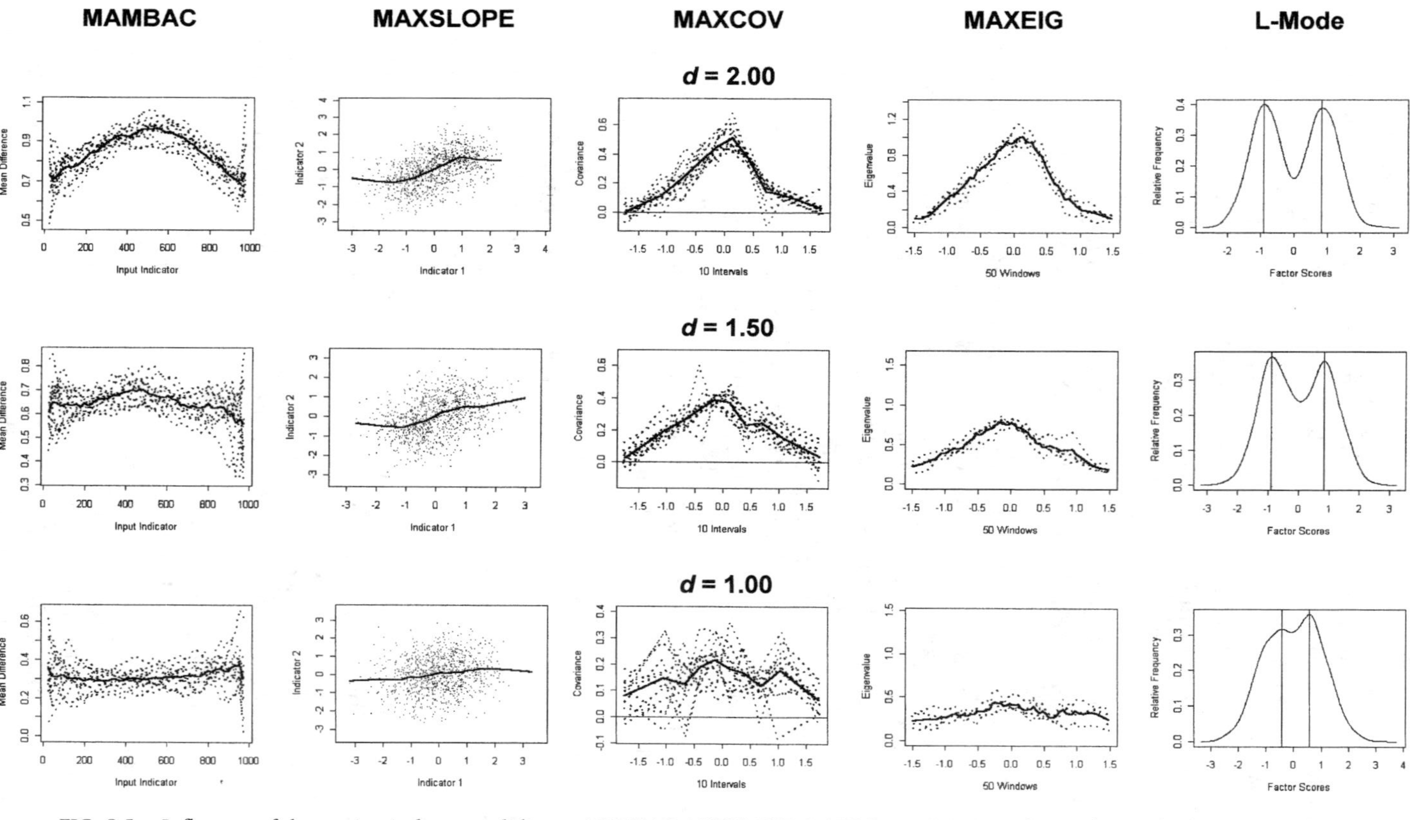

FIG. 8.5. Influence of decreasing indicator validity on MAMBAC, MAXSLOPE, MAXCOV, MAXEIG, and L-Mode results for taxonic data. Each taxonic sample was created by generating four normally distributed indicators in a sample of $N = 1,000$, with no within-group correlations and a taxon base rate of $P = .50$. The first MAXSLOPE curve from each panel of 12 curves is shown, and each MAMBAC, MAXCOV, and MAXEIG graph shows the full panel of curves (dotted lines) plus the averaged curve (solid line).

of indicators within groups has predictable—and sometimes substantial—effects on taxometric curve shapes. For example, MAXSLOPE curves for taxonic data will only be discernibly S-shaped to the extent that the local regression is relatively flat at both ends; this flatness requires that indicators be independent of one another within the taxon and complement. With problematically large within-group correlations, the MAXSLOPE curves for taxonic data can appear linear, increasing the likelihood of incorrect inferences of dimensional structure. When within-group correlations are low, variation around taxon and complement mean scores tends to cancel out when individuals' scores on the first factor are estimated in an L-Mode analysis. However, larger within-group correlations disperse the factor scores for each group around its mean, which makes it more difficult to distinguish taxonic from dimensional structure using L-Mode.

MAMBAC and MAXCOV/MAXEIG curves can likewise be distorted by large within-group correlations. Figure 8.6 presents results for taxonic data sets with four normally distributed indicators, $N = 1{,}000$, $P = .25$, $d = 2.00$, and varying levels of within-group correlations. In some cases, these correlations were equal within the taxon and complement (either both *r*s = .00 or both *r*s = .50); in others, correlations were large in one group ($r = .50$), but not the other ($r = .00$). The taxonic nature of these data sets was more difficult to discern when there were substantial within-group correlations in one—and especially in both—of the groups. Although these illustrative analyses involve data with large within-group correlations, they provide a sense for the way in which results would be expected to degrade with more moderate within-group correlations. Notably, had we started with data possessing less favorable parameters (e.g., a smaller sample size, more extreme taxon base rate, lower indicator validity, some degree of indicator skew within groups), it is likely that some or all of the taxometric procedures would have yielded equally or more ambiguous and misleading results with smaller within-group correlations.

These results are not meant to suggest that one taxometric procedure is typically more robust to low indicator validity or high within-group correlations than another; this may be true, but more systematic study is needed before such general conclusions can be drawn. Our point instead is that any taxometric procedure can yield ambiguous or misleading results when data characteristics deviate from ideal values. Until more is known about these influences under a broader range of conditions, investigators are encouraged to conduct and interpret taxometric analyses with appropriate caution. In particular, we urge researchers to make use of indicator selection and construction strategies that maximize indicator validity and minimize within-group correlations (see chap. 4), and to empirically evaluate the possibility that seemingly dimensional results may stem from poor validity or large within-group correlations.

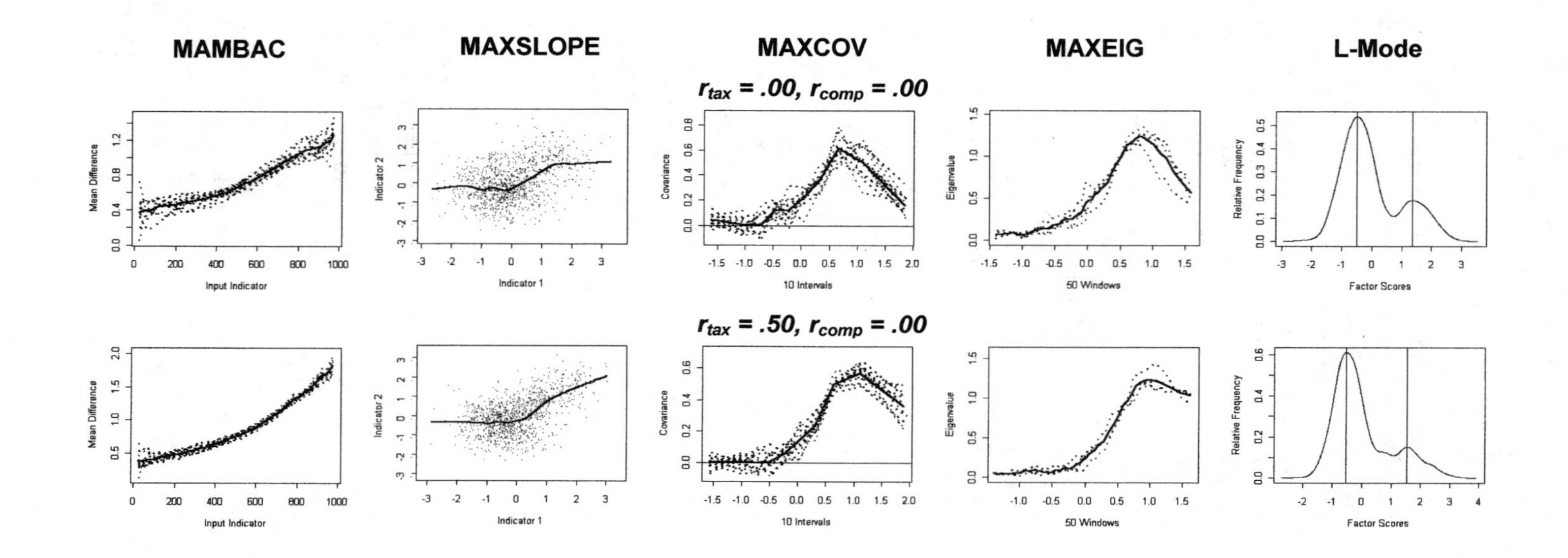

MAMBAC
MAXSLOPE
MAXCOV
MAXEIG
L-Mode
r_{tax} = .00, r_{comp} = .00
r_{tax} = .50, r_{comp} = .00
Mean Difference
Input Indicator
Indicator 2
Indicator 1
Covariance
10 Intervals
Eigenvalue
50 Windows
Relative Frequency
Factor Scores

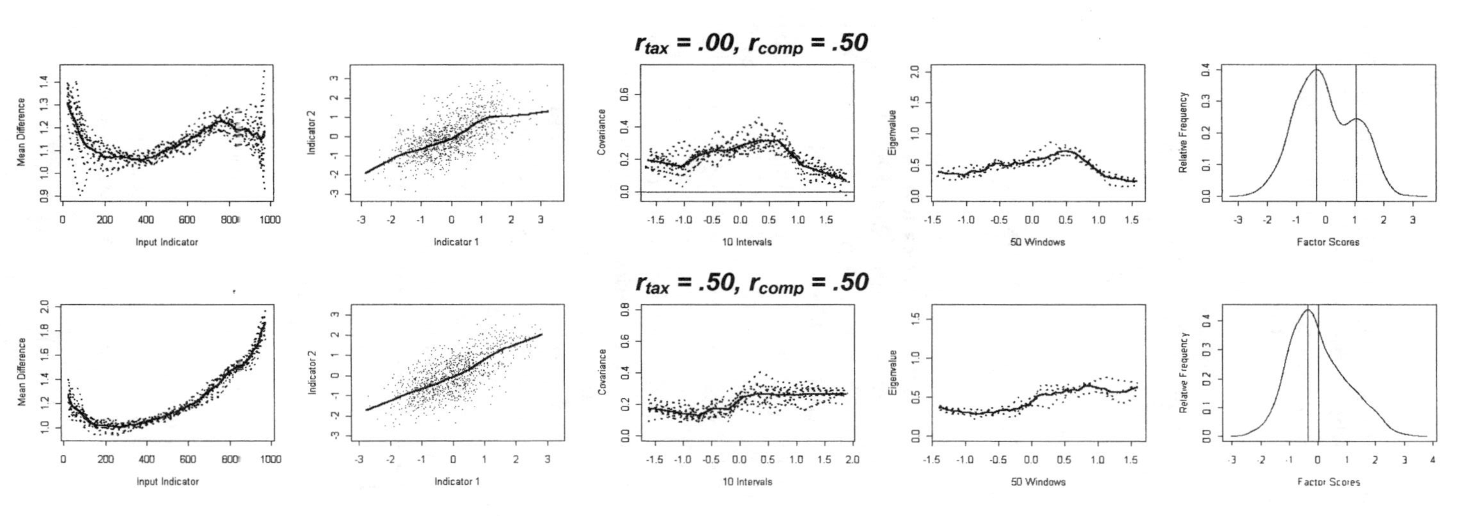

FIG. 8.6. Influence of within-group correlations on MAMBAC, MAXSLOPE, MAXCOV, MAXEIG, and L-Mode results for taxonic data. Each taxonic sample was created by generating four indicators in a sample of N = 1,000; each indicator separated the taxon (P = .25) and complement with substantial validity (d = 2.00). Within each group, indicators were correlated with one another through shared loadings on a latent dimension. The first MAXSLOPE curve from each panel of 12 curves is shown, and each MAMBAC, MAXCOV, and MAXEIG graph shows the full panel of curves (dotted lines) plus the averaged curve (solid line).

Similarly, these illustrative analyses are not intended to suggest that there exist any thresholds above which indicator validities (or below which within-group correlations) are acceptable. Not only would such thresholds likely vary across taxometric procedures, they would also likely depend on other aspects of the data. As noted in chapter 4, we recommend against evaluating data characteristics in isolation, preferring instead to evaluate them jointly by generating empirical sampling distributions of results for taxonic and dimensional comparison data. We revisit this approach in the procedural safeguards to which we turn next.

INTERPRETATIONAL SAFEGUARDS

The centerpiece of a taxometric investigation involves the interpretation of graphical results that are obtained from the application of taxometric procedures. The fact that the taxometric method does not rely on tests of statistical significance is a mixed blessing. On the one hand, reaching accurate structural inferences requires a solid understanding of the relevant analytic procedures, including their susceptibility to providing ambiguous or misleading results under a variety of conditions. It would doubtless be easier to read a p value and arrive at a structural conclusion. On the other hand, evaluating the consistency of results—with the patterns expected for taxonic and dimensional data, with the output of all analytic procedures, and with the findings observed in empirical sampling distributions—offers distinct advantages over a rote interpretational process hinging largely on a p value. There are ample opportunities for the informed investigator to detect results that are (in any of the ways described earlier) inconsistent, and thereby to prevent misinterpretations of the data. In the remainder of this chapter, we discuss procedural safeguards that researchers can use to enhance accurate interpretation of their taxometric results, and therefore to increase the rigor with which they test between taxonic and dimensional structural models.

Presenting as Many Taxometric Curves as Possible

In a taxometric investigation, structural inferences are derived primarily from interpretation of the full panel of individual curves yielded by all taxometric analyses. However, because taxometric analyses often yield more results—particularly graphical results—than can readily be presented in a standard research article, taxometric reports seldom include all of the curves produced in a given investigation. Indeed some researchers have downplayed graphical material in their reports, highlighting instead the percentage of obtained curves appearing taxonic and dimensional (and sometimes also noting the percentage of curves that appear ambiguous).

Given the central role of taxometric curves in the interpretational process, we believe that every effort should be made to present as many taxometric curves as possible in published reports, even if the curves must be reduced in size or combined to fit in the available space. As noted and illustrated throughout this chapter, there are multiple ways in which researchers can aggregate curves to conserve space, with one particularly compact option involving plotting multiple curves on a single graph (see Figs. 8.2–8.6 for examples). At a minimum, either averaged curves (if these adequately represent the full panels of curves) or a random, representative subset of the curves should be presented. Such presentations not only facilitate the interpretational process for researchers, but also ensure that readers, editors, and reviewers are provided with sufficient graphical material to personally determine the strength of the evidence for the researchers' structural conclusions. Although summary percentages may be quite useful, we regard them as a supplement to the core graphical material essential to a taxometric report.

Using Independent Raters

Regardless of how graphical results are summarized or presented, we recommend that researchers solicit judgments from independent raters to inform the interpretation process when there is any ambiguity in the results. All too often, taxometric investigators authoritatively state the percentages of curves that they judged to be taxonic, dimensional, or ambiguous, with no assurance provided that others would interpret the curves in the same manner. Interpreting taxometric curves is an inherently subjective process. At times this process is relatively straightforward and clear; at others, conclusions may be based on relatively subtle features of the curves. When conclusions depend so heavily on subjective judgments, it is valuable to demonstrate an acceptable level of interrater agreement in support of a particular structural conclusion.

When soliciting raters' judgments of taxometric results, a number of procedural decisions should be considered and reported. First, will raters be blind to the structural hypothesis of the researchers? Our reading of taxometric reports suggests that investigators often begin their research with the expectation of obtaining results supporting a particular structure. Often this appropriately reflects the use of the taxometric method as a theory-driven, hypothesis-testing enterprise (e.g., Lenzenweger, 2004; Meehl, 1992, 2004; Meehl & Golden, 1982; Waller & Meehl, 1998), rather than an exploratory tool (but see Meehl, 1999). However, it is well known that in many types of research, investigators' preconceptions may bias their interpretation of results. This may be a particularly relevant concern in taxometric research, which involves a great deal of judgment in planning and implementing analyses and interpreting their results. In cases where these results are not overwhelmingly clear—and we would classify most taxometric studies into this category—it would be helpful to have multiple blind raters independently evaluate the results to increase the likelihood of unbiased interpretation.

Second, will the obtained curves be interpreted relative to those produced by fairly idealized data in published Monte Carlo studies, relative to empirical sampling distributions of taxometric curves yielded by analyses of bootstrap samples of taxonic and dimensional comparison data, or relative to some other standard? As noted later, we believe that comparing curves to the results of comparison data can greatly reduce unwarranted or erroneous structural inferences. Unwarranted inferences—those based on results that do not support one structure over another—are diminished by refraining from reaching any structural conclusion when the results of taxonic and dimensional data cannot be distinguished from one another by the taxometric analyses (as discussed in chap. 4, we do not regard ambiguous results as either a strike against taxonic structure or as support for dimensional structure). Some erroneous inferences that occur when results are systematically misleading in a way that a rater does not recognize can be avoided by comparing the obtained curves to empirical sampling distributions. However, comparing the obtained results to those in published Monte Carlo studies may contribute to inferential mistakes because many factors other than latent structure may influence the extent to which the research curves resemble prototypical taxonic or dimensional curves. As we discussed earlier, if the indicators are insufficiently valid or there are substantial within-group correlations, taxonic data tend to yield results appearing more consistent with dimensional structure. Likewise, as has been demonstrated several times, skewed indicators of a dimensional construct can yield results appearing more consistent with a small taxon. Thus, we believe that fewer unwarranted or erroneous structural conclusions will be reached when raters evaluate curves relative to empirical sampling distributions—an approach we illustrate in the next section.

Third, what background knowledge should appropriate raters have, and what instructions should they be given for the rating task? If Monte Carlo results are to form the basis for interpretations, raters should have extensive experience with taxometric analysis and be familiar with the known influences of data characteristics and procedural implementation on taxometric results. If experienced users of taxometrics are not available to serve as raters, adequate instruction should be provided to ensure a thorough understanding of the expected results for each structure under a variety of conditions. Although this is certainly a tall order, raters who possess only a superficial knowledge of taxometrics may be far more susceptible to reaching unwarranted or erroneous structural inferences. If, however, empirical sampling distributions form the basis for interpretations, raters need not have any prior experience with or special knowledge of taxometrics. All that such raters will be asked to do is determine whether the results for the two known structures can be distinguished (suggesting that the data parameters and implementation decisions afford an informative test), and—if so—whether the results for the research data appear more similar to those for the taxonic or dimensional comparison data.

Fourth, how should the rating task be conducted? Should raters judge each individual curve or evaluate a full panel of curves one at a time? Should raters leaf through the full set of curves generated by a procedure (to gain a sense for their range and consistency) before making any ratings? Should raters be allowed to alter previous judgments after inspecting subsequent results? At present, there are no empirical grounds for making these decisions. Until further research is done, we encourage investigators to consider which approach is most likely to lead to accurate conclusions, and to document the approach that was used.

Clearly, there is no presently established procedure for interpreting taxometric results. However, by making use of multiple, independent, unbiased raters, and by providing these raters with useful interpretive aids, we believe that the rigor of the interpretive process—and the accuracy of the resulting conclusions—can be substantially enhanced.

Using Empirical Sampling Distributions as an Interpretive Aid

Since first introducing the approach in chapter 4, we have repeatedly recommended that researchers analyze comparison data for three related pragmatic purposes. The first is to evaluate the appropriateness of a particular data set for a particular analysis by examining the overlap between the sampling distributions of results for taxonic and dimensional comparison data. If these known structures produce distinguishable results, the data are judged to be appropriate for the procedure. The second purpose is to determine how best to implement a particular taxometric procedure or consistency test. Through the process of adaptive calibration (Efron & Tibshirani, 1993), different strategies can be evaluated and the one that yields the most powerful differentiation between taxonic and dimensional structural models retained. The third purpose is to compare the research results to the empirical sampling distributions to facilitate the accurate interpretation of results. To the extent that results for the research data are more similar to those of one known structure than the other, evidence is garnered in favor of the matching structure. In addition to careful indicator selection and attention to important parameters, we view this bootstrapping approach as a key defense against mistaken structural conclusions arising from inadequate data or poorly implemented analyses.

It is important to distinguish between the overarching approach of generating empirical sampling distributions and the specific technique that is used to simulate the necessary taxonic and dimensional comparison data. In Appendix A, we described several techniques for simulating data and drew attention to the strengths and limitations of each. We believe that our own iterative technique (J. Ruscio, Ruscio, & Meron, 2005), grounded in bootstrap methods, may represent a particularly promising way to reproduce both the distributional and correlational properties of a data set when simulating comparison data; however, it is entirely possible that superior simulation strategies will be developed. Our main point is that, although any simulation technique will have strengths and weaknesses that influence the informativeness of its results, the value of any particular technique should be distinguished from the value of the broader approach for generating and working with empirical sampling distributions.

The results of any data analysis—taxometric or otherwise—require an appropriate benchmark for comparison to afford meaningful interpretations. Sampling distributions of results expected for different hypotheses (e.g., a null hypothesis vs. an alternative hypothesis, one structural model vs. another) almost always guide the interpretation of quantitative results, whether these results are presented as graphs, numerical indexes, or *p* values. In taxometric research, interpreting curve shapes or latent parameter estimates based on the results of Monte Carlo studies (e.g., Meehl & Yonce, 1994, 1996) treats their results as the sampling distributions to be expected for taxonic and dimensional structure. However, the data parameters in a particular taxometric investigation may deviate in substantial ways from the conditions included in the Monte Carlo studies, and one might implement a taxometric procedure in a different way than was done in the Monte Carlo research. Thus, we believe that the appropriateness of generating empirical sampling distributions tailored to the unique data set and analytic implementation strategy in a particular taxometric investigation is unassailable. The question of how best to do this—for example, what technique to use for simulating the requisite samples of taxonic and dimensional comparison data, how many samples of each structure are required—remains to be studied. We have presented one approach, but we make no claim that it is optimal. Our own simulation programs have evolved in a number of ways over time and may continue to evolve in the future. We encourage research that assesses the strengths and weaknesses of our technique as well as proposals for better approaches.

Once comparison data sets have been simulated and analyzed, there are several ways in which their results may be graphed for interpretation and presentation. One graphing approach involving two-panel graphs was illustrated in Fig. 7.4, in which results for the research data (dark line) were superimposed on the empirical sampling distributions for simulated taxonic (left) and dimensional (right) data sets. These sampling distributions were summarized by presenting the curves falling ±1 *SD* from the average curve for all samples of each structure; because the *SD* is calculated within a sampling distribution, it represents the familiar *SE*. An alternative graphing approach involves three-panel graphs in which the first panel displays the results for the research data, the second panel for the simulated taxonic data sets, and the third panel for the simulated dimensional data sets. The second and third panels contain multiple curves, including one dotted line per sample plus a solid line depicting the average across all samples.

Figure 8.7 demonstrates both of these graphing formats for the same MAMBAC and MAXEIG analyses of TS#4. This data set was used because, among our eight illustrative samples, TS#4 posed the greatest challenge for many taxometric procedures and consistency tests. In these analyses, empirical sampling distributions were generated using $B = 10$ samples each of taxonic and dimensional comparison data. Although different researchers might choose to include more samples (given sufficient computing power) or might prefer the two- or three-panel graphs, either approach clearly provides a useful interpretive aid that is likely to result in an accurate inference of taxonic structure. Because some researchers might find the two-panel graphs easier to interpret, whereas others might prefer the three-panel graphs, our taxometric programs offer researchers both graphing options.

In addition to plotting empirical sampling distributions of taxometric curves, investigators can also plot such distributions for quantitative results of taxometric analyses. For example, researchers have often examined the *M* and, even more frequently, the *SD* of taxon base rate estimates to inform their structural interpretations. As described in chapter 7, although a lower *SD* (i.e., more consistent base rate estimates) is typically viewed as evidence in favor of taxonic structure, there are a number of circumstances under which dimensional data can produce low *SD*s. Thus, one can profitably use the bootstrapping approach to help interpret the *M* and *SD* of obtained base rate estimates. One frequently useful strategy is to display the *M* and *SD* of base rate estimates yielded by the simulated comparison data alongside those yielded by the research data in a single table. This format facilitates direct comparisons of estimates within each taxometric procedure, determining whether base rates estimated for the research data are more similar in size and variability to those estimated for

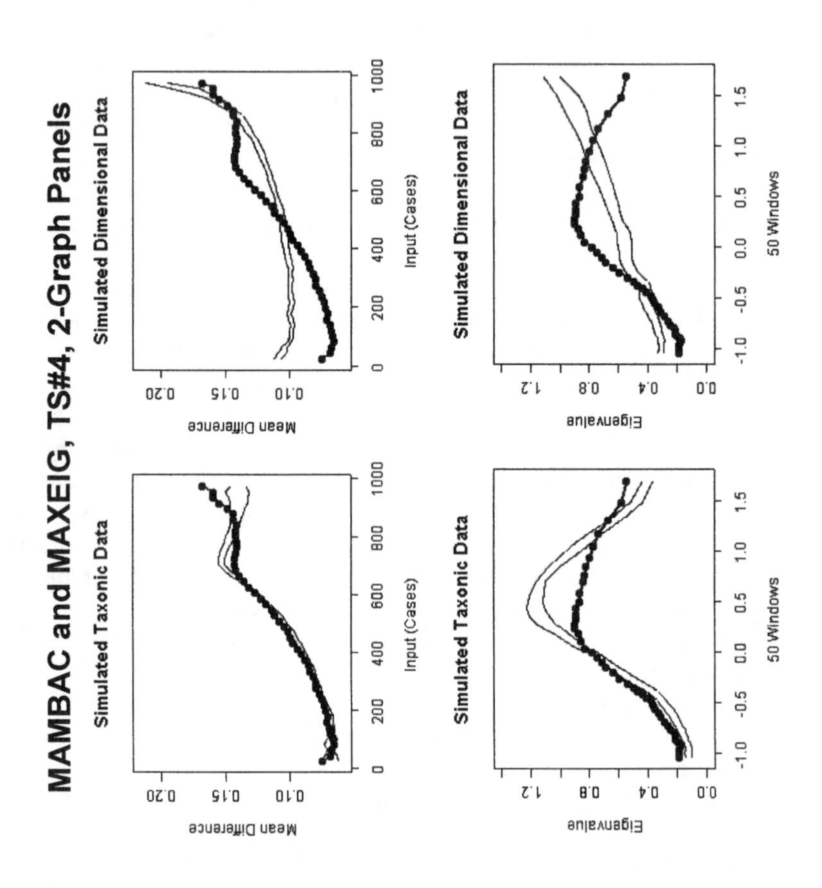
MAMBAC and MAXEIG, TS#4, 2-Graph Panels
Simulated Taxonic Data
Simulated Dimensional Data
Mean Difference
0.10
0.15
0.20
Input (Cases)
0
200
400
600
800
1000
Simulated Taxonic Data
Simulated Dimensional Data
Eigenvalue
0.0
0.4
0.8
1.2
50 Windows
-1.0
-0.5
0.0
0.5
1.0
1.5

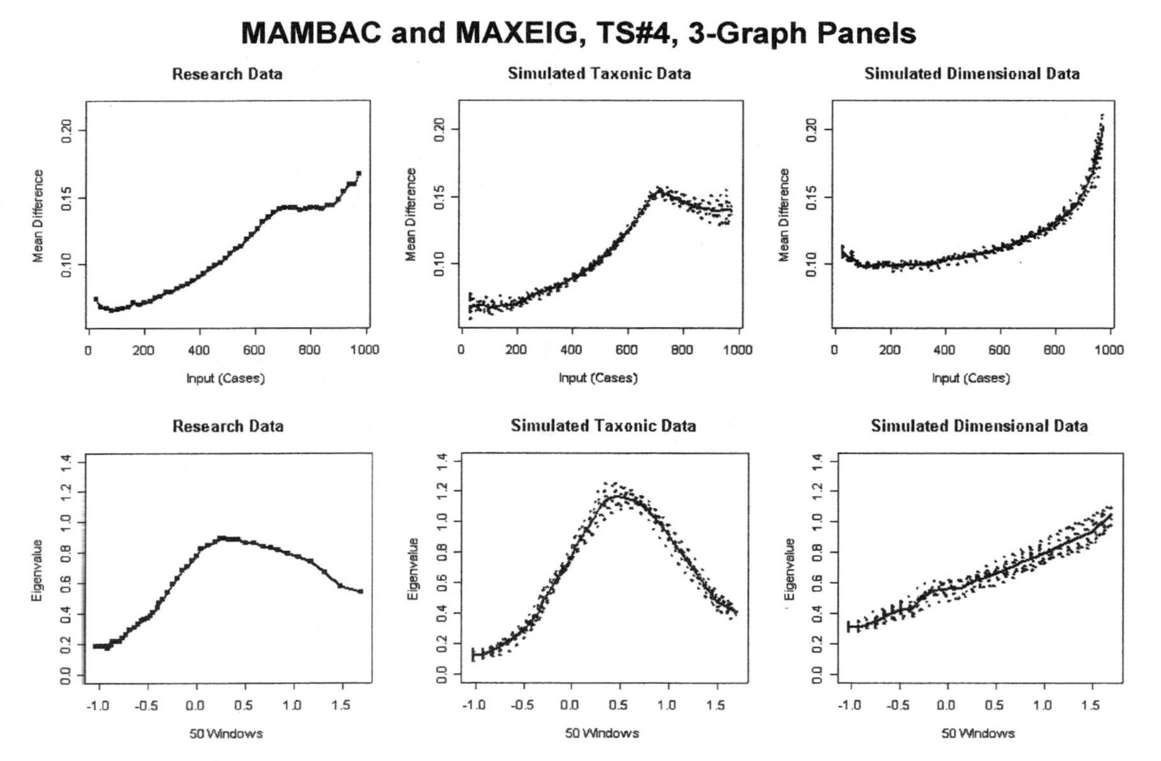

FIG. 8.7. Two ways to present results for research data along with empirical sampling distributions for results provided by analyses of taxonic and dimensional comparison data. In two-graph panels, the averaged curve for the analysis of the research data (dark line with data points) is superimposed first on the sampling distribution of results for simulated taxonic data and then on the sampling distribution of results for simulated dimensional data; each sampling distribution is represented by light lines (without data points) that are calculated as ±1 *SD* from the average of all curves generated within that distribution. In three-graph panels, averaged curves are presented separately for analyses of the research data, simulated taxonic data, and simulated dimensional data; for the latter two panels, dotted lines represent averaged curves for each set of comparison data, and solid lines represent the average of all curves for that structure.

the taxonic or dimensional simulated data. Alternatively, researchers can graph empirical sampling distributions of base rate estimates to facilitate their interpretation. To illustrate this approach, we performed MAMBAC and MAXEIG analyses of TS#3, DS#3, TS#4, and DS#4 to obtain the *M* and *SD* of the taxon base rate estimates for each data set. We then generated empirical sampling distributions of means and standard deviations by analyzing 10 samples each of taxonic and dimensional comparison data. Results are presented in Fig. 8.8, with the empirical sampling distribution for taxonic data sets plotted as a solid curve, the distribution for dimensional data sets plotted as a dashed curve, and results obtained for the research data plotted as a dotted vertical line.

For each analysis, it is easy to gauge the degree of overlap between the two sampling distributions—this overlap is even more apparent with distributions of quantitative results than with distributions of taxometric curves. When the distributions overlap almost entirely, this suggests that the estimate observed for the research data is of little or no probative value in distinguishing taxonic from dimensional structure. In contrast, when the distributions diverge *and* the results for the research data are clearly more consistent with those of one structure than another, this suggests that the estimate observed for the research data can be taken to support one structure relative to the other. Inspection of Fig. 8.8 reveals instances in which *M* base rate estimates discriminate between taxonic and dimensional structure relatively well (e.g., MAXEIG analysis of TS#4) or poorly (e.g., MAMBAC analysis of TS#4), as well as instances in which *SD*s discriminate relatively well (e.g., MAXEIG analysis of DS#4) or poorly (e.g., MAMBAC analysis of DS#4). In the cases of good discrimination cited earlier, it is also clear that the results for the research data fall more squarely within one sampling distribution than the other, and using these comparisons to guide interpretation would lead to correct structural inferences.

Rather than applying a threshold to the *M* or *SD* of taxon base rate estimates (or other consistency tests), one can use empirical sampling distributions to reach a judgment about the power of that test to distinguish taxonic and dimensional structure in the study at hand (Efron & Tibshirani, 1993). When these distributions do not overlap too much, and when the results for the research data are more consistent with those of one distribution than another, one can reach a structural inference with some assurance that the results are not misleading because of idiosyncratic features of the data or the analysis, key aspects of which are held constant across sampling distributions. When empirical sampling distributions largely overlap or the results for the research data are no more consistent with those of either distribution, one can avoid reaching a potentially erroneous structural inference on the basis of a nondiscriminating

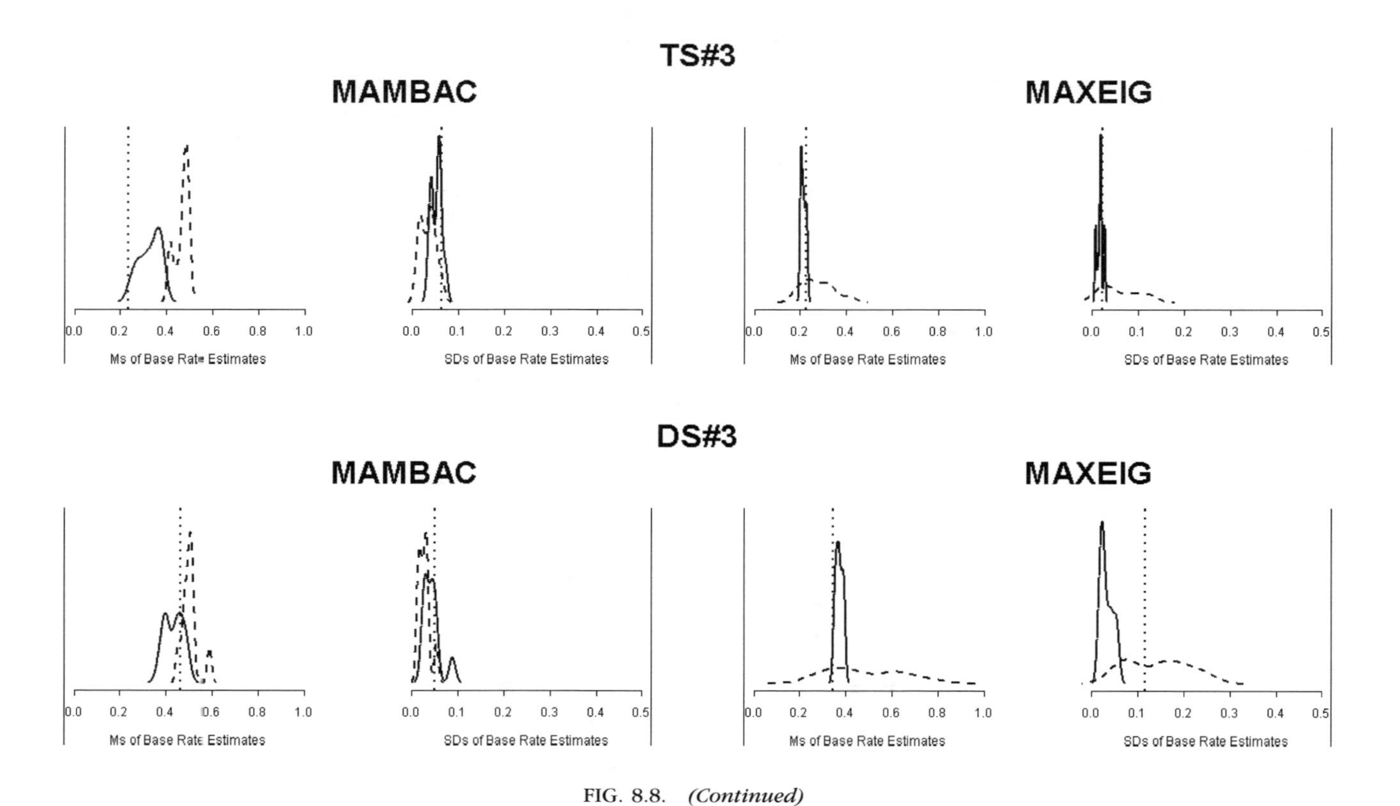

FIG. 8.8. *(Continued)*

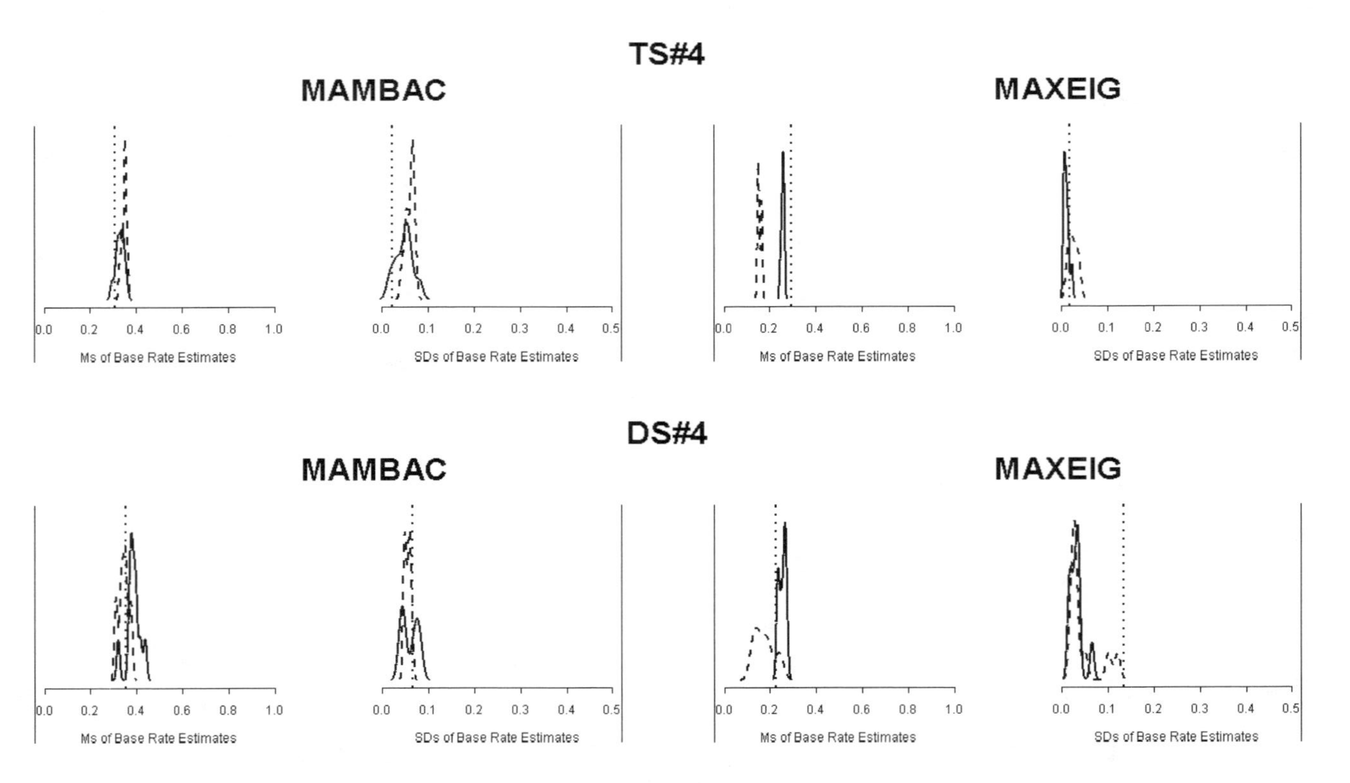

FIG. 8.8. Empirical sampling distributions of *M* and *SD* base rate estimates obtained in analyses of simulated taxonic (solid lines) and dimensional (dashed lines) comparison data sets. Vertical (dotted) lines represent the *M*s and *SD*s obtained in analyses of the research data.

test or ambiguous or misleading results. In summary, generating and inspecting empirical sampling distributions of results for simulated comparison data can be a useful way to determine how much weight to attach to the graphical or quantitative results of any taxometric procedure or consistency test during the interpretation process.

CONCLUSIONS

This chapter has described and illustrated a number of factors that can influence the shape of taxometric curves and the estimates of latent parameters derived from these curves, thereby influencing the validity with which structural inferences are reached using taxometric results. Our goal has been to draw attention to several factors whose effects are fairly well understood, but often overlooked in taxometric investigations. For example, although it has long been known that the manner in which a curve is graphed can influence its apparent shape, taxometric studies often make use of default graphing options and rarely explain the graphing decisions that are made. Our stance is that decisions about graphing—including whether curves will be smoothed and/or averaged for interpretation or presentation—should be made thoughtfully and deliberately, with consideration of the strengths and limitations of each approach.

It is also important that researchers keep in mind the effects of indicator skew, indicator validity, and within-group correlations as they interpret taxometric results. The informed researcher will be aware of the relatively idealized nature of taxometric curves published in Monte Carlo studies and of the limitations of using such curves as a point of comparison for interpreting curves obtained for more complex research data. Given these important limitations, we recommend that researchers bootstrap multiple samples of taxonic and dimensional comparison data, submit these to the taxometric procedures and consistency tests intended for the research data, and use the empirical sampling distributions of results not only to evaluate the appropriateness of the data for each analysis, but also as a point of comparison during the interpretive process. We demonstrated how such sampling distributions can be used to aid the interpretation of quantitative results (such as the *M* or *SD* of taxon base rate estimates) as well as the shapes of taxometric curves. We also suggested that published reports include as many taxometric curves as possible to permit their independent review, and that the full set of curves be judged by multiple, independent raters—preferably those unaware of the structural hypothesis of the study—to help prevent unwarranted or erroneous structural inferences.

The interpretation of taxometric results is largely a subjective enterprise, and much remains to be learned about the factors that can influence this process. It is the responsibility of taxometric researchers to rigorously procure and clearly report the interpretive judgments that inform their structural conclusions, making use of appropriate procedural safeguards to maximize their validity.

CHAPTER NINE

A Taxometric Checklist

In Part I of this book, we raised a number of conceptual issues relevant to the empirical study of latent structure, including the types of structural models that the taxometric method can help to distinguish and the scientific and applied value of making this fundamental distinction. In Part II, we addressed the application of the taxometric method, including the process of determining the adequacy of data for analysis and deciding among the many options for selecting and implementing taxometric procedures and consistency tests. In this chapter, we pull together all of these conceptual and empirical issues into a checklist organized around five broad concerns:

1. Is a taxometric analysis scientifically justified?
2. Are the data appropriate for taxometric analysis?
3. Has a sufficient variety of procedures been implemented properly?
4. Have the results been presented and interpreted appropriately?
5. Are implications of the findings clearly articulated?

Whether one is designing a taxometric study, writing a grant proposal that includes a taxometric analysis, preparing a research report of a taxometric investigation, reviewing a submitted taxometric manuscript, or reading a published taxometric paper, careful consideration of these five concerns can greatly assist in evaluating the relative strengths and limitations of the research. For this reason, we recommend that investigators explicitly address each of these questions when presenting a taxometric study. We elaborate here on these five questions and consider how they may be used to prevent or detect important omissions, questionable techniques, or other weaknesses of a taxometric investigation. We also draw attention to practices that remain controversial, commenting where possible on the extant empirical support for various methodological or analytic strategies. Because this chapter reviews material presented earlier in the book, it contains frequent references to more detailed discussions of the relevant points appearing in previous chapters. Table 9.1 summarizes the proposed checklist in outline form.

QUESTION 1: IS A TAXOMETRIC ANALYSIS SCIENTIFICALLY JUSTIFIED?

A taxometric report must begin by clearly articulating a compelling justification for performing a taxometric investigation. Although a persuasive rationale is important in any research study, it may be especially important in the taxometric literature, where a recent upsurge in interest and published studies (see chap. 10) runs the risk of promoting faddish research

TABLE 9.1
Conceptual and Methodological Checklist of Issues to Consider Carefully in a Taxometric Investigation

1. Is a taxometric analysis scientifically justified?
   A. Theoretical and/or practical implications
   B. Structural question being posed
2. Are the data appropriate for taxometric analysis?
   A. Sampling considerations
      i. Sample size
      ii. Taxon representation (taxon base rate and number of taxon members)
      iii. Population from which sample was drawn
   B. Potentially problematic sampling strategies
      i. Admixed samples
      ii. Splitting sample into subsamples for analysis
      iii. Trimming likely complement members to increase taxon base rate
   C. Selection and construction of indicators
      i. Content coverage
      ii. Indicator validity
      iii. Sensitivity at appropriate trait levels
      iv. Within-group correlations
      v. Multimethod assessment
      vi. Indicator distributions
      vii. Number of indicators
   D. Empirical evaluation of data
      i. A priori versus post hoc estimates of latent parameters
      ii. Generating empirical sampling distributions of taxometric results

(Continued)

TABLE 9.1
(Continued)

3. Has a sufficient variety of procedures been implemented properly?
   A. Taxometric procedures
      i. MAMBAC
      ii. MAXSLOPE
      iii. L-Mode
      iv. MAXCOV/MAXEIG
   B. Consistency tests
      i. Multiple taxometric procedures
      ii. Repeated application of each procedure (different indicator configurations)
      iii. Multiple samples (or subsamples) of data
      iv. Constructing multiple indicator sets
      v. Inchworm consistency test
      vi. Coherence of latent parameter estimates
      vii. Case removal consistency test
      viii. Testing differences between taxon and complement members
      ix. Distribution of Bayesian probabilities of taxon membership
      x. Goodness of fit index (GFI)
      xi. Generating empirical sampling distributions of taxometric results
      xii. Comparison curve fit index (CCFI)
4. Have the results been presented and interpreted appropriately?
   A. Graphing considerations
      i. Constructing the x axis
      ii. Scaling the y axis
      iii. Raw versus smoothed curves (vs. both)
      iv. Full panels versus averaged curves (vs. both)
   B. Interpretational issues
      i. Indicator skew
      ii. Indicator validity
      iii. Within-group correlations
   C. Judging curve shapes and quantitative results
      i. Rating procedure
      ii. Basis for comparison (Monte Carlo studies vs. empirical sampling distributions)
5. Are implications of the findings clearly articulated?
   A. Reiterate original motivation for the study
   B. Describe what knowledge has been gained

that is largely motivated by a desire to use the "flavor of the month" data-analytic technique. To justify a taxometric study, two criteria must be satisfied. First, as with any type of research, there needs to be a scientific rationale for the project. In the case of taxometrics, this means that there must be solid reasons that the distinction between taxonic and dimensional structural models of the target construct is worthy of empirical study. In chapter 2, we considered a number of ways in which taxometric analysis can promote basic and applied science. The presence or absence of a taxonic boundary may reflect a critical assumption of a theory to be tested, may have implications for how a construct should be assessed or how individuals should be classified, or may suggest the most conceptually appropriate and statistically powerful research design for future investigations of the target construct. Lenzenweger (2004) also noted the importance of providing a sound rationale for a taxometric study, emphasizing the need for a theoretical grounding of the structural model being tested.

Second, a taxometric study cannot be justified unless the structural information that is sought falls within the limits of what a taxometric analysis is able to provide. Part I of this book introduced the structural distinction that the taxometric method is designed to make: that between (a) a taxonic model consisting of two latent classes, either of which may contain residual variation along one or more latent dimensions, and (b) a dimensional model consisting of one or more latent dimensions. One point that has been insufficiently stressed in the taxometric literature is that the dimensional model need not be constrained to a single latent dimension. Rather, the taxometric method may be most appropriately conceived as distinguishing between the presence and absence of a putative taxonic boundary against a background of n-dimensional variation, with n unspecified. One can use the taxometric method iteratively to test for additional taxonic boundaries one at a time, using a sample of data and a set of indicators carefully selected to provide a rigorous test of each boundary. However, there remain many interesting and important structural questions that are beyond the scope of taxometric analysis. For example, taxometrics cannot be used to determine the number of, or the relationship among, latent dimensions underlying variation within a dimension or latent class; exploratory or confirmatory factor analysis would be more appropriate for these tasks. Thus, it is important to be aware of the structural models that are (and are not) compared by the taxometric method and to restrict application of the method to those questions that it can appropriately address.

QUESTION 2: ARE THE DATA APPROPRIATE FOR TAXOMETRIC ANALYSIS?

For a taxometric analysis to yield trustworthy results, one must work with data that meet certain conditions. Although there are few hard-and-fast rules for data selection and informed judgments must often be made, one can nonetheless evaluate the strengths and weaknesses of a particular sample of data for taxometric analysis by considering several key factors. These factors were detailed in chapter 4 and are summarized next.

Sample Considerations

Taxometric analysis requires that the overall sample size be sufficiently large and that a putative taxon be adequately represented within the sample. No firm minimum sample size has been established for taxometrics, although $N > 300$ has often been suggested as a rough guideline (e.g., Meehl, 1995a). Likewise, research has not yet established how small a taxon can be detected by any particular taxometric procedure. Whereas a minimum taxon base rate of $P = .10$ has often been suggested (e.g., Meehl, 1995a), there are a number of reasons to regard this guideline with caution. This recommendation appears to have been drawn primarily from two Monte Carlo studies (Meehl & Yonce, 1994, 1996), in which base rates lower than .10 were not studied and in which other aspects of the data were highly favorable (e.g., indicator validity was large, within-group correlations were absent) when a base rate of .10 was evaluated. This leads to two important implications. First, it is unknown whether an even smaller taxon could be detected if other data characteristics were favorable. Second, it is unknown whether a taxon of this size could be detected if other aspects of the data were not so favorable. Two more recent Monte Carlo studies (Beauchaine & Beauchaine, 2002; J. Ruscio, 2005) and one small-scale demonstration (J. Ruscio & Ruscio, 2004a) suggest that some taxometric procedures may be capable of detecting taxa with lower—perhaps much lower—base rates under favorable conditions. A third reason to view with caution the recommended minimum P of .10 is that it fails to take into account the actual number of taxon members in the sample, which research has suggested to be an important factor (J. Ruscio & Ruscio, 2004a). For example, all else being equal, it may be very difficult to detect a taxon with $P = .10$ in a sample of 300 cases, yet fairly easy to detect a taxon with $P = .10$ in a sample of 3,000 cases. One key difference is a tenfold increase in the number of taxon members, which can make the taxon significantly easier to detect.

Unfortunately, we can offer little more guidance than to recommend a careful consideration of the total sample size and the size of a taxon, both in relative (base rate) and absolute (number of taxon members) terms. In making this consideration, it is important to bear in mind the nature of the (mixed-group) population from which cases were originally sampled, which will likely influence the representation of a taxon in the research sample. For example, when studying a psychopathological construct, one would generally expect to find more taxon members in a clinical sample than in an analogue or community sample, and more taxon members in some clinical samples than in others.

Potentially Problematic Sampling Strategies

There are several sampling techniques that can confound inferences drawn from taxometric results. First, often it is best to avoid admixing samples (i.e., pooling samples of selected and unselected individuals for analysis) when possible because this can lead to pseudotaxonic results under some conditions (Grove, 1991a). However, this sampling strategy may pose fewer problems when taxometric analyses in an admixed sample yield dimensional results. Second, although dividing a sample into subsamples can be viewed as a form of consistency testing, it can weaken the power of all taxometric analyses. This sampling practice may be especially problematic when trying to detect a small taxon because it may not be correctly identified in all subsamples, and diminished power may actually prevent the taxon from being detected in *any* subsample. Researchers must judge whether the value of an exact replication is worth the potential loss in statistical power resulting from this approach. If the total sample size and the size (both relative and absolute) of a taxon are especially large, constructing subsamples for analysis may be worthwhile.

A final sampling technique that can complicate the interpretation of results is the practice of trimming likely complement members from the sample. Although some researchers have used or recommended this strategy as a way to increase the base rate of a small taxon, this approach has two potential limitations. First, when a taxon is small, discarding probable complement members ordinarily has little influence on the interpretively critical regions of taxometric curves. For example, if taxon members are overwhelmed by complement members even within high-scoring subsamples of a MAXCOV or MAXEIG analysis, removing low-scoring cases (likely complement members) will not help to reveal the group mixture in the upper region of the curve. Second, depending on how presumed complement members are identified and removed from the sample, a bias can be introduced in favor of taxonic results. As was demonstrated in chapter 4, even the apparently innocuous approach of randomly dropping cases from the full suspected range of complement members can create indicator bimodality. This, in turn, can increase the odds of pseudotaxonic results. Thus, we discourage the removal of complement members unless researchers can build a persuasive case that this is likely to clarify results in an unbiased way.

Selection and Construction of Indicators

In many ways, the selection and construction of indicators is one of the most important factors affecting the quality and interpretability of taxometric results. Researchers must ensure that each set of indicators submitted to analysis adequately represents the target construct and *only* the target construct. If content coverage is poor or if the indicators plausibly represent multiple related constructs, it can be difficult to determine what construct was investigated. The situation is analogous to any other analytic technique involving latent variables (e.g., confirmatory factor analysis, structural equation modeling) in that the extent to which the intended latent variables are actually being studied depends on how well these variables are represented by the manifest indicators.

Among the most critical aspects of a data set for the successful detection of genuine taxa is the validity with which the indicators distinguish taxon and complement members. Taxometric procedures—like other analytic tools for studying latent structure—require a fairly large separation between the indicator scores of the groups before a taxonic boundary between them can be detected. In the taxometric literature, indicator validity is typically scaled in Cohen's d units, although it is important to emphasize that the use of a group difference metric does not necessarily imply a taxonic structure. Regardless of the metric chosen to express indicator validity, the critical task is to determine whether the estimated validity of the indicators is sufficient to detect a taxonic boundary if one exists. Unfortunately, relatively little is presently known about what constitutes minimally acceptable validity, nor about the extent to which high values on some indicators can compensate for low values on others. One widely cited rule of thumb is that indicator validity should be at least $d = 1.25$ (Meehl, 1995a), and one study of MAXCOV classification accuracy provides some support for the desirability of indicator validity of at least this magnitude (Beauchaine & Beauchaine, 2002). Although empirical evidence is limited for specific validity guidelines, it is nevertheless important to estimate indicator validity in some fashion and to present some assurance of its acceptability before proceeding with taxometric analysis (we return shortly to the question of how key model parameters can be estimated). This is especially important if dimensional results are obtained because this raises the possibility that one may have failed to detect a taxonic boundary due to insufficient indicator validity.

Another important data requirement for successful taxometric analysis concerns within-group correlations, or the extent to which indicators covary within groups. When indicators are substantially correlated within groups, some or all taxometric procedures may fail to detect a genuine taxonic boundary. Within-group correlations can stem from shared method variance, from content overlap among indicators (substantive redundancy), or from other sources. Particularly when dimensional results are obtained, it is essential to demonstrate that within-group correlations were sufficiently low to rule out this alternative explanation for nontaxonic findings. In chapter 4, we discussed a number of strategies that can be used to help minimize within-group correlations prior to taxometric analysis. For example, combining items or measures with similar content into composite indicators can reduce within-group correlations. By contrast, assigning every nth item from a scale to each of n indicators virtually guarantees substantial within-group correlations because the resulting indicators are likely to be highly redundant with regard to content. When reasonable steps have been taken to minimize artifactual sources of within-group correlations and indicators nonetheless covary substantially within groups, this provides indirect evidence that the latent structure of the target construct is unlikely to be taxonic.

Various authors have recommended the use of multimethod assessment (e.g., interview, self-report, observational, physiological, and other assessment methods) when collecting data for a taxometric study (e.g., Beauchaine, 2003; Meehl, 1995a). Using the most valid measures of each facet of the target construct should help to maximize indicator validity, and using different methods to assess different facets should help to minimize within-group correlations. As we discuss further in chapter 10, relatively few taxometric investigations have employed multiple methods of data collection. Nonetheless, when appropriate and feasible, multimethod assessment may increase the power with which taxometric analyses can distinguish taxonic and dimensional structures.

Although none of the procedures in the taxometric method involves distributional assumptions, it is the case that curve shapes—and estimates of latent parameters derived from curves—can be influenced by the shape of the indicator distributions. For instance, indicator skew stemming from sources other than group mixture appears to exert a systematic and sometimes dramatic influence on most tests (A. M. Ruscio & Ruscio, 2002; J. Ruscio, Ruscio, & Keane, 2004). However, at present, little else is known about the association between indicator distributions and taxometric results. For example, Monte Carlo studies (e.g., Meehl & Yonce, 1994, 1996) have only examined the performance of taxometric procedures using continuous data even though the vast majority of research data (e.g., binary items, interview ratings or self-report data on Likert-type scales) only roughly approximate continuous distributions. As we discussed in chapters 5 and 6, composite input indicators can be constructed for many taxometric procedures to accommodate ordered categorical data. In addition, we demonstrated that, under certain conditions, MAXEIG may be capable of yielding informative results even when indicators vary across as few as four unique values, although this a rather extreme case that may well represent an exception to generally appropriate analytic practice. The question of when indicators are sufficiently continuous to serve as standalone input indicators—and when they should instead be combined into composites—remains to be addressed empirically.

A final issue to consider when selecting or constructing a set of indicators concerns the total number of indicators to be retained. There can be benefits to having more indicators. For example, all taxometric procedures yield more curves with a larger number of indicators, thereby affording more consistency testing. Multivariate procedures such as MAXEIG and L-Mode can have greater statistical power with larger numbers of indicators; the same may be true of MAXSLOPE, MAMBAC, and MAXCOV when using the composite input indicator technique. However, all of these potential benefits must be weighed against the potential costs of retaining more indicators. Perhaps the greatest possible cost is that when too many indicators are used, there is likely to be substantial content overlap (and perhaps also shared method variance), thereby increasing within-group correlations. This is in contrast to structural equation models, in which increasing numbers of indicators of a latent variable can be useful even if these indicators share content or method variance. It should also be noted that gains in statistical power associated with additional indicators presume that all indicators are comparably valid. Including a larger number of indicators of lower average validity may actually reduce statistical power. Factors influencing the relationship between number of indicators and statistical power have not yet been explored empirically. However, it seems unlikely that a fixed number of indicators would be appropriate or optimal for all taxometric analyses given differences in the constructs being examined and the data and goals of individual studies.

Rather than setting a target number of indicators for a study or attempting to use as many indicators as possible, we strongly suggest approaching indicator selection with an eye toward relevant theory. Specifically, we recommend listing the distinctive facets of the target construct and then using the available data to construct a single indicator for each facet that achieves the greatest validity in separating groups and the lowest content and method overlap with other indicators. Thus, rather than beginning with a certain number of indicators as a goal, we advise researchers to begin with a theoretical model of the target construct and represent this model with indicators that are as valid and nonredundant as possible. Researchers may wish to construct multiple sets of indicators if there are multiple theoretical models for conceptualizing the target construct or multiple ways to distill the available data into theoretically meaningful and potentially informative composites.

Empirical Evaluation of Data Appropriateness

We have stressed the value of estimating key model parameters to help assess the adequacy of research data for taxometric analysis. However, thresholds for acceptable data have yet to be well defined by empirical study, and there is no universally accepted approach to estimating model parameters such as the taxon base rate, indicator validity, and within-group correlations. Thus, determining whether the data requirements reviewed here are met requires a considerable amount of judgment. Estimates of several key parameters can be obtained through a priori or post hoc procedures, each of which is characterized by certain strengths and limitations.

To perform a priori estimation, one might assign all cases to taxon and complement groups using criteria derived from prior theoretical or empirical work in the relevant domain. The taxon base rate can be estimated as the proportion of individuals assigned to the taxon, the validity of each indicator can be estimated by calculating *d* across the two groups, and within-group correlations can be estimated by calculating correlations separately within each group. One advantage of such an approach is that researchers can make an informed judgment about the appropriateness of a data set prior to analysis, saving considerable time and effort by not performing analyses when the data are not deemed adequate for taxometrics. A second advantage is that the estimates are based on a realistic appraisal of all available data because no data are excluded from this analysis. An important drawback is that the estimates are only as trustworthy as the initial assignment of cases to groups. If cases are poorly classified, the estimates may be misleading, perhaps causing the data to appear either more or less favorable for taxometric analysis than they really are.

Another way to perform a priori estimation is to develop an educated guess—based on theory and relevant data—of what the taxon base rate is likely to be, calculate indicator correlations in the full sample as well as in subgroups consisting of likely taxon and complement members, and use the formula provided by Meehl and Yonce (1994; see Eq. A.5 in this book) to estimate indicator validity. This approach requires researchers to specify which cases in the data set are probable taxon and complement members, although how best to do this is not always clear. One way in which this has been done is to sort cases according to their total score on the indicators and to designate the upper quartile of cases as the taxon and the lower quartile as the complement (cf. Meehl & Yonce, 1994). Although this approach can provide a quick and useful sense of important aspects of the data, it also has a few limitations. Using grouping criteria that are too restrictive may lead one or both groups to be represented by very small subsamples that are unrealistically homogeneous, which can yield an underestimate of within-group correlations and an overestimate of indicator validity; using criteria that are too liberal can have the opposite effect. Moreover, when the created subsamples do not collectively include all cases in the sample, the exclusion of some data from the estimation procedure may serve to inflate estimates of indicator validity and deflate estimates of within-group correlations due to range restriction. This approach also averages all correlations (in the full sample and within groups) to yield a single average estimate of indicator validity. The drawback is that it is impossible to determine whether individual indicators are sufficiently valid, and a global average may lead to an overly optimistic or pessimistic assessment of the appropriateness of the data for taxometrics. Finally, when this approach is used, estimates of validity will vary depending on the base rate estimate employed. It is often informative to try multiple, plausible base rate estimates ranging from the most conservative (estimates close to $P = .50$ yield the lowest indicator validity estimates) to the most liberal (estimates close to $P = 0$ or $P = 1$ yield the highest indicator validity estimates) to report the sensitivity of validity estimates to the taxon base rate.

Some researchers prefer to estimate model parameters after taxometric analysis has been performed, and certain taxometric procedures can be used to provide such post hoc output. For example, one can perform a MAXCOV or MAXEIG analysis to procure an estimate of the taxon base rate, assign cases to groups using Bayes' Theorem, and estimate indicator validity and within-group correlations using these groups. An advantage of this approach is that it uses all available data (no cases are excluded). However, a significant limitation is that estimates of latent parameters are not available prior to taxometric analysis to help researchers determine whether to proceed with the analysis. Moreover, estimates obtained in this way may be useful only when taxonic results emerge in the initial taxometric analysis. For example, when a MAXCOV or MAXEIG analysis yields dimensional results, both the estimate of the taxon base rate and the assignment of cases to groups using Bayes' Theorem (which involves, among other things, the base rate estimate) may be of dubious value. Thus, this type of post hoc parameter estimation carries a fairly strong presumption that the data are taxonic, and it may yield ambiguous or misleading results for dimensional data.

When evaluating the appropriateness of data for taxometrics, both the a priori and post hoc approaches described earlier have two important limitations. First, each involves a piecemeal examination of the data. That is, one must reach separate judgments regarding the adequacy of the estimated taxon base rate, indicator validities, and within-group correlations. It may be that minimally acceptable estimates for each of these nonetheless are jointly inadequate, or that some particularly favorable values may compensate for some unfavorable ones. However, almost nothing is known about how the full configuration of data characteristics should be evaluated. Second, these approaches rest on the assumption that all taxometric procedures and consistency tests have the same data requirements—a presumption that is almost certain to be false. For example, multivariate procedures such as MAXEIG may yield informative results with a large number of moderately valid indicators, whereas a procedure performed with only two indicators (e.g., a typical MAXSLOPE or MAMBAC analysis) may require more valid indicators.

A third approach to evaluating the appropriateness of data—generating empirical sampling distributions of taxometric results by analyzing taxonic and dimensional comparison data—addresses both of these limitations. This approach is more holistic in that it examines the joint sufficiency of all relevant aspects of the research data; it is also more specific in that it considers the adequacy of the data in relation to a particular taxometric procedure or consistency test. To the extent that empirical sampling distributions of results for taxonic and dimensional comparison data can be distinguished, the unique data configuration is judged likely to be acceptable for that particular procedure or consistency test. If these distributions overlap substantially, this suggests that one or more characteristics of the data is problematic for the planned analysis. Although the inspection of empirical sampling distributions does not necessarily help to identify what is problematic in such cases, it does strongly caution that the planned analysis is unlikely to yield informative results and either should not be performed or should be interpreted with added caution. It is also possible that these distributions will reveal the data to be acceptable for some of the planned analyses but not others, providing valuable guidance as to which analyses should be pursued and how much weight to assign to the results of each analysis when final conclusions are drawn.

Although there are clear advantages to this approach to evaluating the appropriateness of data, it is not without its own challenges and limitations. Generating empirical sampling distributions requires the simulation of taxonic and dimensional comparison data that reproduce important characteristics of the research data. This is relatively straightforward for dimensional data, wherein one can bootstrap samples of comparison data that reproduce indicator correlations and distributions. Simulating appropriate taxonic data is more complex because this requires a tentative assignment of cases to groups. Empirical sampling distributions of results for taxonic data will therefore be of limited value—and perhaps even systematically misleading—when a questionable case classification is used to simulate the comparison data. We have described several ways to assign cases to groups and recommend that multiple techniques be used to adequately represent all plausible classifications.

QUESTION 3: HAS A SUFFICIENT VARIETY OF PROCEDURES BEEN IMPLEMENTED PROPERLY?

The centerpiece of any taxometric investigation is the implementation of a variety of analytic procedures and consistency tests. Although it is beyond the scope of this chapter to describe taxometric analyses in detail (see chaps. 5–7), we review a number of important implementation decisions that should be considered thoughtfully and justified explicitly in any taxometric study.

Taxometric Procedures

MAXSLOPE and MAMBAC can be performed when as few as two indicators are available for study. MAXCOV, MAXEIG, and L-Mode require at least three indicators. When a larger number of valid and nonredundant indicators are available, any procedure that is implemented in a multivariate manner (e.g., using more than two output indicators in MAXEIG; using composite input indicators in MAXSLOPE, MAMBAC, MAXCOV, or MAXEIG; or providing as many indicators as seems appropriate in L-Mode) may yield increasingly powerful results. Chapter 7 reviewed additional considerations relevant to selecting complementary taxometric procedures for an investigation, such as the degree of mathematical independence of the candidate procedures (e.g., MAMBAC is much more distinct from MAXEIG than is MAXCOV).

Table 9.2 summarizes the implementation decisions that must be made to perform each taxometric procedure. Researchers have often appeared to follow conventions when making these decisions. Chapters 5 and 6 reviewed the relative strengths and weaknesses of a number of available implementation options, including some that researchers have seldom exploited, and offered tentative suggestions for performing each procedure in a manner appropriate to the available data. These suggestions are admittedly quite speculative, based on the limited empirical study of these procedures as well as our own experience in working with them. At least until more extensive Monte Carlo research has been conducted to address and inform these implementation decisions, we recommend the use of empirical sampling distributions as a way to empirically evaluate how best to implement each procedure through the process of adaptive calibration. For example, one can perform a MAXCOV or MAXEIG analysis on simulated comparison data, varying the techniques that are used to assign indicators to input and output roles and to divide cases into subsamples in the analysis. The techniques yielding a clearer differentiation between taxonic and dimensional comparison data can then be employed to maximize the power with which these structural models are distinguished.

Consistency Tests

A hallmark of the taxometric method is the evaluation of consistency of results across multiple nonredundant analyses. A large assortment of consistency tests has been proposed for use in taxometric research, but the extent to which these tests are actually independent of one another or

TABLE 9.2
Summary of Implementation Decisions for Taxometric Procedures

MAXSLOPE

Assigning *k* indicators to input–output roles

- All pairwise combinations; $k(k - 1)$ curves
- Composite input indicator, single output indicator; *k* curves

Calculating the local regression curve

- Smoothing function used to generate curve (e.g., LOWESS)
- Parameters required by smoothing function (e.g., proportion of cases used to estimate each point)

MAMBAC

Assigning *k* indicators to input–output roles

- All pairwise combinations; $k(k - 1)$ curves
- Composite input indicator, single output indicator; *k* curves

Placing cuts along the input indicator

- Each successive case
- Every *n*th case
- Intact scale values
- *SD* units
- Internal replications if cuts may be placed between equal-scoring cases
- Minimal sample size beyond first and last cuts

Graphing and presentation

- Constructing the *x* axis (case numbers vs. input indicator scores)
- Scaling the *y* axis (min/max values, specified min and/or max values, hold constant across all curves in a panel if indicators are standardized)
- Raw curve, smoothed curve, or both
- Full panel of curves, averaged curve, or both

L-Mode

Using *k* indicators in one or more analyses

- Subsets of $k \geq 3$ indicators per analysis
- All *k* indicators used in a single analysis

Locating the latent modes

- Program defaults (i.e., maxima on either side of factor score of $x = 0$)
- Manual override when visually apparent mode is missed

MAXCOV/MAXEIG

Assessing the association among output indicators

- Conditional covariances between two output indicators (MAXCOV)
- Conditional eigenvalues between two or more output indicators (MAXEIG)

Assigning *k* indicators to input–output roles

- All possible triplets; $k(k - 1)(k - 2)/2$ curves
- One input indicator, all others as outputs (MAXEIG only); *k* curves
- Composite input indicator, single output indicator; *k* curves

Dividing cases into subsamples along the input indicator

- Intervals (*SD* units, equal-sized, intact scale values)
- Overlapping windows
- Internal replications if equal-scoring cases may fall in adjacent subsamples
- Minimum subsample size

Graphing and presentation

- Constructing the *x* axis (input indicator scores vs. subsample numbers)
- Scaling the *y* axis (min/max values, specified min and/or max values, hold constant across all curves in a panel if indicators are standardized)
- Raw curve, smoothed curve, or both
- Full panel of curves, averaged curve, or both

contribute incremental validity to inferences of latent structure has not yet been established. We reviewed a number of consistency tests in chapter 7 (see also Table 9.1, Section 3B, for a listing), highlighting several that seem especially promising (e.g., the inchworm consistency test) and cautioning against others that appear to rest on more questionable foundations or that have yielded poor results in empirical trials. We also presented some new results for several of the most popular consistency tests. Unfortunately, even less is known about the performance of most consistency tests under realistic research conditions than is known about the primary taxometric procedures.

Given this paucity of relevant research, we again encourage researchers to evaluate the potential utility of a candidate consistency test for their data by generating empirical sampling distributions of results for simulated taxonic and dimensional comparison data. The comparison curve fit index (CCFI) introduced in chapter 7 can also be considered a consistency test based on empirical sampling distributions of curve shapes, and preliminary evaluations of the CCFI and precursors to it have yielded promising results. For example, J. Ruscio (2004) found that four curve fit indexes performed well with the 700 Meehl–Yonce Monte Carlo samples as well as taxonic and dimensional data sets that varied across broader ranges of a number of data parameters. J. Ruscio, Ruscio, and Meron (2005) found that the CCFI achieved much greater classification accuracy than several widely used consistency tests.

QUESTION 4: HAVE THE RESULTS BEEN PRESENTED AND INTERPRETED APPROPRIATELY?

Chapter 8 raised a number of concerns regarding the presentation and interpretation of taxometric results. In this section, we review several of the key decisions that must be made at this stage of a taxometric investigation. As is discussed further in chapter 10, a review of published taxometric studies suggests that researchers have begun to pay increased attention to these factors in recent years.

Graphing Considerations

Graphing the results of some taxometric procedures involves several choice points. For example, one must decide how to construct the *x* axis and how to scale the *y* axis of MAMBAC, MAXCOV, or MAXEIG graphs. Some of the pros and cons of alternative scaling techniques were discussed in chapters 5 and 6. Here we simply draw attention to the importance of making informed decisions, rather than accepting the default technique of a computer program or automatically following conventional practices.

Other unsettled issues in the realm of graphing include the smoothing and averaging of taxometric curves. There are situations in which smoothing may be argued to be beneficial, such as when the endpoints of a MAMBAC curve are unstable and may yield distorted estimates of the taxon base rate. However, smoothing can be problematic when there are relatively few data points on a curve or when an inappropriate smoothing technique is used. General conditions under which smoothing is advisable and inadvisable—as well as the best way to perform smoothing—remain to be studied. At a minimum, researchers should indicate when, why, and how they have smoothed taxometric curves; they might also consider presenting raw and smoothed curves on the same graph so that readers can evaluate the effects of the smoothing technique.

Taxometric researchers have reached no consensus when to average the many curves generated by a given taxometric procedure. One solution may be to present both a full panel of curves and the average of these curves. Provided that all indicators are standardized prior to analysis, one can plot all of these curves on a single graph so that the appropriateness of averaging can be observed directly. If the curves are too numerous to be plotted on a single graph, they can be aggregated in some other fashion (e.g., organized on separate graphs for each input indicator). Although the CCFI currently requires curves to be averaged, this is not intended as an endorsement of averaging; as noted earlier, we plan to expand this approach for use with a full panel of curves. At present, we suggest that researchers interpret the full panel of curves for each procedure and consider presenting their average only if it fairly and adequately represents the overall trend.

Interpretational Issues

As researchers plan, perform, and interpret taxometric procedures and consistency tests, it is important to be aware of a number of factors that can systematically influence results in ways that are not always fully appreciated. For example, it has long been known that indicator validity and within-group correlations can exert predictable influences on curve shapes (e.g., Meehl & Yonce, 1994, 1996), several of which were illustrated in chapter 8. More recently, the impact of indicator skew has received attention (e.g., A. M. Ruscio & Ruscio, 2002; J. Ruscio, Ruscio, & Keane, 2004). Skewed indicators can tilt MAMBAC, MAXCOV, and MAXEIG curves, making it increasingly difficult to distinguish a small taxon from a dimensional construct. The influence of indicator skew on other procedures (e.g., MAXSLOPE, L-Mode) and consistency tests is even less thoroughly understood, although it is not difficult to envision some of the expected effects. For example, when positively skewed indicators in dimensional data yield rising MAMBAC, MAXCOV, or MAXEIG curves, estimates of the taxon base rate from these curves may be low and quite consistent, which seriously compromises the utility of the base rate consistency test. We urge researchers to consider the potential influence of various characteristics of their data on the obtained results. By adopting a skeptical attitude, constructing alternative explanations for observed patterns in the taxometric output, and subjecting these explanations to empirical testing wherever possible, researchers may prevent mistaken structural inferences or strengthen the evidence in support of a correct conclusion.

Judging Curve Shapes and Quantitative Results

As we note in chapter 10, researchers are becoming more explicit about the way in which they approach the challenging business of interpreting taxometric curves and associated quantitative output. Although curve shapes at times appear quite similar to those presented in Monte Carlo studies of taxometric procedures, very often they do not. This is largely because the curves obtained in Monte Carlo studies are usually based on highly idealized data conditions that may not be representative of the data researchers typically analyze. For example, although Meehl and Yonce (1994, 1996) presented an exhaustive array of figures containing all panels of MAMBAC and MAXCOV curves generated in their analyses, it must be recalled that these procedures can be implemented in different ways, their results can be graphed differently, the studied data configurations may not be representative of the data in a particular research study, and less ideal data than those studied can yield different results. Thus, published curves do not always provide an adequate benchmark for interpreting the results of taxometric studies, and simple comparison with such curves can be supplemented in at least two ways.

First, interpretation can be facilitated by employing knowledgeable raters to judge taxometric results. Chapter 8 described a number of issues to consider when setting up such a rating task. For example, one would ideally use multiple raters who are blind to the hypothesis of the study, empirically examining their agreement to establish that reliable judgments have been made. To the extent that the raters are familiar with the nuances of taxometric analysis, their judgments may be more accurate than those of raters who can only compare curves to those appearing in published Monte Carlo studies without an appreciation for possible complicating or influential factors.

Second, interpretation can be enhanced by generating empirical sampling distributions of taxometric results, creating a comparative benchmark that is tailored to the unique data configuration and procedural implementation of a particular investigation. The results for comparison data can aid in the interpretation of curve shapes and quantitative indexes obtained from any taxometric procedure or consistency test, often providing a highly informative supplement to the results of Monte Carlo studies. Such sampling distributions can be an especially practical and useful tool when the generalizability of Monte Carlo results to one's study is questionable, or when there are simply no results to generalize due to a combination of data characteristics and implementation decisions that have not yet been studied.

Using knowledgeable raters, comparing results to those of Monte Carlo studies, and generating empirical sampling distributions need not be mutually exclusive practices. In fact we advise an approach based on all three practices: Solicit judgments from raters well versed in the relevant Monte Carlo literature and then have them evaluate the obtained results in the context of empirical sampling distributions.

QUESTION 5: ARE IMPLICATIONS OF THE FINDINGS CLEARLY ARTICULATED?

The last of the five concerns addressed by our checklist involves a return to the questions that originally motivated a taxometric investigation. Provided that a satisfactory scientific rationale was initially identified for the study, each of the theoretical and applied reasons for conducting taxometric analysis should be revisited in light of the obtained structural results. As with any type of research, it is incumbent on the investigators to offer a compelling explanation of the knowledge that has been gained, the contribution of this knowledge to the broader literature, the implications of the knowledge for theory and practice, and the questions (structural or otherwise) that remain to be answered. For example, Lenzenweger (2004) posed a number of questions that researchers and readers should ask themselves about taxometric investigations. Many of these questions are echoed throughout this book. Whereas we have focused primarily on methodology, Lenzenweger explores and illustrates fundamental issues in the relationship between one's substantive theory and the testing of structural models using taxometrics. Although Lenzenweger drew most of his illustrations from research in psychopathology, consideration of the types of questions that he posed should improve the contribution of taxometric research to advances throughout the social and behavioral sciences. Ultimately, the critical task at this stage of research is to move beyond the simple reporting of a taxonic or dimensional solution, ensuring that readers of a taxometric study understand why the proferred structural inference matters.

CONCLUSIONS

This chapter concludes Part II of the book by reviewing a number of fundamental conceptual and empirical issues involved in any taxometric investigation. We have presented this material in a checklist format for ease of reference and application. This review models a similar checklist appearing in J. Ruscio and Ruscio (2004b). The present version has been updated to reflect recent developments and has been streamlined to summarize key concerns, with cross-referencing of previous chapters providing greater detail than would be possible in a journal-length survey of taxometrics. We suggest that the scientific contribution of taxometrics could be enhanced if users of the method carefully and habitually think through these five central issues: (a) the scientific justification for a taxometric investigation, (b) the appropriateness of data for taxometric analysis, (c) the sufficiency of the analyses performed, (d) the presentation and interpretation of results, and (e) the articulation of their implications. Likewise, we believe that it is important for readers, reviewers, and editors to familiarize themselves with the central issues of taxometric analysis so that appropriately high standards for taxometric research are established and maintained.

Having described the taxometric method and illustrated its application, we turn now to a discussion of how the method has been—and could be—employed to address important structural questions in the social and behavioral sciences. In the remaining two chapters of this book, we review how and why the taxometric method has previously been used, note emerging trends in the application of taxometrics, and suggest future directions for much-needed substantive and methodological research with this promising analytic tool.

PART THREE

APPLICATIONS AND FUTURE DIRECTIONS

CHAPTER TEN

Applications of the Taxometric Method

The previous section reviewed taxometric methodology and presented a set of analytic approaches, practical considerations, and interpretive guidelines to facilitate its successful implementation by researchers faced with the challenges of real data. Although it is critical that investigators have a firm grasp of the taxometric method, we believe that it is also important for them to be aware of the body of taxometric research that has accumulated over the past two or more decades. This body of work provides an instructive and vital context for future studies. It underscores the specific purpose (distinguishing two particular structural models) and the specific domain (individual differences in personality and psychopathology) for which taxometrics was initially developed. It illustrates the types of research questions that have been and can be addressed using this method. Finally, it shows that taxometrics is currently less standardized in its application than many other data-analytic approaches, and it highlights the importance of making appropriate implementation decisions when performing taxometric analyses.

In this chapter, we review the conceptual context of published taxometric studies performed in each research domain and report the conclu sions drawn by the researchers. Because taxometric investigators have historically been most concerned with psychopathological constructs, a majority of the studies included in the present review focus on these constructs, and we organize these studies according to the broad groupings of disorders listed in Axes I and II of the *DSM*. We subsequently review the smaller body of taxometric research on normal personality, then turn to constructs falling outside the realm of individual differences, a vast and neglected domain that may be an important frontier for taxometric research. We conclude the chapter by reviewing how the taxometric method has been implemented in past research and how applications of the method have changed over time.

This chapter is not intended to provide an exhaustive review of the strengths and limitations of the relevant studies that have been performed to date (for a more detailed and critical survey of this literature, see Haslam, 2003; Haslam & Kim, 2002). For this reason, we do not critique individual studies, comment on the methodology employed in these studies, or offer our impressions of the strength of the evidence or the validity of structural inferences reached in these studies. Instead our aim is to draw on the published taxometric literature to examine where and how the method has been used, thereby setting the stage for the future directions outlined in chapter 11.

THE LATENT STRUCTURE OF PSYCHOPATHOLOGY

Taxometric investigations of psychopathology take place in the context of enduring debates about classification and diagnosis within the mental health professions. Despite the advances of the last quarter century, the taxonomy of mental disorders has yet to reach an acceptable state of professional consensus. One of the most fundamental points of contention in these debates is whether mental disorders are most validly and usefully represented as discrete categories or continuous dimensions. Although the prevailing systems of classification—the *DSM* and the *ICD*—make some allowance for continuous variation, they generally assume the adequacy of a categorical representation. Disorders are laid out as distinct types and diagnosed as present or absent based on the number and nature of presenting symptoms, in a manner that mirrors classification in general medicine.

Critics of categorical diagnosis argue that it embodies a medicalized view of mental disorder and that it reifies disorders (e.g., Jablensky, 1999). Other critics assert that discontinuous boundaries and coherent, mutually distinguishable clinical entities are highly unlikely within the domain of psychopathology (e.g., Carson, 1997). These and other arguments have led some to propose that a dimensional taxonomy be installed for at least some disorders (e.g., Livesley, Schroeder, Jackson, & Jang, 1994).

Preferences for categorical versus dimensional views of mental disorder often coincide with differences of discipline or theoretical orientation. Traditionally, biological psychiatrists have taken a "disease perspective [which] rests on a logic that captures abnormalities within categories" (McHugh & Slavney, 1998, p. 15). Professionals placing greater emphasis on psychosocial factors tend to embrace a dimensional perspective, in which disorder arises out of psychological vulnerabilities and environmental provocations that vary in degree rather than in kind. Compared to their medically trained colleagues, many psychologists tend to favor dimensional models of individual differences for a variety of historical reasons discussed at length by Meehl (1992). Hence, taxonic and dimensional conceptualizations of the nature of mental disorders exist in tension within the mental health professions.

As we have argued in previous chapters, this distinction and the controversies surrounding it are not merely matters of theoretical preference or speculative interest. Numerous clinically vital consequences flow from the distinction, making the latent structure of each disorder an empirical question of considerable importance. The strategies and goals of clinical assessment depend on whether the latent variable to be measured is categorical or continuous. The ways in which classification systems are designed and used—and the decision of who will be diagnosed and thus gain access to services or third-party payments—depend on the structural models of disorder that underlie these systems. Efforts to explain the origin of mental disorders also depend on this issue because different etiological models are best suited to particular latent structures. Multifactorial explanatory models, in which many contributing factors combine additively to the expression or risk of a mental disorder, are likely to produce latent continua, whereas models invoking single dichotomous causal factors (e.g., major genes, specific environmental events, threshold effects) are more likely to produce latent categories. In short, the latent structure of a mental disorder can be pivotal for its investigation, assessment, and treatment, and resolving this structure empirically is of great importance. To this end, we turn now to a review of the published taxometric studies of psychopathology, which are summarized in Table 10.1.

Mood Disorders

The importance of resolving latent structure is well illustrated in the case of mood disorders. A substantial body of taxometric work has accumulated on major depression and its proposed subtypes, although controversy still exists about its conclusions, and much work remains to be done. The importance of the categorical versus dimensional distinction for the study of depression has been reviewed extensively by other authors (Flett, Vredenburg, & Krames, 1997; Solomon, Haaga, & Arnow, 2001). Whether cut scores should be applied to continuous measures of depression to designate the presence of a disorder, whether subthreshold mood conditions should be recognized by classification systems, and whether analogue research on dysphoric undergraduates should be generalized to

TABLE 10.1
Published Taxometric Studies of Psychopathology and Personality

Category and Construct	*Published Study*	*Authors' Conclusion*
Mood disorders		
Dysthymia	Trull et al. (1990)	Dimensional
Depression	J. Ruscio & Ruscio (2000)	Dimensional
Depression	Franklin et al. (2002)	Dimensional
Depression	A. M. Ruscio & Ruscio (2002)	Dimensional
Depression	Beach & Amir (2003)	Dimensional
Depression	Hankin et al. (2005)	Dimensional
Depression	Slade & Andrews (2005)	Dimensional
Nuclear subtype	Grove et al. (1987)	Taxonic
Melancholic subtype	Haslam & Beck (1994)	Taxonic
Endogenous subtype	Keller (2000)	Taxonic
Melancholic subtype	Ambrosini et al. (2002)	Taxonic
Involuntary defeat syndrome	Beach & Amir (2003)	Taxonic
Involuntary defeat syndrome	J. Ruscio et al. (2004)	Dimensional
Involuntary defeat syndrome	Hankin et al. (2005)	Dimensional
Hopelessness subtype	Haslam & Beck (1994)	Dimensional
Hopelessness subtype	Whisman & Pinto (1997)	Dimensional
Hopelessness subtype	Keller (2000)	Taxonic
Autonomous subtype	Haslam & Beck (1994)	Dimensional
Self-critical subtype	Haslam & Beck (1994)	Dimensional
Sociotropic subtype	Haslam & Beck (1994)	Dimensional
Depression proneness	Gibb et al. (2004)	Taxonic
Depression proneness	Strong et al. (2004)	Dimensional
Hypomanic temperament	Meyer & Keller (2003)	Dimensional
Anxiety disorders		
PTSD	A. M. Ruscio et al. (2002)	Dimensional
Pathological worry	A. M. Ruscio et al. (2001)	Dimensional
Anxiety sensitivity	Taylor et al. (1999)	Dimensional
Anxiety sensitivity	Bernstein et al. (2005)	Taxonic
Eating disorders		
Anorexia nervosa	Williamson et al. (2002)	Taxonic
Bulimia nervosa	Gleaves et al. (2000a)	Taxonic
Bulimia nervosa	Gleaves et al. (2000b)	Taxonic
Bulimia nervosa	Williamson et al. (2002)	Taxonic
Eating disturbance	Tylka & Subich (2003)	Dimensional
Binge eating disorder	Williamson et al. (2002)	Taxonic
Dissociative disorders		
Pathological dissociation	Waller et al. (1996)	Taxonic
Pathological dissociation	Waller & Ross (1997)	Taxonic
Dissociative absorption	Waller et al. (1996)	Dimensional
Other Axis I conditions		
Dementia	Golden (1982)	Taxonic
Negative symptom schizophrenia	Blanchard et al. (2005)	Taxonic

(Continued)

TABLE 10.1
(Continued)

Category and Construct	*Published Study*	*Authors' Conclusion*
Abnormal personality		
Schizotypy	Golden & Meehl (1979)	Taxonic
Schizotypy	Erlenmeyer-Kimling et al. (1989)	Taxonic
Perceptual aberration	Lenzenweger & Korfine (1992)	Taxonic
Perceptual aberration	Korfine & Lenzenweger (1995)	Taxonic
Perceptual aberration	Horan et al. (2004)	Taxonic
Schizotypy	Tyrka et al. (1995a, 1995b)	Taxonic
Schizotypy	Lenzenweger (1999)	Taxonic
Social anhedonia	Blanchard et al. (2000)	Taxonic
Social anhedonia	Horan et al. (2004)	Taxonic
Positive schizotypy	Keller et al. (2001)	Dimensional
Negative schizotypy	Keller et al. (2001)	Taxonic
Perceptual aberration	Meyer & Keller (2001)	Taxonic
Physical anhedonia	Meyer & Keller (2001)	Taxonic
Magical ideation	Meyer & Keller (2001)	Dimensional
Magical ideation	Horan et al. (2004)	Dimensional
Psychopathy	Harris et al. (1994)	Taxonic
Psychopathy	Marcus et al. (2004)	Dimensional
Child/adolescent psychopathy	Vasey et al. (2005)	Taxonic
Antisocial PD	Skilling et al. (2001a)	Taxonic
Child antisociality	Skilling et al. (2001b)	Taxonic
Borderline PD	Trull et al. (1990)	Dimensional
Borderline PD	Rothschild et al. (2003)	Dimensional
Broad personality traits		
Jungian preferences	Arnau et al. (2003)	Dimensional
Specific personality traits		
Self-monitoring	Gangestad & Snyder (1985)	Taxonic
Type A	Strube (1989)	Taxonic
Adult attachment styles	Fraley & Waller (1998)	Dimensional
Infant attachment patterns	Fraley & Spieker (2003)	Dimensional
Infant reactivity	Woodward et al. (2000)	Taxonic
Hypnotic susceptibility	Oakman & Woody (1996)	Taxonic
Gender identity	Korfine & Lenzenweger (1995)	Dimensional
Impulsivity	Gangestad & Snyder (1985)	Dimensional
Sexual orientation	Haslam (1997)	Dimensional
Sexual orientation	Gangestad et al. (2000)	Taxonic
Response styles		
Impression management	Strong et al. (1999)	Taxonic
Impression management	Strong et al. (2002)	Taxonic
Self-deceptive positivity	Strong et al. (1999)	Dimensional
Self-deceptive positivity	Strong et al. (2002)	Dimensional
Symptom overreporting	Strong et al. (2000)	Taxonic

clinically depressed populations are among the most important of these implications. Further implications extend beyond those of direct interest to psychologists and psychiatrists. As Kessler (2002b) recently noted, there has been a lasting controversy over the categorical versus dimensional assessment of psychological distress within the fields of sociology and epidemiology. Kessler recommended taxometrics as a means to clarify this debate.

To date, taxometric studies have investigated mood disorders from three angles: the structure of general depression, proposed subtypes of depression, and temperamental vulnerability to mood disorders. Several taxometric studies examining the first of these angles have converged on a dimensional view of depression. The first study was conducted by Trull, Widiger, and Guthrie (1990), who advanced a dimensional model of dysthymia. Six subsequent studies (Beach & Amir, 2003; Franklin, Strong, & Greene, 2002; Hankin, Fraley, Lahey, & Waldman, 2005; A. M. Ruscio & Ruscio, 2002; J. Ruscio & Ruscio, 2000; Slade & Andrews, 2005) replicated this dimensional finding with a variety of measures of depression in clinical and undergraduate samples, and in community samples of adults, children, and adolescents.

Thus, taxometric studies have consistently endorsed a continuum model for general depression. This happy consensus disappears when attention turns to subtypes of depression. Numerous depressive subtypes have been proposed in the psychological and psychiatric literatures, and several have received taxometric attention to date. These include the venerable melancholic or endogenous variant; the autonomous, sociotropic, and hopelessness variants proposed by cognitive theorists (Abramson, Metalsky, & Alloy, 1989; Beck, 1983); a self-critical variant proposed by psychodynamic theorists (Blatt & Homann, 1992); and Involuntary Defeat Syndrome (IDS), a variant derived from an evolutionary perspective (Gilbert, 2000).

The most extensively studied of these proposed subtypes is melancholia. The structure of this subtype, a severe form of depression distinguished by prominent vegetative (e.g., loss of appetite and libido, terminal insomnia) and anhedonic symptoms, has been debated for almost a century (Zimmerman & Spitzer, 1989). Some theorists favor a dimensional position that regards melancholic depression as a more severe variant of milder depression, whereas others maintain that melancholia represents a discrete syndrome. In the first taxometric investigation to address this question, Grove et al. (1987) presented evidence of a discrete nuclear depressive subtype characterized by many classic melancholic or endogenous features. Haslam and Beck (1994) also reported a taxonic finding for melancholia in depressed adults, a finding later replicated by Ambrosini, Bennett, Cleland, and Haslam (2002) in an adolescent sample. More recently, Beach and Amir (2003) reached a taxonic conclusion for IDS, a form of depression whose predominantly somatic symptoms are similar to those of the melancholia subtype. However, J. Ruscio, Ruscio, and Keane (2004) reached a dimensional conclusion in an exact replication of their study, as did Hankin et al. (2005) in a study of child and adolescent depression.

Taxometric investigation of other proposed subtypes of depression is less well advanced. Three studies have examined the hopelessness subtype proposed by Abramson et al. (1989); two yielded dimensional interpretations (Haslam & Beck, 1994; Whisman & Pinto, 1997), whereas one yielded ambiguous findings (Keller, 2000). Finally, each of the other three proposed subtypes that have received taxometric attention (sociotropic, autonomous, and self-critical depression) have been examined in only a single study (Haslam & Beck, 1994), which reported dimensional results for each subtype. Thus, the taxometric investigation of depressive subtypes is still in its infancy. However, with the possible exception of melancholia, conclusions suggest that few discrete subtypes of depression exist, consistent with the apparent continuity of general depression.

A few recent studies have investigated vulnerabilities to depression. One study of numerous cognitive vulnerability factors (e.g., depressogenic explanatory style) in an undergraduate sample reported dimensional results (Gibb, Alloy, Abramson, Beevers, & Miller, in press), whereas a study examining such factors among treatment-seeking smokers reported taxonic results (Strong, Brown, Kahler, Lloyd-Richardson, & Niaura, in press). A separate taxometric study examined a proposed diathesis for bipolar disorder (hypomanic temperament) and concluded that it may be dimensional (Meyer & Keller, 2003). Unfortunately, despite controversies surrounding the existence of a bipolar spectrum and subthreshold bipolar disorders (Akiskal, 2003), there have been no other taxometric studies of bipolar conditions.

Anxiety Disorders

In contrast to the mood disorders, the anxiety disorders have only recently begun to receive taxometric attention. A taxometric investigation of posttraumatic stress disorder (PTSD) in a sample of Vietnam-era combat veterans arrived at a dimensional conclusion (A. M. Ruscio, Ruscio, & Keane, 2002), implying that assessment of the disorder should focus on degrees of impairment rather than categorical distinctions among trauma survivors. A taxometric study of worry also reported dimensional results (A. M. Ruscio, Borkovec, & Ruscio, 2001), suggesting that the pathological worry commonly associated with generalized anxiety disorder (GAD) is continuous with normal worry. No taxometric studies of social phobia have been published, although one study reporting a taxon underlying social inhibition in infancy and childhood (Woodward, Lenzenweger, Kagan, Snidman, & Arcus, 2000; see Normal Personality section) makes the taxonicity of social anxiety an interesting prospect. Finally, although the latent structure of panic disorder has not been investigated, two studies of anxiety sensitivity—a vulnerability factor for panic disorder—have been published: one reaching a dimensional conclusion (Taylor, Rabian, & Fedoroff, 1999) and one reaching a taxonic conclusion (Bernstein, Zvolensky, Weems, Stickle, & Leen-Feldner, 2005). In summary, large areas of the anxiety domain are in need of taxometric study, and those areas that have received some attention are in need of replication.

Eating Disorders

Taxometric research on eating disorders has only recently begun. Taxonic conclusions were reached in two studies of bulimia nervosa (BN)—one using an admixed sample of bulimic and normal college women, the second using a larger normal sample (Gleaves, Lowe, Snow, Green, & Murphy-Eberenz, 2000). A follow-up study examined the *DSM–IV* subtypes of anorexia nervosa (AN) and BN in an admixed sample containing normals and a variety of eating disorder patients (Gleaves, Lowe, Green, Cororve, & Williams, 2000). The authors reported taxonic conclusions for the purging and nonpurging subtypes of BN and found only the restricting subtype of AN to be qualitatively distinct from BN, implying that *DSM–IV* may incorrectly locate the AN/BN boundary on the basis of weight, rather than presence versus absence of binging and purging. Williamson et al. (2002) replicated the taxonic finding for AN as well as for BN and binge eating disorder. More recently, however, Tylka and Subich (2003) reached a dimensional conclusion for BN features in an undergraduate sample. Thus, some uncertainty remains over the taxonicity of AN, BN, and their variants.

Dissociative Disorders

Dissociative conditions have been the subject of three taxometric studies. Using an admixed sample of dissociative identity disorder (DID) cases and normal controls, Waller, Putnam, and Carlson (1996) concluded that the features of depersonalization-derealization and amnesia demarcated a pathological dissociation taxon, whereas the feature of absorption and imaginative involvement was better understood as dimensional. Building on these results, the authors generated an abbreviated instrument for assessing pathological dissociation and assigning individuals to the taxon. This taxonic finding for pathological dissociation was later replicated by Green, Gleaves, Dell, and Riley (2000) and Waller and Ross (1997), who found that 3.3% of the general population belonged to the taxon and that younger age and PTSD diagnosis—but not female gender—are associated with taxon membership. The taxon had a large shared environmental component and a negligible genetic component, consistent with traumatic etiological theories. Although taxometric studies of pathological dissociation have consistently arrived at a taxonic conclusion, doubt cast its temporal stability may call its taxonicity into question (Watson, 2003).

Other Axis I Conditions

Although mood, anxiety, eating, and dissociative disorders have been the focus of most taxometric studies of Axis I psychopathology, a few studies have examined other conditions. Three early studies by Golden and colleagues examined the latent structure of organic brain conditions, arriving at taxonic conclusions for dementia (Golden, 1982), tardive dyskinesia (Golden, Campbell, & Perry, 1987), and neonatal brain dysfunction (Golden, Vaughan, Kurtzberg, & McCarton, 1988). A recent study by Blanchard, Horan, and Collins (2005) also obtained taxonic findings for a negative symptom subtype of schizophrenia. Unpublished studies have investigated additional mental disorders, including alcoholism and other substance use disorders (see Haslam & Kim, 2002, for a review).

Personality Disorders

A substantial number of taxometric investigations have focused on the personality disorders (PDs) appearing on Axis II of *DSM–IV*. Perhaps the main reason for this level of attention is the long and intense debate that has waged over the categorical versus dimensional status of the PDs. At present, many psychologists strongly favor a dimensional view of PDs, and there have been several proposals to overhaul the classification of these disorders along dimensional lines (Livesley, Schroeder, Jackson, & Jang, 1994; McCrae, 1994; Widiger & Clark, 2000). This dimensional position is supported by the robust association and apparent continuity of many PDs with dimensions of normal personality such as the five factor model (Costa & Widiger, 2002), as well as by the finding that dimensional measures of PDs commonly predict clinical phenomena better than categorical measures (e.g., Ullrich, Borkenau, & Marneros, 2001). Moreover, it may be that the disturbingly high levels of comorbidity among some PDs are better understood as the overlap of indistinct conditions on common dimensions of abnormal personality rather than as the co-occurrence of discrete diagnostic entities. At the same time, debates over the latent structure of PDs raise an inherent conflict for many psychologists who assume personality to be dimensional, but recognize that some mental disorders may be discrete entities (see Haslam, 2003).

Although the taxometric method is well suited to resolving issues of latent structure in the PD domain, only three PDs have been investigated to date. By far the most extensive body of taxometric research in the realm of abnormal personality has focused on schizotypy. This construct, representing one component of Meehl's (1962) influential theory of vulnerability to schizophrenia, was the subject of the first published taxometric study. In the study, Golden and Meehl (1979) supported Meehl's hypothesis that this liability was taxonic, revealing a discrete class of schizotypes believed to be at risk for developing schizophrenia. Since that study, nine published taxometric investigations have examined the latent structure of schizotypy and the closely related construct of schizotypal PD, all reporting support for taxonic structure. These include high-risk studies of children assessed by standardized cognitive and neuromotor measures (Erlenmeyer-Kimling, Golden, & Cornblatt, 1989) as well as adolescents followed into adulthood and assessed by clinical interviews and behavioral ratings (Tyrka, Cannon, et al., 1995; Tyrka, Haslam, et al., 1995). Studies found that taxon membership conferred a markedly increased risk of lifetime schizophrenia spectrum diagnosis and was plausibly associated with genetic risk. Taxonic results were also reported by the overwhelming majority of taxometric analyses performed in college samples using self-report measures of schizotypy or its hypothesized components (e.g., perceptual aberration, social anhedonia; Blanchard, Gangestad, Brown, & Horan, 2000; Keller, Jahn, & Klein, 2001; Korfine & Lenzenweger, 1995; Lenzenweger, 1999; Lenzenweger & Korfine, 1992; Meyer & Keller, 2001), which consistently obtained base rates of .05 to .10 for the inferred schizotypal taxon. Interestingly, Horan, Blanchard, Gangestad, and Kwapil (2004) found evidence that social anhedonia and perceptual aberration may in fact represent two distinct taxa. Notably, the obtained estimates are considerably higher than the population prevalence of schizotypal PD, suggesting that schizotypal PD may reflect a relatively severe variant within the broader taxon (or taxa).

A second PD that has received taxometric scrutiny is antisocial PD. Three studies have examined antisocial PD, psychopathy, or their childhood antecedents, and all yielded taxonic conclusions. Using a large adult offender sample, Harris, Rice, and Quinsey (1994) reported evidence of a taxon defined by indicators of chronic antisocial behavior beginning in childhood, but not by indicators of criminality. Skilling, Harris, Rice, and Quinsey (2001) repeated these analyses with the inclusion of some additional data and obtained similar results. Two additional studies reported evidence of a psychopathic taxon among children and adolescents (Skilling, Quinsey, & Craig, 2001; Vasey, Kotov, Frick, & Loney, 2005). However, Marcus, John, and Edens (2004) came to a dimensional conclusion in a study of jail and prison inmates. Existing studies suggest that antisocial PD may be best understood as a taxon, although the evidence is somewhat mixed. Additional investigations currently underway may help resolve these discrepancies.

A third PD that has been studied using taxometrics is borderline PD. Both published taxometric studies of this disorder arrived at a dimensional conclusion (Rothschild, Cleland, Haslam, & Zimmerman, 2003; Trull, Widiger, & Guthrie, 1990).

In summary, taxometric investigations do not support a strictly categorical or dimensional position for all forms of personality disturbance. Instead they suggest that different structural models may be appropriate for different disorders. By implication, this research implies that multiple co-occurring PD diagnoses may at times reflect true comorbidity among discrete disorders, rather than mere overlap of dimensional conditions. Thus, taxometric research has the potential to play an important role in nosological revision of the PD domain, informing difficult theoretical debates with relevant and valuable empirical evidence.

NORMAL PERSONALITY

Meehl and his colleagues initially developed the taxometric method to resolve structural problems in the study of psychopathology, and to this day the vast majority of taxometric studies address latent variables in that domain. However, the method can serve equally well as a tool for studying normal personality; over the last two decades, researchers have conducted taxometric investigations of a variety of personality characteristics. The distinction between dimensional and categorical conceptions of individual differences—often described as traits versus types—has long been a focus of attention for personality psychologists (Asendorpf, 2002). The historical and theoretical context of this distinction has been discussed by several authors (Gangestad & Snyder, 1985; Kagan, 1994; Meehl, 1992), who identified reasons that the dimensional view has generally triumphed within psychology, whereas typological views of personality have often been disparaged. For instance, psychology's fundamental concern with quantitative measurement has encouraged a view of traits as smoothly gradated dimensions, whereas the role of dimension-presuming statistical procedures such as factor analysis in the development of personality models also contributed to the entrenchment of a dimensional perspective. As a result, there is now often "a prevailing and rarely questioned assumption that the units of personality are continuous dimensions and an accompanying prejudice against class variables" (Gangestad & Snyder, 1985, p. 317).

Taxometric methodology allows this assumption to be tested. A body of taxometric research now allows some cautious conclusions to be drawn about the existence of latent categories in the domain of normal personality. Next we briefly review taxometric studies of normal personality characteristics, beginning with broad traits and proceeding to more specific personality features.

Broad Traits

Most taxometric studies in the personality domain have examined relatively specific dispositional factors perhaps because taxonic findings appear more plausible among narrow-band characteristics. Mega-traits, such as those of the popular five factor model (FFM), encompass many specific dispositions, suggesting that they may originate from—and be influenced by—a multitude of factors. Specific traits, in contrast, refer to circumscribed forms of behavior whose variation could conceivably be ascribed to the sorts of single dominant causes that can generate taxa. This approach of looking for personality taxa where they are most likely to be found has some merit, but the present neglect of broad personality characteristics is still somewhat surprising, especially given specific taxonic claims made about broad-band traits—such as that there are three broad personality prototypes (e.g., Asendorpf & van Arken, 1999)—which can be tested using taxometrics. Only one published taxometric study has examined broad characteristics (Arnau, Green, Rosen, Gleaves, & Melancon, 2003)—namely, Jung's *preferences* (extraversion vs. introversion, thinking vs. feeling, sensing vs. intuiting, judging vs. perceiving). Contrary to Jung's typological theory, the authors arrived at dimensional conclusions for all four preference variables. No published studies have addressed Big Five or other broad personality traits, despite the central importance of these characteristics in contemporary personality theory.

Specific Traits

Several taxometric studies have examined more specific personality traits. In the first such study, Gangestad and Snyder (1985) examined self-monitoring, the dispositional monitoring and regulation of expressive behavior and self-presentation. They reported a taxonic finding and speculated on the possible processes of divergent causality operating over the course of social development that may account for the taxon. This finding was buttressed by a recent replication using latent class analysis (von Davier & Rost, 1997).

The second normal personality trait to be examined by taxometrics was the Type A construct—a complex of impatience, hostility, competitiveness, and achievement striving that was conceptualized by its originators as typological. Consistent with this conceptualization, Strube (1989) reported taxonic results, underscoring the need for explanatory models of this personality type that invoke causal processes capable of generating taxa.

A third set of specific traits to receive taxometric attention are those related to attachment style. Adult attachment style has attracted enormous attention from researchers in the past decade, representing a fruitful extension of concepts developed to explain infant and child behavior. In the original child domain, categorical models of attachment have long been favored, with infants typically classified into one of several distinct attachment styles. Inconsistent with these models, Fraley and Waller (1998) concluded that adult attachment styles were dimensional—a finding that was subsequently replicated in a study of infant attachment behavior based on the Strange Situation task (Fraley & Spieker, 2003). These conclusions suggest that theorists move away from single-cause models of adult attachment and give greater attention to multiple-factor models that may be more likely to produce attachment dimensions.

Another taxometric study of developmental phenomena tested Kagan's (1994) prediction that there exists a highly reactive class of infants, characterized by high levels of distress and motor activity, who tend to develop into behaviorally inhibited children. Using systematic observational ratings of 4-month-old infants, Woodward, Lenzenweger, Kagan, Snidman, and Arcus (2000) reported evidence of an infant reactivity taxon with a base rate of .10, replicating the findings of a mixture model analysis by the same research group (Stern, Arcus, Kagan, Rubin, & Snidman, 1995).

Taxometric investigation of a fifth personality trait, hypnotic ability, was motivated by disagreement in the literature over whether it is meaningful to identify a discrete class of highly hypnotizable individuals (or hypnotic "virtuosi"). Oakman and Woody (1996) concluded that such a taxon exists, calling for a change in measurement strategy and in the search for explanatory theories of hypnotic susceptibility.

Two further specific traits have been examined using taxometrics, although only as presumptively dimensional control variables in studies of other constructs. Gangestad and Snyder (1985) reported the expected dimensional result for the trait of impulsivity, as did Korfine and Lenzenweger (1995) for the construct of gender identity.

Sexual Orientation

The characteristics examined thus far in this section fall within the domain of personality and temperament. Sexual orientation—the tendency to desire sexual partners of one gender or another—is rarely considered to belong in this domain, but unquestionably involves behavioral and emotional dispositions. The question of latent structure is especially important in the case of sexual orientation because it has been a focus of pronounced theoretical disagreement. Many laypeople hold a taxonic view of this construct, wherein homosexual and heterosexual people are believed to differ in kind. This position was challenged by Kinsey, Pomeroy, and Martin (1948), who claimed that sexual orientations differ by degree alone and can be represented by a dimensional scale. In the first taxometric examination of sexual orientation, Haslam (1997) employed items from the MMPI–2 *Mf* scale in a representative sample of American men and reached a dimensional conclusion. Conversely, Gangestad, Bailey, and Martin (2000) used measures of sexual preference, childhood gender nonconformity, and adult gender identity in a large sample of twins and concluded that both male and female sexual orientation were taxonic. Further studies are needed to clarify these discrepant findings.

OTHER LATENT VARIABLES

Response Styles and Cognitive Variation

In addition to studying enduring, deeply rooted dispositions, the taxometric method has been used to examine transient behavioral strategies that people adopt in particular contexts and one form of cognitive variability. Strong and his colleagues used taxometrics in this fashion to examine the structure of response sets or styles on the MMPI–2. In one study exploring the underreporting of problems and symptoms by child custody litigants (Strong, Greene, Hoppe, Johnston, & Olesen, 1999), the authors reported taxonic results for validity scales assessing impression management (the deliberate attempt to present a positive social image), but dimensional results for scales assessing unintentional or self-deceptive positivity. A second study (Strong, Greene, & Schinka, 2000) examined the construct of overreporting in two clinical samples and arrived at a taxonic conclusion. Both studies used this structural information in practically useful ways, specifying optimally discriminating items and scale cut scores to improve the detection of invalid test profiles. In a final study of response styles, Arnau, Thompson, and Cook (2001) examined whether different Web-based survey formats (continuous slider bars vs. Likert-style buttons) influenced the latent structure of survey responses. Analysis of survey data concerning perceptions of library quality led to dimensional conclusions for both response formats. Finally, in the first study to employ taxometrics to study individual differences in cognition, Dollaghan (2004) found no evidence for a taxon underlying specific language impairment in 3- and 4-year-old children.

Variables Outside the Realm of Individual Differences

Researchers have traditionally used the taxometric method to examine questions about individual differences, asking in effect whether people fall into discrete categories or differ quantitatively from one another. However, there is no reason that taxometrics should be restricted to this domain. The versatility of taxometric procedures permits their application in many research contexts where the categorical versus dimensional status of a construct is in question and where adequate samples and indicators of the relevant entities can be obtained.

To date, only four taxometric studies have been published in which the person was not the primary unit of analysis. Two of these studies (Haslam, 1994, 1999) examined a theory of social relationships proposed by Fiske (1992), who argued that human sociality is organized by four discrete relational models. Both studies reported support for this taxonic hypothesis in a representative sample of social relationships. Dyadic relationships were also the unit of analysis for a study by Beach, Fincham, Amir, and Leonard (2005), which found evidence for a taxon of marital discordance characterizing about 20% of a sample of couples. The fourth study examined the structure of emotion episodes involving rivalry and concluded that envy and jealousy represent categorically distinct but frequently co-occurring emotions (Haslam & Bornstein, 1996). Studies of this kind have the potential to inform debates over the so-called *basic* emotions, over the distinctness of particular emotions, and over dimensional, categorical, and prototype-based accounts of the emotions.

OVERVIEW OF THE SUBSTANTIVE FINDINGS

Although our review of published taxometric investigations has been brief, we hope that we have conveyed a sense for what has been done and what has yet to be done using the taxometric method. In addition, we hope to draw from this review a few conclusions about existing work that might usefully guide future taxometric endeavors. First, as can be seen in Table 10.1, different studies of particular latent variables tended to yield convergent conclusions, providing increasing confidence in the reliability of the structural inferences drawn from taxometric results. Second, although taxometric studies are being published at an escalating rate, large regions of the psychopathology and personality domains have yet to be explored, few findings have been well replicated, and many conclusions could be challenged on methodological grounds. We identify a number of especially neglected areas—and future research frontiers—in the next chapter. However, the taxometric evidence already at hand allows a few tentative conclusions to be drawn.

In regard to the latent structure of psychopathology, existing research does not support dogmatic positions on either side of the categories versus continua debate, with taxa and dimensions both appearing to be widely distributed in the psychopathology domain. Consistent with the earlier review of Haslam and Kim (2002), roughly half of the psychopathological constructs evaluated in published taxometric studies have produced dimensional conclusions. This suggests that, contrary to occasionally voiced criticism, the taxometric method does not seem to have an obvious bias toward—or against—the detection of taxa. It also suggests that attempts to bring a single, inflexible position to the conceptualization or classification of mental disorders may inadequately capture the latent structure of at least some conditions. The finding that categorical and dimensional models each appear to describe some conditions better than others supports the empiricist stance so resolutely espoused by Meehl: Claims about latent structure should be based not on theoretical fiat, but on concerted structural investigation.

At the same time, while taxometric studies support taxonic and dimensional models to roughly equal degrees in the broad domain of psychopathology, our review also indicates that certain models tend to be favored in particular regions within this domain. Dimensional interpretations have predominated in the traditional "neurotic" spectrum, particularly among the mood and anxiety disorders. However, this conclusion may yet be qualified, as melancholic depression may prove to be better understood as taxonic and there are intimations of a possible anxiety sensitivity taxon. Moreover, plausibly taxonic conditions such as the bipolar disorders and obsessive–compulsive disorder have yet to be studied using taxometrics. At present, categorical models do appear to be more common outside the traditionally neurotic realm, with several taxa reported among the eating, dissociative, and psychotic disorders and a few reported among the personality disorders.

Several studies have shown that taxometric research is capable of challenging and refining accepted structural conceptualizations of psychopathology. For example, on the basis of their taxometric conclusions, Gleaves, Lowe, Snow, et al. (2000) argued that the *DSM–IV* weight-based distinction between AN and BN represents an invalid structural boundary, that a more valid boundary is designated by the presence versus absence of binging and purging, and that the binge/purge subtype of AN would most appropriately be classified with BN. Along similar lines, Waller et al. (1996) usefully parsed the symptoms of dissociation into pathological (taxonic) and nonpathological (dimensional) components in a way that had not hitherto been done. Both studies challenged existing models of psychopathology on empirical grounds, demonstrating that taxometric research can make constructive contributions to nosological revision that go beyond the simple testing of categorical or dimensional assumptions.

Several tentative observations can also be offered concerning taxometric investigation of normal personality. As many psychologists might expect, taxonic conclusions have been reached somewhat less often in studies of normal personality than in studies of psychopathology. Although a sufficient number of taxonic results has been reported in the personality domain to challenge the strong antitypological prejudice described by Gangestad and Snyder (1985), most of the normal personality characteristics that have been studied to date yielded dimensional conclusions. Still there appears to be enough evidence for the existence of some taxa to merit further taxometric research in this domain. Research suggests that taxa may be more prevalent among specific or narrow-band personality characteristics and response styles than among broad personality traits. However, it must be noted that the few normal personality characteristics examined thus far may not be representative of the total pool of individual difference variables, and that the taxonic conclusions generated to date have not yet been thoroughly replicated.

IMPLEMENTATION OF THE TAXOMETRIC METHOD

Having reviewed the substantive conclusions of existing taxometric studies, we now turn to a more focused examination of the ways in which these studies were performed. Table 10.2 summarizes a number of important methodological characteristics of the 57 published taxometric studies of psychopathology and personality included in our substantive review. To help depict trends in taxometric practice, Table 10.2 separates these studies, somewhat arbitrarily, into those published before 2000 and those appearing during or after 2000. The fact that the studies appearing in the last few years already greatly outnumber those appearing in the period from 1979 to 1999 underscores the vigorous pace of taxometric research in recent years. Next we briefly review the methodological information reported in these studies and summarized in the table.

TABLE 10.2
Implementation of the Taxometric Method in Published Studies

	Date of Publication		
Variable	*Pre-2000*	*2000 to Present*	*Total*
Number of studies	21	36	57
Procedure			
MAXCOV	16 (76%)	28 (78%)	44 (77%)
MAMBAC	6 (29%)	27 (75%)	33 (58%)
MAXEIG	1 (5%)	10 (28%)	11 (19%)
L-Mode	0 (0%)	10 (28%)	10 (18%)
Latent class method	4 (19%)	0 (0%)	4 (7%)
MAXSLOPE	1 (5%)	0 (0%)	1 (2%)
Mean number of procedures	1.33	2.08	1.81
Data source			
Self-report	14 (67%)	27 (75%)	41 (72%)
Interview	4 (19%)	6 (17%)	10 (18%)
Behavioral ratings	2 (10%)	4 (11%)	6 (11%)
Archival	2 (10%)	1 (3%)	3 (5%)
Cognitive assessment	2 (10%)	1 (3%)	3 (5%)
Indicator type			
Dichotomous	13 (62%)	7 (19%)	20 (35%)
Sample size			
Median N	639	923	809
$N < 300$	3 (14%)	1 (3%)	4 (7%)
Curve presentation			
Smoothed curves	10 (48%)	10 (28%)	20 (35%)
Averaged curves	12 (57%)	18 (50%)	30 (53%)
All curves	6 (43%)	15 (42%)	21 (37%)
Representative curves	1 (5%)	5 (14%)	6 (11%)
Explicit judgment procedure	1 (5%)	11 (31%)	12 (21%)
Parameter estimates			
Taxon base rates	17 (81%)	32 (89%)	49 (86%)
Indicator validities	5 (24%)	20 (56%)	25 (44%)
Within-group correlations	4 (19%)	15 (42%)	19 (33%)
Consistency tests			
Base rate consistency	15 (71%)	26 (72%)	41 (72%)
Simulated comparison data	4 (19%)	17 (47%)	21 (37%)
Bayesian probabilities	11 (52%)	6 (17%)	17 (30%)
Separate subsamples	2 (10%)	10 (28%)	12 (21%)
Goodness of fit statistic	1 (5%)	9 (25%)	10 (18%)
Inchworm consistency test	1 (5%)	4 (11%)	5 (9%)
Case removal	0 (0%)	3 (8%)	3 (5%)
Nontaxometric statistical procedure	4 (19%)	3 (8%)	7 (12%)
Mean number of consistency tests	1.81	2.17	2.04

Note. Percentages are calculated within columns (e.g., 16 of the 21 pre-2000 studies [76%] included MAXCOV analyses).

Taxometric Procedures

Table 10.2 reveals the taxometric procedures that have been most popular in past studies and those that are now used more frequently. MAXCOV remains the most popular taxometric procedure, but it is increasingly being performed along with other procedures and replaced by MAXEIG in some studies. MAMBAC is easily the second most widely used procedure, rivaling MAXCOV in recent years. MAXEIG and L-Mode, introduced relatively recently by Waller and Meehl (1998), rapidly gained in popularity, whereas the early latent class procedure pioneered by Golden (1982) is no longer in use. MAXSLOPE has yet to become widely used. In addition to the trend toward employing more recently developed taxometric procedures, the table highlights the encouraging finding that researchers are relying increasingly on more than one taxometric procedure when conducting a taxometric study.

Data Source

By far the most common source of data for taxometric studies remains self-report inventories. Relatively few studies have made use of alternative data sources such as behavioral ratings, clinical interviews, or cognitive assessments, although the number of studies using interview data has risen. Beauchaine and Waters (2003) found that individuals' expectations concerning taxonicity can influence the structural nature of their ratings, which suggests that raters should be screened for such biases and trained to make ratings as objectively as possible. The present reliance on self-report data is not intrinsically problematic. Ideally, future research will make greater use of multiple data sources (e.g., interviews, behavioral ratings, biological or physiological markers). Such multimethod assessment practices would not only increase confidence in resultant structural inferences, but might also reduce the extent to which studies are compromised by within-group correlations due to shared method variance.

Indicator Type

Given the concerns that have repeatedly been expressed about the use of dichotomous indicators in taxometric analyses (Maraun, Slaney, & Goddyn, 2003; Miller, 1996; J. Ruscio, 2000), it is encouraging to observe that recent studies employ such indicators far less often than earlier investigations. In fact, whereas dichotomous indicators were once the standard in published taxometric research, continuous or quasi-continuous indicators are now the norm—a desirable state of affairs. Factors contributing to this shift may include the spread of reservations about dichotomous indicators (e.g., concerns about heightened risk of pseudotaxonicity), a greater concern with indicator validity, and the abandonment in recent years of Golden's latent class procedure (which required dichotomous indicators) in favor of multivariate procedures (which can be performed with dichotomous indicators only through procedural modifications such as the use of composite input indicators).

Sample Size

Another beneficial change observed over time is the trend toward larger samples. The median sample size of a taxometric investigation has risen to 923 in recently published studies, and the number of studies that violate Meehl's recommended minimum sample size ($N = 300$) has decreased. The increased reliability and validity of structural conclusions and associated parameter estimates afforded by these larger samples, as well as the greater power that they provide for detecting small taxa, make these changes especially important.

Curve Presentation

Taxometric investigations differ considerably in their approach to the graphing and presentation of taxometric curves. We focus on a few key aspects of graphical presentation in Table 10.2. First, our review suggests that curve smoothing—once a mainstay of taxometric practice—has become somewhat less common in recent years, although this practice is still widespread. As noted in chapter 8, smoothing has been somewhat controversial, and researchers appear to be particularly cautious about applying this technique in situations where smoothing could artificially flatten a genuine taxonic peak (e.g., where there are few data points on a curve). Perhaps another factor contributing to the small observed decline in smoothing was the advent of the MAXEIG procedure, whose overlapping windows provide many more data points—and hence smoother curves—than most MAXCOV analyses. Indeed the use of overlapping windows is one technique for smoothing a curve that may obviate the need for further smoothing. Finally, the larger samples employed in more recent studies also allow data points to be calculated with less sampling error, reducing the necessity for smoothing.

A second important change in recent taxometric practice is the increasing tendency to present many individual taxometric curves—including all curves that are generated, or a representative sampling of curves—rather than solely one or more averaged curves, although presenting averaged curves remains the most common practice. A more comprehensive presentation of panels of curves may improve the reliability of curve judgment, provide more opportunities for consistency testing, and give reviewers and readers a greater chance to form their own judgments of the obtained graphical results. However, this practice does consume more journal space than the presentation of averaged curves. As discussed and illustrated in chapter 8, researchers who work with standardized indicators can present both a full panel of curves *and* an averaged curve on a common set of axes—an approach that presents information efficiently.

In line with the increased number of curves presented in more recent taxometric studies—an increase intensified by the growing popularity of analyzing taxonic and dimensional comparison data sets—there has been an encouraging increase in the use of formal procedures for judging curves. Most early studies relied on the self-evident resemblance of one or more curves to prototypic taxonic or dimensional curve shapes, usually judged solely by the researchers. In contrast, studies are increasingly employing multiple judges whose agreement can be calculated, and some studies have made use of naive judges whose formal training process is described in the published report. Moreover, numerous studies refer explicitly to some sort of judgment procedures, such as those in which obtained curves are sorted into taxonic, dimensional, or ambiguous piles and the resulting judgment rates are reported. Although procedures such as these are still far from standard, they point to a desirable examination and formalization of the curve judgment process. We urge researchers performing taxometric investigations to build on the recent trend toward formalizing and describing the process by which they interpret taxometric curves.

Latent Parameter Estimates

Although curve judgments have often constituted the primary output of taxometric studies, many studies have also reported some information about the parameter estimates drawn from taxometric curves. As shown in Table 10.2, base rate estimates have been reported in most taxometric reports perhaps because consistency among base rate estimates within and across procedures has been a popular consistency test as well as because of research interest in the prevalence and characteristics of taxa. However, early taxometric reports rarely gave similar attention to other important model parameters. Estimates of indicator validities were seldom reported, despite the critical influence of validity on the power of taxometric analyses. Estimates of within-group correlations were similarly neglected in published reports of taxometric studies. By contrast, many recent studies have reported estimates of indicator validity and within-group correlations, or at a minimum indicated that these were sufficiently strong or within tolerable limits for analysis. However, important details concerning how these parameters were estimated are seldom reported, and it is often unclear whether the estimates were generated prior to taxometric analyses (to evaluate the appropriateness of data for the planned analyses) or derived from the taxometric curves (to argue against a pseudotaxonic or pseudodimensional conclusion, or to draw out implications for future research). We recommend that researchers report estimates of indicator validity and within-group correlations more routinely and describe how and why they were estimated more explicitly.

Consistency Tests

As emphasized throughout the book, consistency testing is fundamental to taxometric research. However, the concept of consistency testing is somewhat fuzzy, and at times it can be difficult to determine which analyses served as consistency tests (and even which consistency tests were performed) within a given taxometric investigation. Nevertheless, we made an effort to review the consistency tests employed in the 57 published studies, relying in part on the classification of tests outlined in chapter 7. Table 10.2 displays the range of consistency tests employed and reveals trends in their popularity.

The most popular and enduring consistency test has been the evaluation of consistency among multiple base rate estimates. Consistency may be examined across multiple applications of a single taxometric procedure or across estimates drawn from different procedures. Although studies have varied widely in their approach to reporting base rate estimates and in their criteria for judging these estimates to support taxonic or dimensional conclusions, the majority of published studies have used some form of this test. As noted in chapter 7, it is presently unclear whether the popularity of this consistency test is justified because no published studies have validated the assertion that taxonic data tend to yield more coherent base rate estimates than dimensional data, and because there is evidence that dimensional data can yield coherent base rate estimates under certain conditions (e.g., when indicators are positively skewed). A recent factorial Monte Carlo study (J. Ruscio, 2005) gives reason to question whether this consistency test can provide compelling support for taxonic or dimensional structure under many realistic research conditions. Given the popularity of this test, further investigation—as well as due caution in its use—seems warranted.

The use of simulated comparison data as a consistency test has increased considerably in recent years, and it is now the second most commonly used consistency test, if alternative taxometric procedures are not counted as such tests. Although several studies have incorporated the simulation of comparison data in the manner advocated throughout this book to generate empirical sampling distributions, several earlier studies implemented some variation of this technique. Since Gangestad and Snyder

(1985) first introduced the general approach, its implementation has gradually evolved from the creation of a single set of dimensional data (to demonstrate that one's analysis was not biased toward the detection of a taxon) to the generation of sampling distributions through analyses of multiple samples of taxonic and dimensional comparison data that reproduce the indicator distributions and correlations in the research data. Although much remains to be studied with regard to the optimal use of comparison data, the rapid increase in its popularity suggests that contemporary researchers find it to be a valuable tool. The examination of Bayesian posterior probabilities (see chap. 7), which was the second most popular consistency test in earlier research, has declined in popularity and appears to be falling out of favor.

Several other consistency tests described in chapter 7 have also come to be more widely used in recent taxometric investigations. The parallel analysis of two or more subsamples of data has become relatively common practice, with several studies employing two very distinct samples (e.g., clinical and nonclinical). The goodness of fit (GFI) statistic introduced into the taxometrics literature by Waller and Meehl (1998) has been employed in many recent studies, as has the inchworm consistency test also introduced by these authors. Case removal consistency testing, discussed by Meehl and Yonce (1994) and J. Ruscio (2000), has been employed in only two substantive studies to date.

Finally, a few studies have employed nontaxometric statistical procedures alongside taxometric analyses. Although these methodologies are not recognized as taxometric consistency tests, they advance the goals of consistency testing in these studies, offering an alternative evaluation of latent structure whose agreement with taxometric results can be assessed. An assortment of latent class, cluster analytic, and mixture modeling analyses has been employed in this manner, although this practice is relatively rare.

Overall, Table 10.2 indicates that consistency testing is alive and well in taxometric research. A diverse set of tests has been employed, and new technical developments have been incorporated into many recent studies. In addition to the changing landscape of tests employed, more recent studies appear to utilize a somewhat greater number of consistency tests ($M = 2.17$) than older studies ($M = 1.81$). This trend parallels the separate but corresponding increase in the number of taxometric procedures employed (*M*s = 2.08 vs. 1.33). Indeed if one taxometric procedure is taken as the primary analysis and any other procedures are treated as consistency tests for that analysis, the number of consistency tests included in studies published from 2000 onward is substantially larger than the number included in earlier studies (*M*s = 3.25 vs. 2.14). This is despite the likelihood that this tabulation conservatively estimates the actual number of consistency tests employed in taxometric studies as (a) several infrequently used tests do not appear in the table, (b) it is often difficult to determine precisely which elements of a taxometric report are considered consistency checks, and (c) what constitutes a consistency test is somewhat subjective and may change over time. Bearing in mind the subjectivity inherent in tallying such results, the summary of consistency tests appearing in Table 10.2 provides encouraging evidence for the increasing rigor of taxometric investigations.

CONCLUSIONS

This chapter has surveyed the rapidly growing taxometric literature to provide the reader with a sense of the constructs investigated, the findings generated, and the methodological approaches employed. Taxometric researchers have covered a great deal of ground in the study of individual differences, particularly in the domains of normal personality and psychopathology. However, as revealed by this chapter, much work remains to be done. In addition to summarizing all of the substantive taxometric investigations that have been published to date, this chapter reviewed the ways in which the taxometric method has been implemented and considered how procedural implementation has evolved in the quarter century since the publication of the first substantive taxometric report. Although technical refinements have at times been slow to gain a foothold in practice and extensive consistency testing has not always been employed, there is clear evidence that better research practices are increasingly being adopted. The quality of taxometric research—and the resulting confidence that researchers can have in taxometric findings—show strong signs of improvement. On balance, this review gives ample cause for optimism about the future of taxometric research. Some suggestions about where this research might be headed are presented in the next chapter.

CHAPTER ELEVEN

The Future of Taxometrics

The first 10 chapters of this book laid out our view of the present state of taxometric research. We discussed the conceptual foundations of the taxometric method, the scientific questions that it can answer, the central considerations and state-of-the-art methods involved in its implementation, and the knowledge that it has generated about the latent structure of psychopathology, personality, and related constructs. But what about the future of taxometrics? Although it is more than 30 years old, the taxometric method is undergoing rapid growth in popularity and vitality. As summarized in chapter 10, taxometric investigations are appearing in the literature at an ever-increasing rate, with more taxometric studies of personality and psychopathology published since the millennium than in the two preceding decades.

Clearly, taxometric research has gained considerable momentum that should continue to propel it forward. But what might its future look like? How might taxometric research be conducted in the years to come, and how should it be conducted? We conclude this volume with a discussion of the directions in which taxometric research might profitably move over the next decade. This discussion covers four major areas. First, we focus on research domains that have already received some taxometric attention and identify constructs that represent especially promising subjects for new taxometric research, either because they are particularly pressing or because they have yielded ambiguous findings to date. Second, we propose several new research domains that might benefit from taxometric investigation, including constructs within those domains that represent broad new vistas for taxometric research. Third, we examine several research questions with significant theoretical value that can be addressed or substantially informed by taxometric research. Many of these applications of taxometric research have been proposed in past writings, but none has been fully exploited. Finally, we consider several aspects of the method that require further investigation and questions that future Monte Carlo studies of taxometric procedures might usefully consider. Such methodological development and refinement is essential if taxometric research is to advance rather than stagnate. We hope this chapter cultivates a sense of the considerable promise borne by taxometric research, as well as a sense of the important work that remains to be done.

CONSTRUCTS IN TRADITIONAL DOMAINS REQUIRING FURTHER STUDY

Psychopathology

Although psychopathology has been a primary focus of taxometric investigations and more studies have fallen within this domain than in any other, the coverage of these studies has been patchy and uneven. Large regions in this domain have yet to see a taxometric study, and even the most thoroughly studied conditions would benefit from additional investigation or replication. At the same time, certain disorders seem especially in need of taxometric attention, given their prevalence in the population, their priority within the mental health community, or their equivocal results in past taxometric investigations. Because the taxometric conclusions regarding many mental disorders were reviewed in chapter 10, we touch on only the highlights here.

Unipolar mood disorders have received more taxometric scrutiny than almost any other form of psychopathology. Although the continuity of general depression is relatively well studied, there is less consensus about the many depressive subtypes that have been proposed in the clinical literature. Taxometric findings for subtypes have been undermined by considerable methodological limitations, making rigorous taxometric studies of subtypes a research priority. Equally important are taxometric studies of bipolar disorders, as well as replication studies performed with new samples (e.g., inpatients) and new measures (e.g., clinical interviews, behavioral observations) to evaluate the consistency of previous findings. In contrast to the mood disorders, anxiety disorders and other fear-related constructs have received relatively little taxometric attention. These conditions present an important focus for future taxometric research.

Schizophrenia and related disorders represent a crucial frontier for taxometric study. In light of the massive investment of research funding and effort in the study of these disabling conditions, their neglect by taxometric researchers is surprising, even more so given enduring controversy over whether schizophrenia represents one or several related disorders and given substantial efforts to characterize the broad symptom dimensions of this syndrome (e.g., Crow, 1985). Taxometrics can be used to examine not only the structure of schizophrenia, but the structure of its putative subtypes (e.g., paranoid, disorganized, catatonic) to help address questions about the heterogeneity of schizophrenia. The recent work of Blanchard et al. (2005), which found evidence of a negative symptom taxon, is an encouraging first foray into this domain.

One other area in which taxometric research is greatly needed concerns vulnerability factors for mental disorders. Taxometric procedures are at least as useful for testing the structure of diatheses as they are for testing the structure of expressed syndromes, provided that the samples and measures that are used are appropriate and specific to the vulnerability construct. At present, few taxometric studies have been conducted in this area, with the exception of research on the liability for schizophrenia (schizotypy). Even those vulnerability factors that have been studied—including those for panic disorder (Bernstein et al., 2005; Taylor et al., 1999), depression (Gibb et al., in press; Strong et al., in press), bipolar disorder (Meyer & Keller, 2003), and alcoholism (Knowles, 1989)—have been examined only by one to two studies apiece, despite enormous interest in these diatheses. The paucity of taxometric research in this area awaits rectification.

Personality Disorders

Any discussion of vulnerabilities for psychopathology naturally broaches the subject of personality disorders (PDs). Although these disorders and related personality constructs have been the focus of more than 20 taxometric studies dating back to the earliest years of taxometric research (Golden & Meehl, 1979), only 3 of the 10 *DSM–IV* PDs have been investigated: schizotypal PD and schizotypy, antisocial PD (or psychopathy), and borderline PD. This dearth is remarkable when one considers the importance of the continuity/discontinuity debate within the PD domain. For example, a recent review (Endler & Kocovski, 2002) identified the structure of PDs as perhaps the most crucial unresolved issue in the study of these disorders. Given the potential of taxometric research to help resolve this issue, taxometric studies of the seven remaining PDs are urgently needed.

The outcome of this work is likely to have substantial implications for the future conceptualization, classification, and assessment of PDs. If taxometric research yields consistent support for the continuum view of PDs, with schizotypal and perhaps antisocial PDs constituting the only taxonic forms of personality pathology, this would lend support to the current push to develop and implement dimensional systems of disordered personality (e.g., Widiger & Clark, 2000). The end result of these developments might be an economical set of dimensions that avoids many of the current comorbidity problems of *DSM–IV*'s Axis II. However, if research finds further evidence of taxa underlying at least some PDs, making taxonic disorders at least a significant minority on Axis II, it may prove difficult and inappropriate to install a purely dimensional system of diagnosis. Instead, revisions to the current classification system might need to be more modest, perhaps producing a pluralistic or hybrid model that incorporates both categorical and dimensional elements.

As is readily apparent, taxometric studies of PDs are pressing and are likely to be quite influential in—and important to—the field. Although investigating the seven unstudied PDs should be a top priority, it is critical that researchers not neglect the three PDs that have already received some taxometric attention because existing studies (particularly of antisocial and borderline PDs) are few in number and of sufficiently variable quality to be interpreted with caution.

Normal Personality Characteristics

As chapter 10 revealed, taxometric studies of normal personality have examined a few highly specific personality characteristics, and few replications of these studies have been attempted. Surprisingly, there have been no published studies of the predominant contemporary models of personality structure, such as the five-factor model or Eysenck's three factors (Eysenck, 1990). Given the limited application of taxometrics to the domain of normal personality, it is difficult to nominate particular constructs as especially deserving of investigation. However, one somewhat ill-defined region within this domain has completely escaped taxometric attention and might be a valuable focus of future study.

For want of a better description, this region might be described as the area of cognitive personality constructs. Whereas many personality characteristics are conceptualized primarily as behavioral or affective dispositions and have no special relevance to people's distinctive ways of thinking, such ways of thinking are central to constructs that we organize under the *cognitive* rubric. One example of a cognitive personality characteristic is creativity. This elusive construct falls at the boundary of personality and cognition because it appears to combine elements of character and ability, involving some sort of mental fluency accompanied by associated emotional and behavioral traits. As yet there have been no taxometric studies of creativity-related constructs.

Another collection of cognitive personality characteristics involves complexes of beliefs, theories, or schemas. Social psychological research on person theories (Levy, Stroessner, & Dweck, 1998), for example, invokes a variety of domain-specific lay theories regarding the malleability versus fixedness of human attributes. These theories are conceptualized as individual differences in personality and are measured by personality-style instruments, but—unlike most traits—their primary focus is on the content of cognitive structures. The latent structure of such characteristics is interesting because, unlike most other areas of personality theory and research, these constructs are typically assumed to be categorical. For example, researchers commonly dichotomize scores on personal theory scales on the taxonic assumption that entity theorists and incremental theorists represent meaningful groups. Indeed if person theories are really theory-like, individuals might be expected to gravitate toward one theory or another (particularly when the choice is between two antithetical positions) and to maintain this polarized stance by discounting evidence that challenges their theoretical commitment. This presents an area in which there are good conceptual grounds for questioning the dimensional assumption that prevails within the personality domain—and, consequently, a strong theoretical basis for conducting taxometric research.

Similar issues arise with several other personality characteristics of interest to social psychologists. For example, authoritarianism (Adorno, Frenkel-Brunswick, Levinson, & Sanford, 1950) and social dominance orientation (Pratto, Sidanius, Stallworth, & Malle, 1994) have been extensively studied in relation to prejudice and largely refer to political beliefs and allied dispositions. If holding authoritarian attitudes and favoring social hierarchies are akin to political ideologies, we might expect them to reflect latent categories. However, if these tendencies are more akin to standard personality traits such as disagreeableness or hostile dominance, a dimensional interpretation may be more plausible. Taxometric examination seems particularly important for individual difference constructs such as these, whose structure is ambiguous and whose status is a matter of considerable theoretical interest.

NEW DOMAINS FOR TAXOMETRIC INVESTIGATION

We have identified some broad regions and specific constructs within the fields of psychopathology and personality that would benefit from taxometric study, and we are confident that the momentum of taxometric research in these fields will ensure that they are examined in due course. However, we believe that it is at least as important for the continuing viability of taxometric research that it expand into less traditional fields. Indeed there are whole disciplines outside the realm of individual differences that merit taxometric attention but that have been neglected.

Proposing where such an expansion should start is no easy task. The effort to distinguish between categorical and dimensional models of variation is an almost universal scientific endeavor. Historians argue about the existence of discontinuities between temporal periods, sociologists over the reality of discrete social classes, biologists over the nature of differences between species and subspecies, and so forth. Questions of this nature, rarely expressed as clashes between categorical and continuous models but readily framed as such, repeatedly crop up wherever the drawing of boundaries is proposed or challenged—that is, everywhere that people classify. For any number of practical and conceptual reasons, these questions are not always amenable to taxometric study, but in many cases they are. The human individual may not be the unit of analysis in many such studies—the units might instead be animal specimens, cultures, ancient pottery fragments, emotional states, or heavenly bodies—but the versatility of taxometric procedures suggests that this diversity should present no great obstacle to their application. Indeed we envision taxometric research being fruitfully conducted in fields as diverse as economics, political science, sociology, anthropology, medicine, behavioral and public health, education, and countless other disciplines faced with the task of classification.

We leave it to experts in particular domains to gauge the relevance of taxometric analysis to problems within their respective fields and to carry out the pioneering studies that introduce taxometrics to their disciplines. Here we make some more modest suggestions that are perhaps most germane to an audience of behavioral scientists. In particular, we suggest two domains—epidemiology and developmental psychopathology—that comprise new populations and constructs in great need of taxometric study, and two additional domains—human cognition and biology—that would extend taxometric research beyond the realm of individual differences.

Epidemiology

In a recent article, Kessler (2002a) forcefully advocated the use of taxometric analysis in studies of psychiatric epidemiology. Although the applicability of these relatively large-*N* methods to community epidemiological samples might seem obvious, taxometric studies have so far overwhelmingly been performed with clinical or normal convenience (e.g., college student) samples. Applying taxometric procedures to large health-related data sets—particularly representative community samples—has numerous potential benefits, some of which are well summarized by Kessler:

> If the taxometric results obtained in clinical samples could be replicated in representative community samples, parameter estimates could be generated both for the estimated prevalence of the illness (the discrete latent variable) and for the conditional probabilities of individual symptoms separately in the presence and in the absence of the illness. Empirical Bayes' methods could then be used to generate predicted probabilities of illness for individual symptom profiles from these parameter estimates. Variation in the predicted probabilities of being classified as a case depending on these different threshold symptoms could then be studied to help guide decisions regarding the symptom severity and frequency thresholds that should be included in the diagnostic revisions to optimize classification accuracy. (p. 159)

In short, taxometric studies performed in—or generalized to—the epidemiological context could establish more robust prevalence figures than those based on existing diagnostic conventions and could refine diagnostic thresholds, criteria, and case identification.

All of these benefits of taxometric research depend on taxonic findings. However, Kessler (2002a) also argued that dimensional findings may have important implications for epidemiological research. Whereas there has been a tendency among taxometric researchers to regard categorical distinctions superimposed on dimensional variables as "arbitrary" (and hence suspect), Kessler proposed that these distinctions remain important and their placement need not be capricious. For example, even if major depression is dimensional, this finding does not diminish the need to diagnose the condition for purposes of communication and treatment, nor does it imply that any diagnostic threshold is equally appropriate. When dimensional findings are obtained, Kessler proposed that researchers seek inflection points in the relationship of the latent dimension to important external criteria such as impairment, treatment response, and recurrence risk. Optimal diagnostic thresholds can then be located at the point on the underlying dimension where negative consequences begin to accelerate upward. Similar approaches have been applied to epidemiological data to define disorder thresholds for nontaxonic physical conditions such as obesity and hypertension. A dimensional finding is therefore a basis for further research, rather than the end of the story.

In short, taxometric analysis can play an important role in epidemiological research, helping to invoke and inform classificatory revision regardless of whether taxa are found. Where the construct of interest is sufficiently prevalent in the general population to offer an adequate base rate and case frequency of a taxon, taxometric analyses may be performed directly in epidemiological samples. Conversely, for very rare phenomena, taxometric analyses may be performed in subsamples and their results applied to epidemiological samples. After classifying individuals into taxa or locating individuals along the relevant dimension, follow-up analyses can explore the association between taxon membership or dimensional severity and important risk, course, and outcome factors. Given the potentially significant benefits of such efforts, we urge researchers to begin applying the taxometric method to large community samples.

Developmental Psychopathology

The use of taxometrics in epidemiological or other large-scale health-related research represents one frontier for future work. Another largely unexplored frontier concerns not the scale of taxometric studies, but their populations. In particular, taxometric researchers have tended to study adults to the neglect of infants, children, adolescents, and the elderly. Infants have been studied in investigations of reactive temperament (Woodward et al., 2000), attachment behavior (Fraley & Spieker, 2003), and neonatal brain dysfunction (Golden et al., 1988). Preadolescent children have been featured in studies of antisocial behavior (Skilling et al., 2001; Vasey et al., 2005) and vulnerability to schizophrenia (Erlenmeyer-Kimling et al., 1989). Adolescent samples have been examined in studies of schizotypy (Tyrka et al., 1995) and depression (Ambrosini et al., 2002; Hankin et al., 2005; Whisman & Pinto, 1997). There have been no published studies of normal personality structure in children and adolescents, virtually no studies of preadolescent psychopathology, and no studies of any construct in exclusively elderly samples.

In a recent review, Beauchaine (2003) made a strong case for conducting taxometric investigations in the field of developmental psychopathology. Many of the most-studied childhood disorders—including autism, attention-deficit/hyperactivity disorder (ADHD), childhood depression, and oppositional defiant disorder (ODD)—have not been studied using taxometrics. The latent structure of most of these disorders has been the focus of professional disagreement. For example, Asperger's syndrome is now recognized as a milder variant of autism, but there is no consensus as to whether the two conditions are qualitatively distinct or are located along shared continua such as severity or onset. The status of ADHD and its proposed subtypes have also come under debate, with some authors favoring categorical models (e.g., Milich, Balentine, & Lynam, 2001) and others favoring continuum models (e.g., Lahey, 2001; Levy, Hay, McStephen, Wood, & Waldman, 1997), but the taxometric studies that might resolve these disagreements have yet to be conducted. One other potentially valuable role for taxometrics is to help address etiological questions surrounding vulnerability and the longitudinal emergence of disorders that are of concern to developmental psychopathologists, using methods described later in this chapter. For these reasons, we hope that taxometric researchers will begin to devote more attention to the rapidly growing domain of child psychopathology.

Cognition

Parsing cognitive processes is a central concern within cognitive psychology and related disciplines. However, taxometric analysis has not yet been used to directly evaluate cognitive constructs, with the partial exceptions of Golden's (1982) study of dementia, Dollaghan's (2004) study of specific language impairment, and Haslam's (1994) study of relationship schemas. This neglect is unfortunate because many cognitive constructs represent individual difference variables that could be investigated in precisely the same ways as individual differences in the personality and psychopathology domains. A brief review of some of these unexamined constructs is therefore warranted.

Many cognitive theories posit that different people think in different ways or with different degrees of skill and speed. Because virtually all of the human intellectual aptitudes and abilities were identified or validated using factor analysis, these widely studied constructs are explicitly or implicitly believed to represent latent dimensions. No taxometric studies have tested these strongly held but empirically untested assumptions. Although most broad intellectual abilities (like broad personality characteristics) may be expected to be dimensional due to the combined effects of many causal influences, it may be more likely that some more specific or narrow-band abilities are taxonic. Moreover, even if most normal variation in cognitive abilities is dimensional, particular patterns of abnormal variation—such as cognitive deficits, learning disabilities, or forms of mental retardation—probably reflect detectable taxa; the only study conducted to date yielded dimensional findings (Dollaghan, 2004).

Many cognitive constructs reflecting individual differences can be found outside the ability domain. For example, taxometric researchers might fruitfully examine such constructs as learning styles and patterns of academic achievement. Researchers, theorists, and practitioners often proceed from informal taxonomies of these constructs, and taxometric research could play an important role in challenging, validating, or refining these classifications. The present neglect of such constructs reflects the lack of the taxometric method's penetration into educational psychology and education research more generally. We suggest this might be fertile ground for new taxometric investigators.

The cognitive constructs described to this point are all individual difference variables. However, as argued in chapter 10, taxometric analysis need not be confined to constructs of this sort. Indeed many proposed distinctions between phenomena can be tested using taxometrics, and a clear priority for future taxometric work is to expand to research areas in which the appropriate unit of analysis is not an individual person. Distinctions among cognitive functions, mechanisms, or structures, for example, are less distinctions among people than differences within them. If a theorist proposes two distinct kinds of memory, for instance, the taxonicity of the distinction can be examined if a suitable method for sampling the relevant variation in memory phenomena is available.

An illustration of a problem of this nature for which taxometrics may be potentially useful can be found in recent research on visual attention. An article by Wolfe (1998) presented distributions of search slopes (relationships between the number of items in a visual array and the time taken to search this array, summarized as ratios) for over a million trials of assorted visual search experiments. Observing that distributions of search slopes were unimodal, Wolfe concluded that visual search processes vary along a continuum of efficiency, rather than reflecting two discrete kinds of processes. This conclusion challenged the venerable distinction between parallel (fast) and serial (slow) search processes, which are often conventionally distinguished by a 10 ms/item search slope threshold. However, based as it was on the common misperception that discrete categories always generate bimodal manifest distributions, Wolfe's dismissal of the two-process model of visual search—and of the mythical search slope boundary between them—seems premature and potentially erroneous.

To test Wolfe's claim, Haslam, Porter, and Rothschild (2001) reanalyzed his search slope data and found consistent support for the existence of two admixed search slope distributions. Moreover, the optimal cut point between the distributions was strikingly close to the 10 ms/item convention. Although mixture analysis, rather than taxometric analysis, was used in this study (because only one indicator—search slope—was available), the study exemplifies the ways in which empirical evaluations of latent structure might address research problems in cognitive psychology even where individual differences are not at issue. Creative taxometric studies could help test structural propositions in cognitive psychology, and we urge researchers to conduct them.

Biology

Meehl's taxometric method was conceived, nurtured, and sent out into the world by psychologists. It is notable that, decades later, taxometric studies are conducted overwhelmingly by psychologists and published in psychology journals. Psychologists' concerns have largely determined the choice of constructs to be investigated in taxometric studies, and psychologists' assessment tools (e.g., ratings of behavior, attitudes, or subjective experiences) have predominated in these studies. Although the creators of the taxometric method had strong interests in biological psychiatry and behavioral genetics, few applications of the method have appeared in the journals of these disciplines.

This state of affairs is somewhat surprising. As noted earlier, taxometric analysis can be performed in most any research domain where classification matters, using assessments that are appropriate to the subject at hand. Disciplines such as medicine, neuroscience, neuropsychology, and biopsychology represent domains where many of the constructs of interest are biological in nature and where classification research relies on biological data (e.g., electrophysiological measurements, biochemical assays, volumetric properties of brain regions, genetic markers). To date, however, no taxometric study has examined a primarily biological construct or used biological indicators in analyses.

This neglect of biological constructs and indicators by taxometric researchers is regrettable for several reasons. First, it has confined taxometric methodology to one discipline and limited its visibility to others. This has not been the case for other quantitative procedures that test latent structure. Mixture analysis, for example, has been widely used in psychiatric research for some time (Gibbons et al., 1984) and is a statistical staple in biomedical research more broadly (see McLachlan & Peel, 2001, for many illustrations). Second, this neglect may have limited the quality of taxometric investigations. It is often methodologically desirable to employ multiple methods of assessment—including biological measures, which may assess aspects of a construct that are not readily captured by verbal reports or overt behaviors—to improve the likelihood that taxometric results accurately represent the underlying construct. If multimethod assessment is not used, particularly if all indicators are ratings made by a small number of observers, there is always the possibility that taxometric results will merely restate observers' judgments or biases, rather than reflect the structure of phenomenon (Beauchaine & Waters, 2003).

Therefore, we propose that taxometric research begin to address biological constructs, perhaps especially those in the hitherto neglected domains of biological psychiatry and neuropsychology. When the potentially greater costs of these measurements may be offset by worthwhile methodological benefits, researchers might strive to include pertinent biological measurements in taxometric investigations.

NEW RESEARCH QUESTIONS TO BE ADDRESSED USING TAXOMETRICS

We have reviewed an array of constructs that would benefit from taxometric investigation. However, future advances are likely to stem not only from the expansion of constructs that are studied with the method, but from the application of the taxometric method to new kinds of research questions. Until now taxometric researchers have generally been content to focus solely on the question of whether a particular construct is taxonic or dimensional in nature. However, this basic structural conclusion need not—and arguably should not—be the end of the story. Instead, it can serve as a springboard for follow-up investigations of more complex questions. It is important for the development of taxometrics that researchers begin to explore these more complex questions. Doing so would counter the potential criticism that taxometrics is a narrow methodology that resolves a single fundamental question about latent structure, but does not integrate well with other forms of research.

Full Delineation of Latent Structure

As discussed in chapter 1, the latent structure of a construct may be considerably more complex than taxonic (two latent classes) or dimensional (*n* latent factors). This suggests that the initial testing of a taxonic boundary by taxometric analysis should represent only the first step of a program of research designed to flesh out the complete latent structure of the underlying construct. The trajectory of this research program, as well as the analytic techniques used to conduct it, can be guided by a taxometric finding of taxonic or dimensional structure (J. Ruscio & Ruscio, 2004a).

If taxometric analysis reveals that a construct is dimensional, its structure can be further elaborated using tools developed for use with continuous constructs. Researchers can use exploratory or confirmatory factor analysis (e.g., Byrne, 2001; Gorsuch, 1983; Long, 1983) to determine the number of dimensions that underlie the construct, to identify the indicators that most validly assess and differentiate these dimensions, or to search for higher and lower order dimensions. They can also use latent trait analyses such as IRT (Embretson, 1996; Hambleton, Swaminathan, & Rogers, 1991; Lord, 1980) to identify indicators that are appropriate for general assessment or research of the construct (i.e., indicators possessing similar measurement reliability across the full range of trait levels) versus indicators that may be more appropriate for decision-making tasks (those concentrating measurement precision near a particular trait level of the construct).

However, if taxometric analysis reveals that the construct is taxonic, there are a number of promising avenues for additional investigation, any or all of which may be pursued using a variety of complementary analytic techniques. One possibility is to submit the same (or a very similar) set of indicators to additional taxometric analyses to test for further boundaries within the taxon or complement. Whereas taxometric procedures can detect only one boundary (between two groups) at a time, this does not mean that no groups other than this taxon and complement exist.

A second possibility is to submit a *new* set of indicators to additional taxometric analyses to test for hypothesized subtypes within the taxon or complement. At this stage, the analysis is performed in a subsample consisting only of members of one group, using indicators that are believed to distinguish individuals who belong to the subtype from individuals who do not. Because a single indicator set seems unlikely to define multiple subtypes equally well, we recommend that researchers test for one subtype at a time using carefully chosen, nonredundant indicators that together cover all the major facets of that subtype (J. Ruscio & Ruscio, 2004a). Although it often makes sense to search for a type before searching for its subtypes (a top–down approach), it may be necessary to begin by studying the subtypes (a bottom–up approach) if they are thought to represent independent types that have heretofore been misclassified as subtypes, or if they are so heterogeneous that a single indicator set cannot be readily constructed to test the structure of their overarching type.

A third line of investigation that one might pursue after a taxon is detected is to identify the most efficient and least redundant indicators of this taxon. Although factor analysis and IRT serve this purpose well when a construct is dimensional, the indicators of taxonic constructs are more appropriately evaluated by reviewing parameter values (e.g., validity estimates) yielded by the taxometric analysis or by performing latent class analysis (e.g., Green, 1951; Lazarsfeld & Henry, 1968) or latent profile analysis (e.g., Muthén, 2001).

A fourth potential avenue of investigation is to assess each taxon that is revealed (including types and subtypes) for meaningful dimensional variation. Researchers can use procedures such as factor analysis or analyses of internal consistency to examine whether individuals within either the taxon or the complement differ reliably from one another in intensity or severity along one or more dimensions. If reliable residual variation is found, researchers can employ IRT modeling, exploratory or confirmatory factor analysis, or any other strategies appropriate for working with dimensions to elucidate the variation within a latent class.

Construct Validation of Latent Structure

Whether the data suggest a relatively simple or complex structural model, it would be useful to follow up with additional research exploring its construct validity. As Cronbach and Meehl (1955) discussed in their landmark article on construct validation, it is essential to postulate and test linkages between latent constructs. Doing so simultaneously provides evidence of the adequacy of the theoretical model used to generate predictions and the measures with which the model is evaluated. In the context of structural research, construct validation should focus on predictions based on a posited structural model such that corroboration of the hypotheses would lend support to the model, whereas refutation of the hypotheses would call the model into question. For example, in chapter 1, we argued that a taxon should be relatively enduring over an appropriate time frame, so an examination of temporal stability might comprise one component of the construct validation of taxonic results. Watson (2003) performed such an analysis of the pathological dissociation taxon identified by Waller, Putnam, and Carlson (1996) and Waller and Ross (1997). Using longitudinal data, Watson classified individuals as taxon or complement members using the items and scoring algorithm suggested by Waller and Ross (1997). Class membership was highly unstable over time, and Watson argued that this challenges the construct validity of the pathological dissociation taxon.

Examining temporal stability is not necessarily the most important way to explore the construct validity of all proposed taxa, but it represents one potentially fruitful test. Especially for more complex structural models, additional or different types of evidence are required. Beyond suggesting the importance of demonstrating that one's structural model contributes to a theoretical understanding of the target construct as well as related constructs, the current state of theory and research on the construct validation of structural models does not provide detailed guidance on how to undertake this task. We raise this issue only to recommend that investigators attend to construct validity as they would in any other line of psychometric inquiry. Researchers in the social and behavioral sciences are trained in the construct validation of measurement instruments, but the validation of structural models may require some creativity to devise and test hypotheses that subject these models to informative tests.

Comorbidity

A third question that taxometrics can help to clarify involves the nature of *comorbidity*, defined as the co-occurrence of multiple diagnosed disorders in one individual. We prefer to use the more precise term *diagnostic co-occurrence* (Lilienfeld, Waldman, & Israel, 1994) to describe this phenomenon to emphasize that the frequent overlap among disorders is observed at the manifest level, where fallible signs and symptoms define diagnostic categories that may or may not accurately reflect the true nature of psychopathology. In other words, diagnostic co-occurrence may be meaningful, reflecting discrete conditions that are found within the same individual because they are etiologically related (e.g., they share a common cause or one is a risk factor for the other); or co-occurrence may be artifactual, reflecting an overlap that exists only at the manifest level and that could (at least in principle) disappear following changes in sampling, nosology, or measurement. This issue is of considerable clinical importance and has been receiving increased research attention in recent years, in part, because multiple diagnoses can fragment clinical formulations and complicate the planning and conduct of treatment. Widespread diagnostic co-occurrence also raises concerns among professionals that existing classification systems are not parsimonious or they inappropriately fragment clinical phenomena.

Several writers have discussed ways in which taxometric research could clarify and elucidate diagnostic co-occurrence (e.g., Meehl, 2001; J. Ruscio & Ruscio, 2004a; Waldman & Lilienfeld, 2001). All have argued that our understanding of the relationships among disorders and their consequent theoretical and practical implications may be most rapidly advanced by focusing on the latent level of analysis. For example, J. Ruscio and Ruscio (2004a) reviewed the 11 explanatory models of diagnostic co-occurrence described by Klein and Riso (1993) and noted that each model poses a specific, testable prediction about the relation between co-occurring disorders at the latent level. Ruscio and Ruscio further proposed that many of these explanatory models could be tested using the results of taxometric analysis. They recommended that researchers apply taxometrics to each disorder in turn to determine whether it is qualitatively or quantitatively distinct from normality (and/or from subclinical forms of the disorder). Based on the results of each taxometric analysis, researchers could assign individuals to taxa or locate them along latent dimensions for each disorder. Then, using traditional measures of association, they could examine the nature and degree of the relation between the two latent constructs.

It might be asked whether taxometrics could be used to evaluate the boundary between disorders more directly—that is, by conceptualizing one disorder as the taxon and the other as the complement. We would argue that it is not appropriate to use taxometrics in this way (J. Ruscio & Ruscio, 2004a). One reason for this is that diagnostic co-occurrence generates the most interest when it involves disorders whose clinical features partially overlap (e.g., mood and anxiety disorders). In such cases, efforts to select indicators that validly differentiate a taxon (Disorder 1) from its complement (Disorder 2) will be forced to choose between (a) using shared (and thus poorly differentiating) indicators whose low validity may undermine the taxometric analysis, and (b) using only indicators that are unique to each disorder, resulting in such narrowed content coverage that taxometric results may no longer reflect the structure of the intended construct. A second reason is that taxometric analysis presumes that each case is a member of a taxon *or* its complement, *but not both*. One could satisfy this methodological requirement by excluding all cases who are diagnosed with both disorders, but doing so would systematically remove the very co-occurrence that the analysis is attempting to explain. Perhaps the most important reason is that the taxometric method presumes that any groups that it detects are mutually exclusive. In other words, the method does not allow meaningful associations between the taxon and complement at the latent level, making it unsuitable for answering questions about the nature of such associations. For these reasons, we recommend that taxometric analysis serve as the first step of multistage investigations into the boundaries between disorders, rather than evaluating these boundaries directly.

Whenever two disorders that co-occur at the manifest level are found to reliably co-occur at the latent level, we regard this association as important regardless of whether the constituent disorders are taxonic or dimensional (J. Ruscio & Ruscio, 2004a). If two taxa, two dimensions, or a taxon and a dimension are etiologically related—such that class membership or severity status on one is causally linked to status on the other—further explication of this relationship will likely lead to greater understanding of both disorders, improved ability to predict presentation and course, and more rapid development of increasingly effective interventions. Thus, research on diagnostic co-occurrence is urgently needed, and taxometric results may play a valuable early role in illuminating this theoretically and clinically important phenomenon.

Causation

The taxometric method was developed not only to clarify latent structure, but to generate information that would advance causal understanding. Because different causal models are implied by taxonic versus dimensional structure, Meehl and his colleagues intended taxometric studies to have implications for the etiology of mental disorders and related constructs. Indeed the first published taxometric study (Golden & Meehl, 1979) examined a critical component of Meehl's (1962) etiological theory of schizophrenia and found evidence supporting the existence of a taxon predicted by the theory.

Unfortunately, taxometric research conducted since that early study has rarely addressed causal issues directly, and more often than not the causal implications of taxometric results are discussed only in passing. For example, although authors often note that their taxonic findings imply the existence of a dichotomous causal factor of some sort, they rarely attempt to specify more precisely what this factor might be, nor whether a specific etiology (Meehl, 1977) or a threshold effect is involved. However, taxometric studies have the potential to move beyond the basic characterization of latent structures to more aggressively examine the processes that give rise to these structures.

One way in which this could be done is to use the output of a taxometric analysis as input for subsequent etiological research. This possibility is nicely illustrated by Waller and Ross (1997), who, following their detection of a pathological dissociation taxon in taxometric analysis, identified taxon and complement members in a separate sample of twins and performed a behavioral genetic study. This study found that almost half of the variance in pathological dissociation was explained by shared environmental experiences, whereas virtually none of the variance was explained by genetic factors, consistent with trauma-related theories of the etiology of dissociation. This example demonstrates the potential role of taxometric studies as a valuable first step toward causal explanation.

Etiological understanding may also be facilitated by taxometric studies of vulnerability constructs, which not only provide useful causal information, but can also be used to identify high-risk samples for further research. If vulnerability to a disorder is taxonic, it may make little sense to investigate predictors of disorder onset in unselected samples that may dilute or even obscure important effects. Instead it may be more fruitful to assign individuals to the vulnerability taxon using parameter estimates generated by taxometric analyses, and then more powerfully examine triggers within this at-risk group. This research strategy is only one of several ways in which parameter estimates (e.g., base rates and indicator validities) yielded by taxometric procedures can inform causal models.

Another way in which taxometrics can be used to illuminate etiology is through longitudinal research of vulnerability constructs. To date, almost no research has tracked individuals over time to determine whether the latent structure of these constructs remains stable across assessments (see Tyrka, Cannon, et al., 1995; Watson, 2003, for rare exceptions). Research of this kind has the potential to make several important contributions. If a vulnerability taxon with stable membership is consistently obtained across several longitudinal waves, this depicts the vulnerability characteristic as an enduring predisposition for the related disorder. Such a finding could facilitate the early detection of vulnerable individuals and help to identify the changing phenotypic characteristics associated with vulnerability and pathology. It would also make it possible to rule out risk factors appearing at later assessment waves as the sole causes of vulnerability, and to identify vulnerable individuals for more focused research into the factors that trigger expression of a disorder among at-risk cases. If, however, a taxon is consistently detected, but its composition shifts over time, it may be most accurately conceived as a transitory vulnerability state. Such a result could prompt research into the nature of this state and the factors that predict its onset, which may in turn inform efforts to develop interventions for reducing vulnerability.

Another scenario that has particularly interesting implications for the causal understanding of psychopathology may arise in longitudinal studies of the disorder taxon. This is when later waves in a longitudinal study reveal a taxon, but earlier waves do not. Although it would be important to rule out insufficient indicator validity at earlier waves as the reason for this inconsistency, such a finding might indicate the point at which a psychopathology taxon emerges. Beauchaine (2003) referred to this possibility as the location of a bifurcation point at which an abnormal trajectory splits off. Locating such a point has obvious practical implications, both for discovering decisive influences on taxon membership and for the timing of prevention efforts. The scientific potential of this research strategy, however, has yet to be exploited.

Finally, creative use of taxometrics might help to test between specific etiology and threshold accounts of taxonic findings in psychopathology research. These two causal possibilities imply different relationships between vulnerability variables and clinical syndromes. Under the specific etiology model, any variable that underlies or is strongly associated with the pathological taxon should also be taxonic. Under the threshold model, the vulnerability variable that underlies or is associated with the disorder is dimensional, and the disorder only emerges as a taxon when some critical level of vulnerability or severity is reached. Thus, taxometric research could test between threshold and specific etiology models by conducting analyses of vulnerability variables and the corresponding syndrome. Dimensional findings for the former, combined with taxonic findings for the latter, would support a threshold model. Follow-up research examining the relation between the vulnerability dimension and syndrome taxon at the latent level could help to identify the location of the threshold and hone in on the factors that operate at this threshold to fuel the transition from normality to pathology.

In summary, we believe that there are many ways in which the taxometric method could be employed to address questions of causation. We hope that taxometric researchers will make use of these opportunities and contribute to this critical endeavor. More generally, we hope that researchers will more regularly follow up taxonic findings with research and theorizing about what the taxon in question might actually be.

METHODOLOGICAL ISSUES FOR FURTHER STUDY

As previously noted, taxometric practice has matured in the decades since the method was first introduced. A sizable body of taxometric research has accumulated, produced by a growing community of investigators. The range of constructs that have been studied has expanded steadily, moving beyond an exclusive focus on psychopathology. Monte Carlo evidence for the validity and boundary conditions of taxometric procedures has grown, although there remain enormous gaps in this knowledge base.

Taxometric research has not grown in breadth and quantity alone, however. A number of trends toward greater methodological sophistication and care have also emerged. As noted in chapter 10, researchers are increasingly performing multiple taxometric procedures in larger samples. Moreover, established taxometric procedures are being improved and refined. Such developments provide cause for optimism about the increasing robustness of taxometric conclusions.

Despite these positive developments, however, taxometric investigation is still a relatively new enterprise, and several significant methodological issues have yet to be resolved. In this final section, we lay out some of these issues and describe the major challenges that remain.

Sources of Erroneous Structural Inferences

One methodological issue in need of further work involves locating possible causes of pseudotaxonicity and safeguarding taxometric studies against them. At present we know a few of these causes. First, the possibility of *institutional pseudotaxonicity*, obtaining taxonic findings produced as an artifact of selecting cases based on a high threshold on one or more—but not all—indicators, is reasonably well recognized as a result of Grove's (1991a) work. Second, pseudotaxonic peaks in MAXCOV curves were identified as a potential consequence of the use of dichotomous indicators by Miller (1996) and J. Ruscio (2000), who offered suggestions for guarding against misinterpretation. Partly as a result, taxometric studies are now using dichotomous indicators less frequently than before. Third, recent studies have demonstrated that indicator skew in dimensional data sets can produce taxometric curves that not only appear similar to those traditionally associated with a small taxon, but also yield the consistent base rate estimates that would be expected from taxonic results, thereby passing a widely used consistency test (A. M. Ruscio & Ruscio, 2002; J. Ruscio, 2005; J. Ruscio, Ruscio, & Keane, 2004). This finding provides a strong rationale for generating empirical sampling distributions of taxometric results through analyses of taxonic and dimensional comparison data (see chap. 8).

More recently, a fourth potential source of pseudotaxonicity was identified. Beauchaine and Waters (2003) found that when raters were told that target individuals came from discrete categories, their ratings of these individuals yielded taxonic MAMBAC and MAXCOV curves. In a comparison condition that did not include this experimental manipulation, participants' ratings of the same target individuals did not produce taxonic curves. The implication of this research is that preexisting beliefs about the nature of the construct being rated can bias raters to generate data that conform to this expectation. It is not yet known whether dimensional expectations might lead taxa to be mistakenly rated in a way that yields nontaxonic findings. Nevertheless, this finding suggests that when rating data (e.g., interviewer ratings, behavioral observations, perhaps even self-report data) are used in taxometric research, it is important to consider whether those providing the ratings possess a structural bias that might influence the ensuing results.

One way to limit this cause of pseudotaxonicity is to employ indicators of varying types from multiple sources and to avoid using rating data drawn from a single source, particularly a small and highly homogeneous set of raters. However, rating data are so prevalent in psychological research, and so crucial to the assessment of many personality and psychopathology constructs, that dispensing with them may be unrealistic, if not unnecessary. Research is needed to examine the strength of the pseudotaxonic bias in rating data and to specify the circumstances under which it is most likely to be problematic. For example, people do appear to have organized beliefs about the ontological status of some forms of human variation, including their likely structure (Haslam & Ernst, 2002; Haslam, Rothschild, & Ernst, 2000), and it is conceivable that these beliefs could influence their ratings of self or others. However, it is also possible that laypersons' beliefs about the constructs typically examined in taxometric studies are too weak or inconsistent across participants to plausibly distort taxometric findings. Similarly, it is unknown to what extent trained observers, such as diagnostic interviewers or behavioral coders, hold sufficiently strong and consistent beliefs about the structure of the phenomena that they rate to bias their ratings, nor whether certain training procedures can reduce existing biases. Finally, it may be that pseudotaxonicity is less probable when ratings are provided by a large number of raters who do not share a single, pronounced structural bias for the construct (such as naive research participants completing self-report measures)—or by raters who have different structural biases (such as a mix of psychologist and psychiatrist interviewers)—rather than similarly trained raters with comparable professional socialization and salient shared beliefs about the construct in question. These questions await investigation.

Research is also needed to determine whether rating-related factors *other* than beliefs about the nature of a construct can lead to pseudotaxonic findings. For example, clinical interviewers who find evidence of one symptom during an assessment might more thoroughly or forcefully seek evidence of associated symptoms, or might subtly lower their threshold for declaring other symptoms to be present. Such ordinary processes of information gathering may polarize diagnostic judgments and perhaps lead to mistakenly taxonic results.

Our discussion to this point has been predicated on the common perception of pseudotaxonicity as a particularly serious problem for taxometric researchers. However, it is important to emphasize that pseudodimensionality poses no less a problem for taxometrics, despite the dearth of explicit attention it has received. We believe that it is just as critical to avoid missing actual taxa as it is to avoid inferring the existence of taxa that do not exist. Both mistakes represent inferential errors to be minimized by applying the taxometric method evenhandedly without an a priori bias for or against the existence of taxa. Although pseudotaxonic inferences are sometimes presented as somehow riskier or more problematic than pseudodimensional inferences—akin to making a Type I (vs. a Type II) error—we noted in chapter 4 why we regard this as a false analogy. Ideally, taxometric analyses should offer strong and consistent support for either taxonic or dimensional structure; if they do not, we suggest that the results be judged uncertain, rather than presumed to be nontaxonic. In short, we urge researchers to guard against the possibility of pseudodimensionality and to be keenly aware of the factors that can contribute to it—notably poor indicator validity, substantial within-group correlations, and small samples.

As Meehl repeatedly stressed (e.g., Meehl, 1999), *no statistic is self-interpreting*. This point has been reinforced by studies documenting a variety of ways in which data may give rise to mistaken structural conclusions. Indeed we suspect that still more sources of pseudotaxonicity and pseudodimensionality will be revealed by future research. Fortunately, the taxometric method is a dynamic system that is continually being scrutinized, developed, and refined. As the strengths and limitations of the method become better understood, standards for what constitutes rigorous taxometric studies change accordingly. This ongoing process of methodological refinement is by no means unique to taxometrics, but is particularly visible in this area given the method's relative youth and the growth stimulated by its recent popularity.

The result is that this is a truly exciting time to become involved in taxometric research, but that such research should not be undertaken lightly. Researchers wishing to conduct informative taxometric investigations with a high likelihood of staying power are encouraged to: (a) design their studies with an eye toward avoiding known contributors to pseudotaxonicity, pseudodimensionality, and other taxometric pitfalls identified by Monte Carlo research; (b) generate empirical sampling distributions to control for unique characteristics of the research data that may lead to misinterpretations of taxometric results; (c) perform risky tests of the obtained structural solution by conducting extensive replications and consistency tests across multiple samples, measures, data-collection methods, and taxometric procedures; and (d) avoid definitive or overconfident conclusions about latent structure, maintaining an openness to new methodological developments and new conclusions that may challenge earlier findings.

Monte Carlo Elaborations

A second methodological challenge facing taxometrics researchers is the need to expand the existing body of simulation research on the taxometric method. Although the appropriateness of data for taxometric analysis and the accuracy of structural inferences in any individual study can be assessed using empirical sampling distributions (as discussed throughout Part II of this book), it is still desirable for researchers to have a general understanding of how best to implement the taxometric method and how accurate their inferences are likely to be under particular data conditions.

The extant body of Monte Carlo research (e.g., Beauchaine & Beauchaine, 2002; Meehl & Yonce, 1994, 1996) already offers some guidance on these matters. However, this research has a number of significant limitations. First, almost all Monte Carlo research on taxometrics has been conducted with data sets composed of normally distributed variables (in the full sample for dimensional data, within groups for taxonic data), an idealized case that is probably unrepresentative of most empirical data sets (Micceri, 1989). Although two small studies have examined the influence of indicator skew on taxometric curves (Cleland & Haslam, 1996; Haslam & Cleland, 1996) and a larger study of taxon base rate estimates included indicator skew as one factor in its design (J. Ruscio, 2005), no studies have examined the effects of deviations from distributional normality on curve shapes. Likewise, little is known about the ability of taxometric procedures to correctly detect taxa with base rates more extreme than .10 or .90. It will be important for future Monte Carlo studies to be performed with data that more closely reflect the characteristics of research data typically submitted to taxometric analysis, including indicators that are not continuously distributed, very small taxa, and indicators whose within-group distributions and correlations are more realistic.

Second, most Monte Carlo studies have investigated individual taxometric procedures in isolation, often implemented in only one fashion. Although this approach provides a useful sense for the utility of a taxometric procedure performed in a particular way under the data conditions studied, it does not accurately reflect typical taxometric practice, which should (and almost always does) involve multiple procedures that are implemented in a manner that accommodates the data at hand. Because structural inferences based on agreement across several procedures should be more accurate than those based on a single procedure, it is likely that existing Monte Carlo studies underestimate the accuracy, power, and robustness of taxometric analysis. Indeed the only Monte Carlo study (Cleland, Rothschild, & Haslam, 2000) that addressed this issue found that joint consideration of MAXCOV and MAMBAC results led to more accurate detection of taxa than either alone. It would be valuable for simulation studies to evaluate taxometric analysis as it is usually practiced, complete with data-driven implementation decisions, multiple taxometric procedures, and consistency tests. Such research would provide a clearer sense of the boundary conditions of successful taxometric analysis and of the incremental benefits of particular analyses than is offered by the currently available literature.

A third limitation of existing Monte Carlo research is that it has not kept pace with the development and refinement of new taxometric procedures. A considerable amount of published simulation research has examined MAXCOV, and a bit has examined MAMBAC, but almost no systematic attention has been paid to MAXSLOPE or the newer MAXEIG and L-Mode procedures. Although these newer procedures may afford advantages over their predecessors that will ultimately strengthen Monte Carlo support for taxometrics, the extent of any overall improvement remains to be determined. Similarly, simulation studies have yet to test whether new procedures or practices substantially resolve problems with earlier approaches that they were designed to overcome.

A fourth limitation of Monte Carlo research is that it has not involved the kinds of simulations needed to inform implementation decisions in taxometric research. As emphasized throughout Part II of this book, taxometric procedures can be implemented in many different ways, and currently there is no empirical basis for deciding between these options. For example, for MAXCOV or MAXEIG analyses, there are no guidelines to help researchers determine how best to assign indicators to input and output roles, what the minimal acceptable number of cases should be for each interval or window, how axes should be scaled to best facilitate accurate curve interpretation, when curves should be smoothed or averaged, and so forth. Monte Carlo research addressing such questions would go a long way toward setting appropriate standards for taxometric investigations and reducing the likelihood of unreliable or misinterpreted taxometric results.

A final limitation of existing simulation research is that very few studies have been comparative in nature. Monte Carlo studies are needed not only to establish the strengths and limits of taxometric procedures, but also to test the performance of these procedures against the available alternatives. As discussed in chapter 3, classification researchers have a choice among several quantitative methods that were designed to evaluate latent structure, and their choice in either general or particular circumstances would ideally be made on empirical grounds. At present, many forms of cluster analysis, latent class analysis, and mixture analysis are available and widely used, and it is fair to ask when taxometric procedures are preferable to these methods. However, only two studies have tackled this sort of comparison. Cleland, Rothschild, and Haslam (2002) compared the accuracy of structural conclusions for MAXCOV and MAMBAC with those for one mixture modeling technique and one clustering procedure, whereas Beauchaine and Beauchaine (2002) compared the classification performance of MAXCOV with a clustering algorithm. Although the taxometric procedures performed well in both studies, they were disadvantaged by being examined in isolation and by the fact that the simulated taxonic data sets were created out of normal mixtures. This represents a disadvantage because taxometric procedures would be expected to perform better in combination with one another (as they are employed in empirical studies) and may be less susceptible to departures from normality than procedures such as mixture analysis (which make distributional assumptions). In short, there is a serious paucity of simulation research on the relative merits of taxometric and alternative procedures. Provided that comparative studies are conducted under realistic research conditions, results supportive of the taxometric method may help to promote its use by skeptical yet open-minded researchers who currently favor other data-analytic procedures.

We have argued that a great deal of simulation work remains to be done in establishing the merits and limits of taxometric analysis, and this work is likely to make a substantial contribution to future taxometric research. We would like to end, however, with two caveats. First, although we focused our critique on Monte Carlo studies, it is important to acknowledge that some methodological issues pertaining to taxometrics will not be appropriately addressed by simulation studies and will require investigation using other analytical approaches. One example of this is Beauchaine and Waters' (2003) experiment testing the influence of prior expectations on the results of taxometric analyses of rating data. Second, although the information provided by Monte Carlo studies is essential for building broad knowledge about taxometric procedures, this information should be applied cautiously—not rigidly—to the data analyzed in individual research studies, especially when the characteristics of the research data are quite different than those included in Monte Carlo research. The most defensible approach may be to utilize Monte Carlo results as general guiding principles when planning and designing a taxometric study, then supplement these with empirical sampling distributions when determining whether one's data are appropriate for taxometric analysis and interpreting the results of taxometric analyses.

CONCLUSIONS

As we contend throughout this book, the taxometric enterprise is very much alive and growing. A multitude of empirical frontiers are open to energetic and creative researchers, with many substantive and methodological questions awaiting their attention. There is truly no shortage of work to be done and no shortage of researchers capable of bringing a taxometric approach to their work. We hope that this book contributes to the education of the research community and brings attention to some of the possibilities that lie ahead in this exciting area. Besides publicizing taxometrics, however, we hope we have convinced you, the reader, that taxometric research is not only interesting, important, and timely, but also something you can do rigorously and well.

APPENDIX A

Simulating Taxonic and Dimensional Comparison Data

The easiest way to simulate taxonic and dimensional data sets, and the one used in two foundational Monte Carlo studies of taxometric procedures, was described by Meehl and Yonce (1994, 1996) and briefly explained in chapter 3 when our illustrative data sets were introduced. In what follows, we describe this technique in greater detail. First, consider the properties that characterize taxonic data: Each indicator separates two groups of specified sizes by a desired amount. To achieve this end, one begins by drawing a vector of N random values from a unit normal distribution (i.e., $M = 0$, $SD = 1$) to represent each indicator; most software packages generate random values from such a distribution. Then to incorporate taxonic separation, a constant d is added to a proportion P of cases that will constitute the taxon. For example, if two points were added to each indicator score for 25% of cases in the sample, this would yield a taxonic data set in which the taxon base rate is $P = .25$ and the validity of each indicator is $d = 2.00$. The first three taxonic data sets introduced in chapter 3 (TS#1 to TS#3) were generated in this way using the aforementioned values of P and d, with the added step of restandardizing each indicator after taxonic separation was achieved to equate their means and variances within and between taxonic and dimensional data sets.

Next, consider the properties that characterize dimensional data in which each indicator loads onto a single latent factor. To achieve this goal, one begins by drawing one vector of N random unit normal values that will be used to establish the shared variance of each indicator on the latent factor (call this vector $\boldsymbol{S}$). One also draws several additional vectors of N random unit normal values that will be used to establish the unique variance of each indicator (call these vectors $U_1, U_2, U_3, \ldots, U_k$, where k is the number of indicators). Each simulated indicator is then generated by combining the shared vector S with the unique vector U_i, applying appropriately chosen weights (or, in the language of factor analysis, *loadings*) to each vector to produce a desired correlation between the indicators. For example, to create a dimensional data set in which the indicators correlate with one another at $r = .25$, one would weight the shared vector S by a loading of .50 and the unique vector U_i by a loading of .87 for each indicator. In general, the loadings for the shared and unique vectors that will reproduce a specified correlation r can be determined in the following way:

$$loading_{shared} = \sqrt{r} \quad \text{(A.1)}$$

$$loading_{unique} = \sqrt{1-r} \quad \text{(A.2)}$$

Thus, one computes each indicator (from $i = 1$ to k) according to the following expression:

$$Indicator_i = loading_{shared} \times S + loading_{unique} \times U_i \quad \text{(A.3)}$$

The first three dimensional data sets introduced in chapter 3 (DS#1 to DS#3) were generated in this way, again with the added step of restandardizing each indicator after the shared and unique loadings were applied.

When data are generated using this simple approach, each indicator distinguishes the taxon and complement with validity d (in taxonic data sets) or exhibit a correlation r with the other indicators (in dimensional data sets), plus or minus normal sampling error. Meehl and Yonce (1994, 1996) have shown that one can convert back and forth between the indicator correlation r observed for dimensional data and the taxon base rate P, complement base rate $Q = 1 - P$, and indicator validity d observed for taxonic data:

$$r = \frac{PQd^2}{PQd^2 + 1} \quad \text{(A.4)}$$

$$d = \sqrt{\frac{r}{PQ(1-r)}} \quad \text{(A.5)}$$

Thus, taxonic and dimensional data sets simulated using this approach can be made highly similar to one another in their full-sample indicator correlation matrixes, although the indicator score distributions will differ to some extent because taxonic separation causes a deviation from normality that will not occur for dimensional data (distributions for taxonic data will be platykurtic, tending toward bimodality with sufficiently large indicator validities, positively skewed when $P < .50$ and negatively skewed when $P > .50$). The resulting indicators will vary along continuous scales; in dimensional data sets, the distributions will be normal, whereas in taxonic data sets they will be normal only within groups.

Most Monte Carlo research has generated data in this fashion because it allows many of the key data characteristics to be varied systematically while yielding otherwise highly similar full-sample indicator correlations and distributions. It also allows the addition of within-group correlations to simulated taxonic data by first creating the taxon and complement groups separately, following the procedure for generating dimensional data (with the desired magnitude of within-group indicator correlations introduced in each group), then merging these two subsamples and adding a constant d to each indicator for taxon members. However, as noted in chapter 4, such data sets differ from actual research data in several important ways, including noncontinuous, non-normal score distributions and unequal indicator validities and correlations. Moreover, dimensional data need not be constrained to load onto a single latent factor. Thus, it would be desirable to have a technique that can be used to simulate taxonic and dimensional comparison data sets that more faithfully reproduce a broader array of data parameters. Next we describe a technique that builds on the method used by Meehl and Yonce (1994, 1996) by applying a bootstrap method in an iterative manner.

When applying the bootstrap, the generation of appropriate univariate distributions is straightforward: From a sample of N scores on a single variable, one randomly resamples, with replacement, N values (Efron & Tibshirani, 1993). This technique uses the sample distribution as an unbiased estimate of the population distribution, from which one then draws B random samples, each of size N. Bootstrapping multivariate distributions involves a careful consideration of the appropriate model to guide the process by which the dependence among variables is reproduced.

For example, to empirically estimate a 95% confidence interval for the correlation between variables x and y, one can randomly resample (with replacement) N pairs of x–y values to generate each of B bootstrap samples of bivariate data, calculate the correlation for each of the bootstrap samples, and select the 2.5th and 97.5th percentiles within the sorted distribution of these B correlations as the lower and upper limits of the confidence interval (see Efron & Tibshirani, 1993, for details on this and other ways to construct bootstrap confidence intervals). This technique treats the observed distribution of bivariate scores as an unbiased estimate of the population distribution and draws bivariate samples accordingly; a sufficiently large value of B (e.g., ≥ 1000) is required to obtain stable estimates of the tails of the empirical sampling distribution. The empirically derived confidence interval (as opposed to one that is derived analytically) makes no assumptions about the univariate or bivariate distributions of x and y. Instead it reproduces the distributions observed in the research data by resampling from an unbiased estimate of the population distribution. If one's objective is, instead, to test the statistical significance of the correlation in the research data, one can generate bootstrap samples by randomly resampling (with replacement) x and y values independently, rather than in pairs (Lee & Rodgers, 1998). Doing so models the null hypothesis of $\rho = 0$ while reproducing the univariate distributions observed in the research data, once again resampling from unbiased estimates of population distributions. Lee and Rodgers found that, across varying sample sizes, population correlations, and population distributions, superior hypothesis test results were obtained using univariate rather than bivariate bootstrap samples. The distinction between bivariate and univariate applications of the bootstrap in this context demonstrates that the choice of a resampling strategy should be guided by one's research goals.

To apply the bootstrap to taxometrics, one might extend the bivariate resampling technique described earlier to the multivariate case, treating the multivariate distribution of indicator scores as an unbiased estimate of the population distribution and randomly resampling (with replacement) from this distribution to generate bootstrap samples of comparison data. However, such bootstrap samples are not model-based and cannot be used to generate sampling distribution of taxometric results for known taxonic or dimensional latent structures. Instead one must bootstrap samples of comparison data using models representing the latent structures that the taxometric method is designed to distinguish.

Generating a bootstrap sample of dimensional comparison data represents the most fundamental challenge because there is an apparent chicken-and-egg problem that one must solve to simultaneously reproduce indicator distributions and correlations using the common factor model. If this can be achieved, then repeated application of the algorithm would allow one to reproduce these data characteristics within two separate groups, and hence in a full sample of taxonic data. Whereas it is relatively simple to reproduce either indicator distributions or correlations, reproducing both is more complex. Any time one transforms indicator distributions (e.g., by applying cutoff scores to convert a continuum to a series of ordered categorical values), one also alters the indicator correlations. Similarly, any time one alters indicator correlations (e.g., by applying shared and unique factor loadings to reproduce observed correlations), one alters the distributions of the reproduced indicators.

Algorithms used to simulate data for Monte Carlo studies yield samples drawn from specified population parameters, but they are not adaptable by the researcher who wishes to reproduce features of a unique sample of data. For example, the Meehl–Yonce algorithm (described earlier) generates continuous, normally distributed indicators whose expected correlations are identical and cannot generate ordered categorical, non-normal indicators whose correlations with one another vary. Waller, Underhill, and Kaiser (1999) developed a sophisticated algorithm that allows researchers to draw samples from populations that differ in many potentially important ways, including provisions for non-normal distributions and varying indicator correlations. Their Monte program is intended for use in Monte Carlo studies of classification procedures, although it can also be used to generate dimensional comparison data when all indicators are continuously distributed. However, because the version of this algorithm that can be used to reproduce ordered categorical distributions modifies indicator distributions after having reproduced their correlations, the latter will be attenuated. Thus, the Monte program is not intended for the investigator who wishes to reproduce the ordered categorical distributions observed in a sample of research data (N. G. Waller, personal communication, June 9, 2004).

Our solution to the chicken-and-egg problem is to use an iterative technique that combines the common factor model with univariate bootstrapped indicator distributions (J. Ruscio, Ruscio, & Meron, 2005). This technique first reproduces the desired indicator correlation matrix by applying factor loadings to normally distributed, continuous variables and then substitutes distributions bootstrapped from the observed indicator distributions. Because the substituted bootstrap distributions will usually be non-normal and often vary across ordered categories, rather than a smooth continuum, imposing them will alter the correlations that have been reproduced. Therefore, the technique iteratively updates the target correlation matrix and calculates new factor loadings, retaining the data set that best reproduces the observed indicator correlations when the bootstrap indicator distributions are imposed. The following is a step-by-step description of our algorithm for generating a bootstrap sample of dimensional comparison data in this iterative manner. The program "Dim-Sample" on the accompanying Web site carries out this algorithm:

1. Read and store the sample size, number of indicators, and indicator correlation matrix.

2. Generate and store a bootstrap score distribution for each indicator by resampling, with replacement, N scores from each observed score distribution.

3. Store a copy of the target correlation matrix as the *desired* correlation matrix, which will be updated as the procedure iterates.

4. Perform a factor analysis of the target data and record the number of factors with eigenvalues > 1. This determines the number of latent factors that will be used to reproduce the indicator correlation matrix. The liberal minimum eigenvalue criterion is used so that potentially useful latent factors are not missed.

5. Create vectors of N random unit normal values that will be partly shared by all indicators. One such vector of shared values is needed per factor.

6. Create additional vectors of N random unit normal values that will contribute the unique, or error, component for each indicator.

7. Determine the weights (loadings) for the shared and unique components of each indicator that will reproduce the desired correlation matrix by performing a factor analysis of that matrix.

8. Calculate each simulated indicator as a new vector of N values by weighting the shared and unique components for each by the appropriate factor loadings.

9. To reproduce the indicators' estimated population distributions, replace each simulated indicator's score distribution with the bootstrap score distribution generated earlier. This key step replaces the artificially normal, continuous simulated distribution with a bootstrap sample from the observed distribution that was used as the best estimate of the population distribution.

10. Calculate the correlation matrix in the simulated data and then compute a matrix of residual correlations (the target correlation matrix minus the reproduced correlation matrix). Particularly in early iterations of the procedure, these residuals may be substantial because the indicator distributions were altered after the correlations were reproduced.

11. Check whether, after substituting the bootstrap distributions, the current iteration achieves the best correlational reproduction so far by calculating the root mean square residual (RMSR) correlation:

$$RMSR = \sqrt{\frac{\sum r_{residual}^2}{n_{residual}}}, \tag{A.6}$$

where each $r_{residual}$ represents a value below the diagonal in the residual correlation matrix and $n_{residual}$ represents the number of these values. If the RMSR correlation is the lowest so far (because on the first iteration it automatically is), store a copy of the desired correlation matrix as the *best* correlation matrix, store the current RMSR correlation, and begin (or reset) a counter c at 0; the counter is used to allow a finite number of additional iterations to attempt to achieve a better correlational reproduction.

Empirical results reveal rapidly diminishing returns beyond five additional iterations, so that is the default; users can specify an alternative value. Finally, update the desired correlation matrix by adding to it the errors contained in the residual correlation matrix to attempt to achieve a better reproduction on the next iteration and return to Step 7.

12. If the current iteration has not achieved the best correlational reproduction thus far, increment the counter (i.e., $c = c + 1$). If this marks the fifth iteration (or an alternative value specified by the user), proceed to Step 13; otherwise, create a new desired correlation matrix by adding a fraction of the residual correlation matrix to the best correlation matrix and return to Step 7. The fraction of the residual matrix to be added is calculated as $1/(2^c)$, where c is the counter. Thus, our algorithm allows for five iterations to attempt to improve the reproduction of correlations. On the first try, the residual correlations are added in their entirety to the best correlation matrix (i.e., the step size multiplier = 1.00). If this first try fails to improve the reproduction of correlations, a second try is made using a smaller fraction of the residual correlations (i.e., the step size multiplier = .50). This repeats, if necessary, for a third, fourth, and fifth try (step size multipliers = .25, .125, and .0625, respectively). If even the smallest step from the best correlation matrix fails to yield improvement, the iterative routine terminates. If any of these trials improves correlational reproduction, the counter is reset to allow up to five trials starting from a new correlation matrix.

13. Construct indicators using the factor loadings that generated correlations which, when altered by substituting the bootstrap distributions, best reproduced the target correlation matrix. These loadings are derived from a factor analysis of the best correlation matrix.

14. Report the RMSR correlation and return the bootstrap sample of dimensional comparison data.

Although the early stages of this procedure work with multivariate normal data, the procedure nonetheless yields a final sample in which the indicators are distributed as in the research data due to the substitution, in Step 9, of the bootstrap score distributions generated in Step 2. Because the simulated data are generated at random, sampling error places a limit on how well indicator distributions and correlations are reproduced. Use of the common factor model introduces additional error to the extent that it does not fit the data, as will approximations inherent in the iterative step size function of the simulation program. Indicator variability places a further limit on correlational reproduction. Whereas correlations among indicators that vary along continuous scales can be reproduced most precisely, the comparatively crude distinctions along ordered categorical scales provide less opportunity to reproduce correlations with high precision. For example, all else being equal, correlations among a set of dichotomous indicators are more difficult to reproduce than are correlations among an equal number of indicators constructed as composites of dichotomous items. Analyses presented in J. Ruscio, Ruscio, and Meron (2005) suggested that even when indicators vary across a small number of ordered categories, indicator correlations and distributions can be reproduced with good precision and negligible bias.

As outlined earlier, this technique creates dimensional comparison data on the basis of the common factor model (i.e., one latent class, with variation due to shared loadings on one or more latent factors). The technique is easily extended to create taxonic comparison data on the basis of a two-class taxonic structural model. To do so, one must supply a criterion variable that denotes class membership. The procedure can then be executed twice to simulate data representing each class by following the steps listed earlier. When data for each class have been simulated, these subsamples are merged to form the full sample. The final taxonic comparison data set will match the sizes of the full sample and each class, and the indicator distributions and correlations within each class will be reproduced with good precision and negligible bias. Consequently, full-sample indicator distributions and correlations, as well as between-group indicator validities, should also be reproduced well (evidence presented in J. Ruscio, Ruscio, & Meron, 2005, supports this). The program "TaxSample" on the accompanying Web site generates taxonic comparison data by following this procedure.

There are many ways that one can split a sample of research data into taxon and complement classes. One might assign cases to groups based on *DSM* diagnoses or another established classification scheme, or by applying a cutting score on a widely used and well-validated measure. Alternatively, one can estimate the taxon base rate and assign a proportion *P* of cases with the highest total score on all available indicators to the taxon. The utility of taxonic comparison data depends in part on the justification that can be provided for the process used to assign cases to the taxon and complement classes. If the objective of the taxometric study is to evaluate the validity of a particular classification scheme, this should be used to assign cases. If, however, no particular classification system is being evaluated, cases could be assigned to groups using multiple strategies. For example, using the base rate classification method described before, one might supply a range of plausible taxon base rates to the simulation procedure, generating separate taxonic comparison data sets for each base rate estimate. If subsequent analyses reveal that the empirical sampling distribution for the dimensional comparison data could be distinguished from the sampling distributions for each type of taxonic comparison data (i.e., those generated using the full range of plausible taxon base rates or different classification schemes), this offers strong evidence of the adequacy of the data for these analyses. If, however, the empirical sampling distributions can only be distinguished when taxonic comparison data are simulated using select base rate estimates or classification schemes, this limits the structural hypotheses that can be tested—and, ultimately, the confidence that can be held in the structural solution. For instance, if taxonic and dimensional structures can only be distinguished with taxon base rates of at least .25 in the comparison data sets, and taxometric analyses of the research data suggest a dimensional solution, one could not rule out the possibility that the analyses failed to detect an existing taxon whose base rate is less than .25. Likewise, if empirical sampling distributions for taxonic and dimensional structure can be distinguished when cases are classified using *DSM* diagnoses, but not when an alternative classification system is used, one could not rule out the existence of a taxon defined by the alternative system.

Regardless of the method used to assign cases to groups, we recommend bootstrapping and analyzing multiple sets of taxonic and dimensional comparison data to determine whether their results can consistently be distinguished. Conventions in the bootstrapping literature suggest that at least $B = 1{,}000$ bootstrap samples be used when constructing empirical confidence intervals, whereas $B = 25$ to 200 bootstrap samples are used to estimate standard errors (Efron & Tibshirani, 1993; Mooney & Duval, 1993). The approach that we describe here corresponds much more closely to the estimation of a statistic's standard error, yet even fewer than 25 bootstrap samples may suffice when generating empirical sampling distributions of taxometric results because of the large sample sizes ordinarily employed in this type of research. With large samples, each sampling distribution is likely to be fairly homogeneous, with relatively little difference observed between the results of each successive set of comparison data. With smaller samples, each sampling distribution is likely to be more heterogeneous, with greater differences observed between the results of multiple sets of comparison data. Given the rule of thumb that $N > 300$ in taxometric investigations, homogeneous sampling distributions should be the norm; this expectation has been borne out in our experience with this approach. In fact we have found that setting $B = 10$ often provides highly informative empirical sampling distributions. The iterative nature of our preferred simulation algorithm and the moderately computationally intensive nature of most taxometric procedures place limits on the value of B that one can feasibly use. Continued increases in computing power should render this practical constraint moot in the near future.

APPENDIX B

Estimating Latent Parameters and Classifying Cases Using MAXCOV

Given that the product of the indicator validities is treated as a constant (K) across intervals, and provided that within-group covariance can be presumed to be zero, Eq. 6.3 shows that covariance is determined by the taxon and complement base rates. Substituting $1 - p_i$ for q_i, writing out the multiplicative expression, and rearranging the terms yields the following quadratic equation:

$$0 = Kp_i^2 - Kp_i + \text{cov}_i(xy) \tag{B.1}$$

Because $\text{cov}_i(xy)$ is the data point plotted for an interval on the MAXCOV curve and K is constant across intervals, one can solve for each interval's p_i using the quadratic formula, where a = K, b = $-K$, and c = $\text{cov}_i(xy)$:

$$p_i = \frac{K \pm \sqrt{K^2 - 4K\,\text{cov}_i(xy)}}{2K} \tag{B.2}$$

The smaller of the two roots is used for intervals below the hitmax cut, whereas the larger of the two roots is used for intervals above the hitmax cut. In the event that a negative covariance value is obtained, p_i is set to 0 (if below the hitmax cut) or 1 (if above the hitmax cut). Performing this calculation for each interval yields an estimate of the proportion of taxon members contained within that interval. These values can then be combined with the sample size of each interval (n_i) to estimate the proportion of taxon members in the full sample or the overall taxon base rate:

$$\hat{P} = \frac{\sum n_i p_i}{N} \tag{B.3}$$

Estimates of p_i can also be used to estimate the valid and false positive rates achieved by the hitmax cut on the input indicator (Meehl & Yonce, 1996):

$$VP = \frac{.5n_b + \sum_{i=b}^{\infty} p_i n_i}{PN}, \text{ and} \tag{B.4}$$

$$FP = \frac{.5n_b + \sum_{i=b}^{\infty} q_i n_i}{QN}, \tag{B.5}$$

where b is the hitmax interval.

Given an estimate of the overall taxon base rate P and complement base rate Q, as well as the valid positive (VP) and false positive (FP) rates estimated for each indicator, one can use Bayes' Theorem to calculate the probability that a given individual is a member of the taxon (Meehl & Yonce, 1996):

$$prob(taxon) = \frac{P\prod_{j=1}^{k} VP_j^{\theta}\,(1 - VP_j)^{1-\theta}}{P\prod_{j=1}^{k} VP_j^{\theta}\,(1 - VP_j)^{1-\theta} + Q\prod_{j=1}^{k} FP_j^{\theta}\,(1 - FP_j)^{1-\theta}}, \tag{B.6}$$

where Π is the cumulative product operator (i.e., for four values of x_i = 6, 3, 4, 2: $\prod_{i=1}^{4} x_i = 6 \times 3 \times 4 \times 2 = 144$), k is the number of indicators, VP_j and FP_j are the valid and false positive rates achieved by the jth indicator, and $\theta = 1$ for a response above the hitmax cut, 0 otherwise.

Because $prob(complement) = 1 - prob(taxon)$ for each individual, it is a simple matter to assign each individual to the more probable class. Once cases are assigned to groups, estimates of indicator distributions and correlations can be obtained within each group, and estimates of the validity with which each indicator distinguishes groups can be calculated. When estimating indicator validity, the raw mean differences (e.g., $D = M_t - M_c$) can be expressed in the more familiar metric of Cohen's d, the mean difference standardized by the pooled within-group variances, using Eq. 4.1.

Estimating the Taxon Base Rate Using MAXEIG

Because MAXEIG involves the same assumption of negligible nuisance covariance as MAXCOV, one can use Eq. 6.2 to represent each element in the modified indicator covariance matrix as a function of the taxon and complement base rates and the indicator validities. For example, the modified covariance matrix C for a MAXEIG analysis with k indicators (x_1, x_2, x_3, . . . x_k), $k - 1$ of which serve as output indicators (x_1 *to* x_{k-1}) for a particular MAXEIG curve, is as follows:

$$C = \begin{bmatrix} 0 & PQD_1D_2 & PQD_1D_3 & \cdots & PQD_1D_{k-1} \\ PQD_2D_1 & 0 & PQD_2D_3 & \cdots & PQD_2D_{k-1} \\ PQD_3D_1 & PQD_3D_2 & 0 & \cdots & PQD_3D_{k-1} \\ \vdots & \vdots & \vdots & \ddots & \vdots \\ PQD_{k-1}D_1 & PQD_{k-1}D_2 & PQD_{k-1}D_3 & \cdots & 0 \end{bmatrix} \qquad (C.1)$$

The matrix in Eq. C.1 can be simplified by factoring out PQ and—just as we used $K = D_xD_y$ to represent the constant product of two indicators' validities that do not vary across intervals in MAXCOV—by representing each constant product of two indicators' validities in MAXEIG as $K_{x,y} = D_xD_y$:

$$C = PQ\begin{bmatrix} 0 & K_{1,2} & K_{1,3} & \cdots & K_{1,k-1} \\ K_{2,1} & 0 & K_{2,3} & \cdots & K_{2,k-1} \\ K_{3,1} & K_{3,2} & 0 & \cdots & K_{3,k-1} \\ \vdots & \vdots & \vdots & \ddots & \vdots \\ K_{k-1,1} & K_{k-1,2} & K_{k-1,3} & \cdots & 0 \end{bmatrix} \qquad (C.2)$$

Because all entries in this matrix remain constant across MAXEIG windows (just as K remains constant across MAXCOV intervals), the product of the taxon and complement base rates (PQ) will be the only cause of systematic variation in $\boldsymbol{C}$ across windows. Hence, for MAXEIG analyses, we can represent the matrix in Eq. C.2 as $\boldsymbol{K}$, making this equation directly parallel to Eq. 6.3 within each subsample w:

$$\boldsymbol{C}_w = p_w q_w \boldsymbol{K}, \tag{C.3}$$

where p_w and q_w again represent the taxon and complement base rates within a particular subsample (here, a window rather than an interval). MAXEIG analysis is performed by calculating and plotting the first (largest) eigenvalue of the modified covariance matrix $\boldsymbol{C}_w$ within each window, so we use the term ***eigen*** to denote this first eigenvalue only. Because $\text{eigen}_w(p_w q_w \boldsymbol{K}) = p_w q_w \times \text{eigen}(\boldsymbol{K})$, one can relate the observed eigenvalue for each window to a constant of $K_{MAXEIG} = \text{eigen}(\boldsymbol{K})$:

$$\text{eigen}_w(\boldsymbol{C}) = p_w q_w K_{MAXEIG} \tag{C.4}$$

For the same reason that $K = 4\text{cov}_{\max}(xy)$ in a MAXCOV analysis (the covariance peaks in the interval of maximal group mixture, where $p_i = q_i = .5$), $K_{MAXEIG} = 4\text{eigen}_{\max}(\boldsymbol{C})$ in a MAXEIG analysis. Thus, on the basis of Eq. C.4, one can proceed to estimate the proportion of taxon members within each MAXEIG window just as is done for each MAXCOV interval. The relevant quadratic equation is:

$$0 = (K_{MAXEIG})(p_w^2) - (K_{MAXEIG})(p_w) + \text{eigen}_w(\boldsymbol{C}) \tag{C.5}$$

The quadratic formula to be solved is therefore:

$$p_w = \frac{K_{MAXEIG} \pm \sqrt{K_{MAXEIG}^2 - 4K_{MAXEIG}\, eigen(\boldsymbol{C}_w)}}{2K_{MAXEIG}} \tag{C.6}$$

As before, the smaller root is used for windows below the hitmax cut, whereas the larger root is used for windows above the hitmax cut. Although negative covariances are possible in a MAXCOV analysis, the largest eigenvalue calculated within each window of a MAXEIG analysis will always be positive, so negative values need not be accommodated. Because MAXEIG windows contain equal numbers of cases (n_w remains constant across all windows), there is also no need to weight the estimated proportion of taxon members within each window by its corresponding sample size. Instead, the overall taxon base rate P can be estimated simply by taking the mean p_w across W windows:

$$\hat{P} = \frac{\sum p_w}{W} \tag{C.7}$$

In summary, Eqs. C.1–C.7 show how one can estimate the taxon base rate by estimating and aggregating the proportion of taxon members within each subsample in a MAXEIG analysis; this represents an adaptation of the same basic process that was presented for MAXCOV in Appendix B.

References

Abramson, L. Y., Metalsky, G. I., & Alloy, L. B. (1989). Hopelessness depression: A theory-based subtype of depression. *Psychological Review, 96*, 358–372.

Adorno, T. W., Frenkel-Brunswick, E., Levinson, D., & Sanford, N. (1950). *The authoritarian personality*. New York: Harper.

Akiskal, H. S. (2003). Validating "hard" and "soft" phenotypes within the bipolar spectrum: Continuity or discontinuity? *Journal of Affective Disorders, 73*, 1–5.

Ambrosini, P., Bennett, D., Cleland, C., & Haslam, N. (2002). Taxonicity of adolescent melancholia. *Journal of Psychiatric Research, 36*, 247–256.

American Psychiatric Association. (1994). *Diagnostic and statistical manual of mental disorders* (4th ed.). Washington, DC: Author.

Arabie, P., Hubert, L. J., & DeSoete, G. (1996). *Clustering and classification*. River Edge, NJ: World Scientific Publishing.

Arnau, R. C., Green, B. A., Rosen, D. H., Gleaves, D. H., & Melancon, J. G. (2003). Are Jungian preferences really categorical? An empirical investigation using taxometric analysis. *Personality and Individual Differences, 34*, 233–251.

Arnau, R. C., Thompson, R. L., & Cook, C. (2001). Do different response formats change the latent structure of responses? An empirical investigation using taxometric analysis. *Educational and Psychological Measurement, 61*, 23–44.

Asendorpf, J. B. (2002). Editorial: The puzzle of personality types. *European Journal of Personality, 16*, S1–S5.

Asendorpf, J. B., & van Arken, M. A. G. (1999). Resilient, overcontrolled, and undercontrolled personality prototypes in childhood: Replicability, predictive power, and the trait-type issue. *Journal of Personality and Social Psychology, 77*, 815–832.

Bartholomew, D. J. (1987). *Latent variable models and factor analysis*. New York: Oxford University Press.

Bartlett, M. S. (1937). The statistical conception of mental factors. *British Journal of Psychology, 28*, 97–104.

Beach, S. R. H., & Amir, N. (2003). Is depression taxonic, dimensional, or both? *Journal of Abnormal Psychology, 112*, 228–236.

Beach, S. R. H., Fincham, F. D., Amir, N., & Leonard, K. E. (2005). The taxometrics of marriage: Is marital discord categorical? *Journal of Family Psychology, 19*, 276–285.

Beauchaine, T. P. (2003). Taxometrics and developmental psychopathology. *Development and Psychopathology, 15*, 501–527.

Beauchaine, T. P., & Beauchaine, R. J. (2002). A comparison of maximum covariance and *k*-means cluster analysis in classifying cases into known taxon groups. *Psychological Methods, 7*, 245–261.

Beauchaine, T. P., & Beauchaine, R. J. (2003). *A comparison of measures of fit for inferring taxonic structure for maximum covariance analysis*. Unpublished manuscript.

Beauchaine, T. P., & Waters, E. (2003). Pseudotaxonicity in MAMBAC and MAXCOV analyses of rating scale data: Turning continua into classes by manipulating observers' expectations. *Psychological Methods, 8*, 3–15.

Beck, A. T. (1983). Cognitive therapy of depression: New perspectives. In P. J. Clayton & J. E. Barrett (Eds.), *Treatment of depression: Old controversies and new approaches* (pp. 265–290). New York: Raven.

Beck, A. T., Ward, C. H., Mendelson, M., Mock, J., & Erlbaugh, J. (1961). An inventory for measuring depression. *Archives of General Psychiatry, 4*, 561–571.

Bernstein, A., Zvolensky, M. J., Weems, C., Stickle, T., & Leen-Feldner, E. W. (2005). Taxonicity of anxiety sensitivity: An empirical test among youth. *Behaviour Research and Therapy, 43*, 1131–1155.

Blanchard, J. J., Gangestad, S. W., Brown, S. A., & Horan, W. P. (2000). Hedonic capacity and schizotypy revisited: A taxometric analysis of social anhedonia. *Journal of Abnormal Psychology, 109*, 87–95.

Blanchard, J. J., Horan, W. P., & Collins, L. M. (2005). Examining the latent structure of negative symptoms: Is there a distinct subtype of negative symptom schizophrenia? *Schizophrenia Research, 77*, 151–165.

Blatt, S. J., & Homann, E. (1992). Parent–child interaction in the etiology of dependent and self-critical depression. *Clinical Psychology Review, 12*, 47–91.

Bock, H. (1996). Probability models and hypothesis testing in partitioning cluster analysis. In P. Arabie, L. J. Hubert, & G. DeSoete (Eds.), *Clustering and classification* (pp. 377–453). River Edge, NJ: World Scientific Publishing.

Bux, D. A., Jr. (1999). *The critical evaluation of a dichotomous approach to classifying alcoholics*. Unpublished doctoral dissertation, Rutgers University.

Byrne, B. (2001). *Structural equation modeling with AMOS: Basic concepts, applications, and programming*. Mahwah, NJ: Lawrence Erlbaum Associates.

Carson, R. C. (1997). Costly compromises: A critique of the Diagnostic and Statistical Manual of Mental Disorders. In S. Fisher & R. P. Greenberg (Eds.), *From placebo to panacea: Putting psychiatric drugs to the test* (pp. 98–112). New York: Wiley.

Cleland, C., & Haslam, N. (1996). Robustness of taxometric analysis with skewed indicators: I. A Monte Carlo study of the MAMBAC procedure. *Psychological Reports, 79*, 243–248.

Cleland, C., Rothschild, L., & Haslam, N. (2000). Detecting latent taxa: Monte Carlo comparison of taxometric, mixture and clustering methods. *Psychological Reports, 87*, 37–47.

Cleveland, W. S. (1979). Robust locally-weighted regression and smoothing scatterplots. *Journal of the American Statistical Association, 74*, 829–836.

Cohen, J. (1962). The statistical power of abnormal-social psychological research: A review. *Journal of Abnormal and Social Psychology, 65*, 145–153.

Cohen, J. (1983). The cost of dichotomization. *Applied Psychological Measurement, 7*, 249–253.

Costa, P. T., & Widiger, T. A. (Eds.). (2002). *Personality disorders and the five-factor model of personality* (2nd ed.). Washington, DC: American Psychological Association.

Coyne, J. C. (1994). Self-reported distress: Analogue or ersatz depression? *Psychological Bulletin, 116*, 29–45.

Cronbach, L. J., & Meehl, P. E. (1955). Construct validity in psychological tests. *Psychological Bulletin, 52*, 281–302.
Crow, T. J. (1985). The two-syndrome concept: Origins and current status. *Schizophrenia Bulletin, 11*, 471–485.
DeBoeck, P., Wilson, M., & Acton, G. S. (2005). A conceptual and psychometric framework for distinguishing categories and dimensions. *Psychological Review, 112*, 129–158.
Dollaghan, C. A. (2004). Taxometric analyses of specific language impairment in 3- and 4-year-old children. *Journal of Speech, Language, and Hearing Research, 47*, 464–475.
Efron, B., & Tibshirani, R. J. (1993). *An introduction to the bootstrap*. San Francisco: Chapman & Hall.
Embretson, S. E. (1996). The new rules of measurement. *Psychological Assessment, 8*, 341–349.
Endler, N. S., & Kocovski, N. L. (2002). Personality disorders at the crossroads. *Journal of Personality Disorders, 16*, 487–502.
Erlenmeyer-Kimling, L., Golden, R. R., & Cornblatt, B. A. (1989). A taxonometric analysis of cognitive and neuromotor variables in children at risk for schizophrenia. *Journal of Abnormal Psychology, 98*, 203–208.
Everitt, B. S. (1993). *Cluster analysis* (3rd ed.). London: Edward Arnold.
Everitt, B. S., & Hand, D. J. (1981). *Finite mixture distributions*. New York: Chapman & Hall.
Eysenck, H. J. (1990). Biological dimensions of personality. In L. A. Pervin (Ed.), *Handbook of personality: Theory and research* (pp. 244–276). New York: Guilford.
Fiske, A. P. (1992). The four elementary forms of sociality: Framework for a unified theory of social relations. *Psychological Review, 99*, 689–723.
Flett, G. L., Vredenburg, K., & Krames, L. (1997). The continuity of depression in clinical and nonclinical samples. *Psychological Bulletin, 121*, 395–416.
Fraley, R. C., & Spieker, S. J. (2003). Are infant attachment patterns continuously or categorically distributed? A taxometric analysis of strange situation behavior. *Developmental Psychology, 39*, 387–404.
Fraley, R. C., & Waller, N. G. (1998). Adult attachment patterns: A test of the typological model. In J. A. Simpson & W. S. Rholes (Eds.), *Attachment theory and close relationships* (pp. 77–114). New York: Guilford.
Franklin, C. L., Strong, D. R., & Greene, R. L. (2002). A taxometric analysis of the MMPI-2 depression scales. *Journal of Personality Assessment, 79*, 110–121.
Gangestad, S. W., Bailey, J. M., & Martin, N. G. (2000). Taxometric analyses of sexual orientation and gender identity. *Journal of Personality and Social Psychology, 78*, 1109–1121.
Gangestad, S. W., & Snyder, M. (1985). "To carve nature at its joints": On the existence of discrete classes in personality. *Psychological Review, 92*, 317–349.
Gelman, S. A. (2003). *The essential child: Origins of essentialism in everyday thought*. Oxford: Oxford University Press.
Gelman, S. A., & Heyman, G. D. (1999). Carrot-eaters and creature-believers: The effects of lexicalization on children's inferences about social categories. *Psychological Science, 10*, 489–493.
Gibb, B. E., Alloy, L. B., Abramson, L. Y., Beevers, C. G., & Miller, I. W. (in press). Cognitive vulnerability to depression: A taxometric analysis. *Journal of Abnormal Psychology*.
Gibbons, R. D., Dorus, E., Ostrow, D. G., Pandey, G. N., Davis, J. M., & Levy, D. L. (1984). Mixture distributions in psychiatric research. *Biological Psychiatry, 19*, 935–961.
Gigerenzer, G., Todd, P. M., & the ABC Research Group. (Eds.). (1999). *Simple heuristics that make us smart*. New York: Oxford University Press.
Gilbert, P. (1992). *Depression: The evolution of powerlessness*. New York: Guilford.
Gilbert, P. (2000). Varieties of submissive behavior as forms of social defense: Their evolution and role in depression. In L. Sloman & P. Gilbert (Eds.), *Subordination and defeat* (pp. 3–45). Mahwah, NJ: Lawrence Erlbaum Associates.

Giles, J. W. (2003). Children's essentialist beliefs about aggression. *Developmental Review, 23*, 413–443.

Gleaves, D. H., Lowe, M. R., Green, B. A., Cororve, M. B., & Williams, T. L. (2000). Do anorexia and bulimia nervosa occur on a continuum? A taxometric analysis. *Behavior Therapy, 31*, 195–219.

Gleaves, D. H., Lowe, M. R., Snow, A. C., Green, B. A., & Murphy-Eberenz, K. P. (2000). Continuity and discontinuity models of bulimia nervosa: A taxometric investigation. *Journal of Abnormal Psychology, 109*, 56–68.

Golden, R. R. (1982). A taxometric model for the detection of a conjectured latent taxon. *Multivariate Behavioral Research, 17*, 389–416.

Golden, R. R. (1991). Bootstrapping taxometrics: On the development of a method for detection of a single major gene. In D. Cicchetti & W. M. Grove (Eds.), *Thinking clearly about psychology* (Vol. 2, pp. 259–294). Minneapolis, MN: University of Minnesota Press.

Golden, R. R., Campbell, M., & Perry, R. (1987). A taxometric method for diagnosis of tardive dyskinesia. *Journal of Psychiatric Research, 21*, 233–241.

Golden, R. R., & Meehl, P. E. (1973). Detecting latent clinical taxa: IV. An empirical study of the maximum covariance method and the normal minimum chi-square method using three MMPI keys to identify the sexes. *Reports from the research laboratories of the Department of Psychiatry, University of Minnesota* (Report No. PR-73-2).

Golden, R. R., & Meehl, P. E. (1979). Detection of the schizoid taxon with MMPI indicators. *Journal of Abnormal Psychology, 88*, 217–233.

Golden, R. R., Tyan, S. H., & Meehl, P. E. (1974). Detecting latent clinical taxa: VI. Analytical development and empirical trials of the consistency hurdles theory. *Reports from the research laboratories of the Department of Psychiatry, University of Minnesota* (Report No. PR-74-4).

Golden, R. R., Vaughan, H. G., Jr., Kurtzberg, D., & McCarton, C. M. (1988). Detection of neonatal brain dysfunction with neurobehavioral indicators: The statistical problem. In P. Vietze & H. G. Vaughan, Jr. (Eds.), *Early detection of infants at risk for mental retardation* (pp. 71–95). Orlando, FL: Grune & Stratton.

Gorsuch, R. L. (1983). *Factor analysis* (2nd ed.). Hillsdale, NJ: Lawrence Erlbaum Associates.

Grayson, D. A. (1987). Can categorical and dimensional views of psychiatric illness be distinguished? *British Journal of Psychiatry, 151*, 355–361.

Green, B. A., Gleaves, D. H., Dell, P. F., & Riley, K. C. (2000, August). *Taxometric analysis of the DES and the QED.* Poster presented at the annual meeting of the American Psychological Association, Washington, DC.

Green, B. F. (1951). A general solution for the latent class model of latent structure analysis. *Psychometrika, 16*, 151–166.

Grove, W. M. (1991a). The validity of taxometric inferences based on cluster analysis stopping rules. In W. M. Grove & D. Cicchetti (Eds.), *Thinking clearly about psychology* (Vol. 2, pp. 313–329). Minneapolis, MN: University of Minnesota Press.

Grove, W. M. (1991b). When is a diagnosis worth making? A statistical comparison of two prediction strategies. *Psychological Reports, 68*, 3–17.

Grove, W. M. (2004). The MAXSLOPE taxometric procedure: Mathematical derivation, parameter estimation, consistency tests. *Psychological Reports, 95*, 517–550.

Grove, W. M., Andreasen, N. C., Young, M., Endicott, J., Keller, M. B., Hirschfeld, R. M. A., & Reich, T. (1987). Isolation and characterization of a nuclear depressive syndrome. *Psychological Medicine, 17*, 471–484.

Grove, W. M., & Meehl, P. E. (1993). Simple regression-based procedures for taxometric investigations. *Psychological Reports, 73*, 707–737.

Hambleton, R. K., Swaminathan, H. S., & Rogers, H. J. (1991). *Fundamentals of item response theory*. Newbury Park, CA: Sage.

Hankin, B. L., Fraley, R. C., Lahey, B. B., & Waldman, I. D. (2005). Is depression best viewed as a continuum or discrete category? A taxometric analysis of childhood and adolescent depression in a population-based sample. *Journal of Abnormal Psychology, 114*, 96–110.

Harding, J. P. (1949). The use of probability paper for the graphical analysis of polymodal frequency distributions. *Journal of the Marine Biological Association, 28*, 141–153.

Harris, G. T., Rice, M. E., & Quinsey, V. L. (1994). Psychopathy as a taxon: Evidence that psychopaths are a discrete class. *Journal of Consulting and Clinical Psychology, 62*, 387–397.

Haslam, N. (1994). Categories of social relationship. *Cognition, 53*, 59–90.

Haslam, N. (1997). Evidence that male sexual orientation is a matter of degree. *Journal of Personality and Social Psychology, 73*, 862–870.

Haslam, N. (1999). Taxometric and related methods in relationships research. *Personal Relationships, 6*, 519–534.

Haslam, N. (2002). Kinds of kinds: A taxonomy of psychiatric categories. *Philosophy, Psychiatry, & Psychology, 9*, 203–217.

Haslam, N. (2003). The dimensional view of personality disorders: A review of the taxometric evidence. *Clinical Psychology Review, 23*, 75–93.

Haslam, N., & Beck, A. T. (1994). Subtyping major depression: A taxometric analysis. *Journal of Abnormal Psychology, 103*, 686–692.

Haslam, N., & Bornstein, B. (1996). Envy and jealousy as discrete emotions: A taxometric analysis. *Motivation and Emotion, 20*, 255–272.

Haslam, N., & Cleland, C. (1996). Robustness of taxometric analysis with skewed indicators: II. A Monte Carlo study of the MAXCOV procedure. *Psychological Reports, 79*, 1035–1039.

Haslam, N., & Cleland, C. (2002). Taxometric analysis of fuzzy categories: A Monte Carlo study. *Psychological Reports, 90*, 401–404.

Haslam, N., & Ernst, D. (2002). Essentialist beliefs about mental disorders. *Journal of Social and Clinical Psychology, 21*, 628–644.

Haslam, N., & Kim, H. C. (2002). Categories and continua: A review of taxometric research. *Genetic, Social and General Psychology Monographs, 128*, 271–320.

Haslam, N., Porter, M., & Rothschild, L. (2001). The parallel/serial distinction in visual search: A matter of degree or a matter of kind? *Psychonomic Bulletin and Review, 8*, 742–746.

Haslam, N., Rothschild, L., & Ernst, D. (2000). Essentialist beliefs about social categories. *British Journal of Social Psychology, 39*, 113–127.

Horan, W. P., Blanchard, J. J., Gangestad, S. W., & Kwapil, T. R. (2004). The psychometric detection of schizotypy: Do putative schizotypy indicators identify the same latent class? *Journal of Abnormal Psychology, 113*, 339–357.

Jablensky, A. (1999). The nature of psychiatric classification: Issues beyond ICD-10 and *DSM–IV. Australian and New Zealand Journal of Psychiatry, 33*, 137–144.

Joreskog, K. G., & Sorbom, D. (2001). *LISREL 8.51 user's manual*. Lincolnwood, IL: Scientific Software.

Kagan, J. (1994). *Galen's prophecy*. New York: Basic.

Keller, F. (2000). Methoden zur unterscheidung dimensionaler und kategorialer latenter strukturen in der depressionsdiagnostik [Methods for understanding dimensional and categorical latent structure in the diagnosis of depression]. In J. Reinecke & C. Tarnai (Eds.), *Angewandte klassifikationsanalyse in den sozialwissenschaften* [*Applied classification analysis in the social sciences*] (pp. 178–207). Münster: Waxmann.

Keller, F., Jahn, T., & Klein, C. (2001). Anwendung von taxometrischen methoden und von mischverteilungsmodellen zur erfassung von schizotypie [Application of taxometric methods and mixture distribution models for the assessment of schizotypy]. In B.

Andresen & R. Mass (Eds.), *Schizotypie. Psychometrische entwicklungen und biopsychologische forschungsansätze* (pp. 391–412). Göttingen: Hogrefe.

Kessler, R. C. (2002a). Epidemiological perspectives for the development of future diagnostic systems. *Psychopathology, 35*, 158–161.

Kessler, R. C. (2002b). The categorical versus dimensional assessment controversy in the sociology of mental illness. *Journal of Health and Social Behavior, 43*, 171–188.

Kim, H. (2001). *The threshold model of melancholia: A taxometric analysis*. Unpublished master's thesis, New School University.

Kinsey, A. C., Pomeroy, W. B., & Martin, C. E. (1948). *Sexual behavior in the human male*. Philadelphia: W. B. Saunders.

Klein, D. N., & Riso, L. P. (1993). Psychiatric disorders: Problems of boundaries and comorbidity. In C. G. Costello (Ed.), *Basic issues in psychopathology* (pp. 19–66). New York: Guilford.

Knowles, E. E. (1989). *Detection of a latent alcohoid taxon: A taxometric search*. Unpublished doctoral dissertation, University of Arkansas.

Korfine, L., & Lenzenweger, M. F. (1995). The taxonicity of schizotypy: A replication. *Journal of Abnormal Psychology, 104*, 26–31.

Kraemer, H. C., Noda, A., & O'Hara, R. (2004). Categorical versus dimensional approaches to diagnosis: Methodological challenges. *Journal of Psychiatric Research, 38*, 17–25.

Kripke, S. (1980). *Naming and necessity*. Cambridge, MA: Harvard University Press.

Lahey, B. B. (2001). Should the combined and predominantly inattentive types of ADHD be considered distinct and unrelated disorders? Not now, at least. *Clinical Psychology: Science and Practice, 8*, 494–497.

Lazarsfeld, P. F., & Henry, N. W. (1968). *Latent structure analysis*. Boston: Houghton Mifflin.

Lee, W.-C., & Rodgers, J. L. (1998). Bootstrapping correlation coefficients using univariate and bivariate sampling. *Psychological Methods, 3*, 91–103.

Lenzenweger, M. F. (1999). Deeper into the schizotypy taxon: On the robust nature of maximum covariance analysis. *Journal of Abnormal Psychology, 108*, 182–187.

Lenzenweger, M. F. (2004). Consideration of the challenges, complications, and pitfalls of taxometric analysis. *Journal of Abnormal Psychology, 113*, 10–23.

Lenzenweger, M. F., & Korfine, L. (1992). Confirming the latent structure and base rate of schizotypy: A taxometric analysis. *Journal of Abnormal Psychology, 101*, 567–571.

Levy, F., Hay, D. A., McStephen, M., Wood, C., & Waldman, I. D. (1997). Attention-deficit hyperactivity disorder: A category or a continuum? Genetic analysis of a large-scale twin study. *Journal of the American Academy of Child and Adolescent Psychiatry, 36*, 737–744.

Levy, S., Stroessner, S. J., & Dweck, C. S. (1998). Stereotype formation and endorsement: The role of implicit theories. *Journal of Personality and Social Psychology, 74*, 1421–1436.

Lilienfeld, S. O., Waldman, I. D., & Israel, A. C. (1994). A critical examination of the use of the term "comorbidity" in psychopathology research. *Clinical Psychology: Science and Practice, 1*, 71–83.

Livesley, W. J., Schroeder, M. L., Jackson, D. N., & Jang, K. L. (1994). Categorical distinctions in the study of personality disorder: Implications for classification. *Journal of Abnormal Psychology, 103*, 6–17.

Long, J. S. (1983). *Confirmatory factor analysis*. Beverly Hills, CA: Sage.

Lord, F. M. (1980). *Applications of item response theory to practical testing problems*. Hillsdale, NJ: Lawrence Erlbaum Associates.

Lorr, M. (1994). Cluster analysis: Aims, methods, and problems. In S. Strack & M. Lorr (Eds.), *Differentiating normal and abnormal personality* (pp. 179–195). New York: Springer.

Lykken, D. T., McGue, M., Tellegen, A., & Bouchard, T. J. (1992). Emergenesis: Genetic traits that may not run in families. *American Psychologist, 47*, 1565–1577.

MacCallum, R. C., Zhang, S., Preacher, K. J., & Rucker, D. D. (2002). On the practice of dichotomization of quantitative variables. *Psychological Methods, 7*, 19–40.

Maraun, M. D., Slaney, K., & Goddyn, L. (2003). An analysis of Meehl's MAXCOV-HITMAX procedure for the case of dichotomous indicators. *Multivariate Behavioral Research, 38*, 81–112.

Marcus, D. K., John, S. L., & Edens, J. F. (2004). A taxometric analysis of psychopathic personality. *Journal of Abnormal Psychology, 113*, 626–635.

Maxwell, S. E. (2004). The persistence of underpowered studies in psychological research: Causes, consequences, and remedies. *Psychological Methods, 9*, 147–163.

McCrae, R. R. (1994). A reformulation of axis II: Personality and personality related problems. In P. T. Costa & T. A. Widiger (Eds.), *Personality disorders and the five-factor model of personality* (pp. 303–310). Washington, DC: American Psychological Association.

McHugh, P. R., & Slavney, P. R. (1998). *The perspectives of psychiatry* (2nd ed.). Baltimore, MD: Johns Hopkins University Press.

McLachlan, G. J., & Basford, K. E. (1988). *Mixture models: Inference and applications to clustering*. New York: Marcel Dekker.

McLaughlin, G., & Peel, D. (2001). *Finite mixture models*. New York: Wiley-Interscience.

Meehl, P. E. (1962). Schizotaxia, schizotypy, schizophrenia. *American Psychologist, 17*, 827–838.

Meehl, P. E. (1967). Theory testing in psychology and in physics: A methodological paradox. *Philosophy of Science, 34*, 103–115.

Meehl, P. E. (1968). Detecting latent clinical taxa: II. A simplified procedure, some additional hitmax cut locators, a single-indicator method, and miscellaneous theorems. *Reports from the research laboratories of the Department of Psychiatry, University of Minnesota* (Report No. PR-68-4).

Meehl, P. E. (1973). MAXCOV-HITMAX: A taxonomic search method for loose genetic syndromes. In P. E. Meehl (Ed.), *Psychodiagnosis: Selected papers* (pp. 200–224). Minneapolis: University of Minnesota Press.

Meehl, P. E. (1977). Specific etiology and other forms of strong influence: Some quantitative meanings. *The Journal of Medicine and Philosophy, 2*, 33–53.

Meehl, P. E. (1990). Toward an integrated theory of schizotaxia, schizotypy, and schizophrenia. *Journal of Personality Disorders, 4*, 1–99.

Meehl, P. E. (1992). Factors and taxa, traits and types, differences of degree and differences in kind. *Journal of Personality, 60*, 117–174.

Meehl, P. E. (1995a). Bootstraps taxometrics: Solving the classification problem in psychopathology. *American Psychologist, 50*, 266–274.

Meehl, P. E. (1995b). Extension of the MAXCOV-HITMAX taxometric procedure to situations of sizable nuisance covariance. In D. Lubinski & R. Dawis (Eds.), *Assessing individual differences in human behavior: New concepts, methods, and findings* (pp. 81–92). Palo Alto, CA: Consulting Psychologists Press.

Meehl, P. E. (1999). Clarifications about taxometric method. *Applied and Preventive Psychology, 8*, 165–174.

Meehl, P. E. (2001). Comorbidity and taxometrics. *Clinical Psychology: Science and Practice, 8*, 507–519.

Meehl, P. E. (2004). What's in a taxon? *Journal of Abnormal Psychology, 113*, 39–43.

Meehl, P. E., & Golden, R. R. (1982). Taxometric methods. In P. C. Kendall & J. N. Butcher (Eds.), *Handbook of research methods in clinical psychology* (pp. 127–181). New York: Wiley.

Meehl, P. E., Lykken, D. T., Burdick, M. R., & Schoener, G. R. (1969). Identifying latent clinical taxa: III. An empirical trial of the normal single-indicator method, using MMPI Scale 5 to identify the sexes. *Reports from the research laboratories of the Department of Psychiatry, University of Minnesota* (Report No. PR-69-4).

Meehl, P. E., & Rosen, A. (1955). Antecedent probability and the efficiency of psychometric signs, patterns, or cutting scores. *Psychological Bulletin, 52*, 194–216.

Meehl, P. E., & Yonce, L. J. (1994). Taxometric analysis: I. Detecting taxonicity with two quantitative indicators using means above and below a sliding cut (MAMBAC procedure). *Psychological Reports, 74*, 1059–1274.

Meehl, P. E., & Yonce, L. J. (1996). Taxometric analysis: II. Detecting taxonicity using covariance of two quantitative indicators in successive intervals of a third indicator (MAXCOV procedure). *Psychological Reports, 78*, 1091–1227.

Meyer, T., & Keller, F. (2001). Exploring the latent structure of the Perceptual Aberration, Magical Ideation and Physical Anhedonia Scales in a German sample—a partial replication. *Journal of Personality Disorders, 15*, 521–535.

Meyer, T., & Keller, F. (2003). Is there evidence for a latent class called "hypomanic temperament"? *Journal of Affective Disorders, 75*, 259–267.

Micceri, T. (1989). The unicorn, the normal curve, and other improbable creatures. *Psychological Bulletin, 105*, 156–166.

Milich, R., Balentine, A. C., & Lynam, D. R. (2001). ADHD combined subtype and ADHD predominantly inattentive subtype are distinct and unrelated disorders. *Clinical Psychology: Science and Practice, 7*, 463–488.

Miller, M. B. (1996). Limitations of Meehl's MAXCOV-HITMAX procedure. *American Psychologist, 51*, 554–556.

Milligan, G. W. (1996). Clustering validation: Results and applications for applied analyses. In P. Arabie, L. J. Hubert, & G. DeSoete (Eds.), *Clustering and classification* (pp. 341–375). River Edge, NJ: World Scientific Publishing.

Milligan, G. W., & Cooper, M. C. (1985). An examination of procedures for determining the number of clusters in a data set. *Psychometrika, 50*, 159–179.

Mooney, C. Z., & Duval, R. D. (1993). *Bootstrapping: A nonparametric approach to statistical inference*. Newbury Park, CA: Sage.

Murphy, E. A. (1964). One cause? Many causes? The argument from the bimodal distribution. *Journal of Chronic Disease, 17*, 301–324.

Muthén, B. O. (2001). Latent variable mixture modeling. In G. A. Marcoulides & R. E. Schumacker (Eds.), *New developments and techniques in structural equation modeling* (pp. 1–33). Mahwah, NJ: Lawrence Erlbaum Associates.

Oakman, J. M., & Woody, E. Z. (1996). A taxometric analysis of hypnotic susceptibility. *Journal of Personality and Social Psychology, 71*, 980–991.

Pratto, F., Sidanius, J., Stallworth, L., & Malle, B. (1994). Social dominance orientation: A personality variable predicting social and political attitudes. *Journal of Personality and Social Psychology, 67*, 741–763.

Rothbart, M., & Taylor, M. (1992). Category labels and social reality: Do we view social categories as natural kinds? In G. R. Semin & K. Fiedler (Eds.), *Language and social cognition* (pp. 11 36). London: Sage.

Rothschild, L., Cleland, C. M., Haslam, N., & Zimmerman, M. (2003). Taxometric analysis of borderline personality disorder. *Journal of Abnormal Psychology, 112*, 657–666.

Ruscio, A. M., Borkovec, T. D., & Ruscio, J. P. (2001). A taxometric investigation of the latent structure of worry. *Journal of Abnormal Psychology, 110*, 413–422.

Ruscio, A. M., & Ruscio, J. (2002). The latent structure of analogue depression: Should the BDI be used to classify groups? *Psychological Assessment, 14*, 135–145.

Ruscio, A. M., Ruscio, J., & Keane, T. M. (2002). The latent structure of post-traumatic stress disorder: A taxometric investigation of reactions to extreme stress. *Journal of Abnormal Psychology, 111*, 290–301.

Ruscio, J. (2000). Taxometric analysis with dichotomous indicators: The modified MAXCOV procedure and a case removal consistency test. *Psychological Reports, 87*, 929–939.

Ruscio, J. (2004, May). *Bootstrapping sampling distributions to quantify the fit of taxometric curves*. Paper presented at the annual meeting of the American Psychological Society, Chicago, IL.

Ruscio, J. (2005). *Estimating taxon base rates with the MAXCOV, MAXEIG, and MAMBAC taxometric procedures: A Monte Carlo study*. Manuscript submitted for publication.

Ruscio, J., & Ruscio, A. M. (2000). Informing the continuity controversy: A taxometric analysis of depression. *Journal of Abnormal Psychology, 109*, 473–487.

Ruscio, J., & Ruscio, A. M. (2002). A structure-based approach to psychological assessment: Matching measurement models to latent structure. *Assessment, 9*, 4–16.

Ruscio, J., & Ruscio, A. M. (2004a). Clarifying boundary issues in psychopathology: The role of taxometrics in a comprehensive program of structural research. *Journal of Abnormal Psychology, 113*, 24–38.

Ruscio, J., & Ruscio, A. M. (2004b). A conceptual and methodological checklist for conducting a taxometric investigation. *Behavior Therapy, 35*, 403–447.

Ruscio, J., & Ruscio, A. M. (2004c). A nontechnical introduction to the taxometric method. *Understanding Statistics, 3*, 151–193.

Ruscio, J., Ruscio, A. M., & Keane, T. M. (2004). Using taxometric analysis to distinguish a small latent taxon from a latent dimension with positively skewed indicators: The case of Involuntary Defeat Syndrome. *Journal of Abnormal Psychology, 113*, 145–154.

Ruscio, J., Ruscio, A. M., & Meron, M. (2005). *Applying the bootstrap to taxometric analysis: Generating empirical sampling distributions to interpret results.* Manuscript submitted for publication.

Schmidt, N. B., Kotov, R., & Joiner, T. E. (2004). *Taxometrics: Toward a new diagnostic scheme for psychopathology*. Washington, DC: American Psychological Association.

Schotte, C. K. W., Maes, M., Cluydts, R., & Cosyns, P. (1997). Cluster analytic validation of the DSM melancholic depression. The threshold model: Integration of quantitative and qualitative distinctions between unipolar depressive subtypes. *Psychiatry Research, 71*, 181–195.

Sedlmeier, P., & Gigerenzer, G. (1989). Do studies of statistical power have an effect on the power of studies? *Psychological Bulletin, 105*, 309–316.

Simpson, W. B. (1994). *Borderline personality disorder: Dimension or category? A maximum covariance analysis*. Unpublished doctoral dissertation, Boston University.

Skilling, T. A., Harris, G. T., Rice, M. T., & Quinsey, V. L. (2001). Identifying persistently antisocial offenders using the Hare Psychopathy Checklist and *DSM–IV* antisocial personality disorder criteria. *Psychological Assessment, 14*, 27–38.

Skilling, T. A., Quinsey, V. L., & Craig, W. M. (2001). Evidence of a taxon underlying serious antisocial behavior in boys. *Criminal Justice and Behavior, 28*, 450–470.

Slade, T., & Andrews, G. (2005). Latent structure of depression in a community sample: A taxometric analysis. *Psychological Medicine, 35*, 489–497.

Sokal, R. R., & Michener, C. D. (1958). A statistical method for evaluating systematic relationships. *University of Kansas Scientific Bulletin, 38*, 1409–1438.

Solomon, A., Haaga, D. A. F., & Arnow, B. A. (2001). Is clinical depression distinct from subthreshold depressive symptoms? A review of the continuity issue in depression research. *Journal of Nervous and Mental Disease, 189*, 498–506.

Sonuga-Barke, E. J. S. (1998). Categorical models of childhood disorder: A conceptual and empirical analysis. *Journal of Child Psychology and Psychiatry, 39*, 115–133.

Stern, H. S., Arcus, D., Kagan, J., Rubin, D. B., & Snidman, N. (1995). Using mixture models in temperament research. *International Journal of Behavioral Development, 18*, 407–423.

Stone, C. A. (2000). Monte Carlo based null distributions for an alternative goodness-of-fit test statistic in IRT models. *Journal of Educational Measurement, 37*, 58–75.

Strong, D. R., Brown, R. A., Kahler, C. W., Lloyd-Richardson, E. E., & Niaura, R. (in press). Depression proneness in treatment-seeking smokers: A taxometric analysis. *Personality and Individual Differences*.

Strong, D. R., Greene, R. L., Hoppe, C., Johnston, T., & Olesen, N. (1999). Taxometric analysis of impression management and self-deception on the MMPI-2 in child-custody litigants. *Journal of Personality Assessment, 73*, 1–18.

Strong, D. R., Greene, R. L., & Schinka, J. A. (2000). Taxometric analysis of the MMPI-2 Infrequency scales [F and F(p)] in clinical settings. *Psychological Assessment, 12*, 166–173.

Strube, M. J. (1989). Evidence for the type in type A behavior: A taxometric analysis. *Journal of Personality and Social Psychology, 56*, 972–987.

Swets, J. A., Dawes, R. M., & Monahan, J. (2000). Psychological science can improve diagnostic decisions. *Psychological Science in the Public Interest, 1*, 1–26.

Taylor, S., Rabian, B., & Fedoroff, I. C. (1999). Anxiety sensitivity: Progress, prospects, and challenges. In S. Taylor (Ed.), *Anxiety sensitivity: Theory, research, and treatment of the fear of anxiety* (pp. 339–353). Mahwah, NJ: Lawrence Erlbaum Associates.

Thurstone, L. L. (1935). *The vectors of mind*. Chicago: University of Chicago Press.

Thurstone, L. L. (1947). *Multiple factor analysis*. Chicago: University of Chicago Press.

Trull, T. J., Widiger, T. A., & Guthrie, P. (1990). Categorical versus dimensional status of borderline personality disorder. *Journal of Abnormal Psychology, 99*, 40–48.

Tukey, J. W. (1977). *Exploratory data analysis*. Reading, MA: Addison-Wesley.

Tylka, T. L., & Subich, L. M. (2003). Revisiting the latent structure of eating disorders: Taxometric analyses with nonbehavioral indicators. *Journal of Counseling Psychology, 50*, 276–286.

Tyrka, A., Cannon, T. D., Haslam, N., Mednick, S. A., Schulsinger, F., Schulsinger, H., & Parnas, J. (1995). The latent structure of schizotypy: I. Premorbid indicators of a taxon of individuals at risk for schizophrenia-spectrum disorders. *Journal of Abnormal Psychology, 104*, 173–183.

Tyrka, A., Haslam, N., & Cannon, T. D. (1995). Detection of a longitudinally-stable taxon of individuals at risk for schizophrenia spectrum disorders. In A. Raine, T. Lencz, & S. A. Mednick (Eds.), *Schizotypal personality disorder* (pp. 168–191). New York: Cambridge University Press.

Uebersax, J. S. (1999). Probit latent class analysis with dichotomous or ordered category measures: Conditional independence/dependence models. *Applied Psychological Measurement, 23*, 283–297.

Ullrich, S., Borkenau, P., & Marneros, A. (2001). Personality disorders in offenders: Categorical versus dimensional approaches. *Journal of Personality Disorders, 15*, 442–449.

Vasey, M. W., Kotov, R., Frick, P. J., & Loney, B. R. (2005). The latent structure of psychopathy in youth: A taxometric investigation. *Journal of Abnormal Child Psychology, 33*, 411–429.

von Davier, M., & Rost, J. (1997). Self-monitoring—A class variable? In J. Rost & R. Langenheine (Eds.), *Applications of latent trait and latent class models in the social sciences* (pp. 296–304). Münster, New York, München, Berlin: Waxmann.

Waldman, I. D., & Lilienfeld, S. O. (2001). Applications of taxometric methods to problems of comorbidity: Perspectives and challenges. *Clinical Psychology: Science and Practice, 8*, 520–527.

Waller, N. G., & Meehl, P. E. (1998). *Multivariate taxometric procedures: Distinguishing types from continua*. Thousand Oaks, CA: Sage.

Waller, N. G., Putnam, F. W., & Carlson, E. B. (1996). Types of dissociation and dissociative types: A taxometric analysis of dissociative experiences. *Psychological Methods, 3*, 300–321.

Waller, N. G., & Ross, C. A. (1997). The prevalence and biometric structure of pathological dissociation in the general population: Taxometric and behavior genetic findings. *Journal of Abnormal Psychology, 106*, 499–510.

Waller, N. G., Underhill, J. M., & Kaiser, H. A. (1999). A method for generating simulating plasmodes and artificial test clusters with user-defined shape, size, and orientation. *Multivariate Behavioral Research, 34*, 123–142.

Ward, J. H. (1963). Hierarchical grouping to optimize an objective function. *Journal of the American Statistical Association, 58*, 236–244.

Watson, D. (2003). Investigating the construct validity of the dissociative taxon: Stability analyses of normal and pathological dissociation. *Journal of Abnormal Psychology, 112*, 298–305.

Whisman, M. A., & Pinto, A. (1997). Hopelessness depression in depressed inpatient adolescents. *Cognitive Therapy and Research, 21*, 345–358.

Widiger, T. A. (2001). What can be learned from taxometric analyses? *Clinical Psychology: Science and Practice, 8*, 528–533.

Widiger, T. A., & Clark, L. A. (2000). Toward *DSM–V* and the classification of psychopathology. *Psychological Bulletin, 126*, 946–963.

Williamson, D. A., Womble, L. G., Smeets, M. A. M., Netemeyer, R. G., Thaw, J., Kutlesic, V., & Gleaves, D. H. (2002). The latent structure of eating disorder symptoms: A factor analytic and taxometric investigation. *American Journal of Psychiatry, 159*, 412–418.

Wolfe, J. M. (1998). What can 1 million trials tell us about visual search? *Psychological Science, 9*, 33–39.

Woodward, S. A., Lenzenweger, M. F., Kagan, J., Snidman, N., & Arcus, D. (2000). Taxonic structure of infant reactivity: Evidence from a taxometric perspective. *Psychological Science, 11*, 296–301.

World Health Organization. (1993). *The ICD-10 classification of mental and behavioral disorders: Diagnostic criteria for research*. Geneva, Switzerland: World Health Organization.

Zachar, P. (2000). Psychiatric disorders are not natural kinds. *Philosophy, Psychiatry & Psychology, 7*, 167–194.

Zadeh, L. A. (1965). Fuzzy sets. *Information and Control, 8*, 338–353.

Zimmerman, M., & Spitzer, R. (1989). Melancholia: From *DSM–III* to *DSM–III–R*. *American Journal of Psychiatry, 146*, 20–28.

Author Index

A

ABC Research Group, 146
Abramson, L. Y., 268, 269, 289
Acton, G. S., 50
Adorno, T. W., 291
Akiskal, H. S., 269
Alloy, L. B., 268, 269, 289
Ambrosini, P., 268, 294
American Psychiatric Association, 22
Amir, N., 268, 269, 277
Andreasen, N. C., 268
Andrews, G., 268
Arabie, P., 47
Arcus, D., 270, 275, 294
Arnau, R. C., 274, 276
Arnow, B. A., 265
Asendorpf, J. B., 273, 274

B

Bailey, J. M., 276
Balentine, A. C., 294
Bartholomew, D. J., 40
Bartlett, M. S., 116
Basford, K. E., 45, 46
Beach, S. R. H., 268, 269, 277
Beauchaine, R. J., 59, 67, 135, 148, 151, 183, 187, 197, 245, 247, 308, 310
Beauchaine, T. P., 29, 32, 38, 59, 67, 135, 148, 151, 183, 187, 197, 245, 247, 248, 281, 294, 297, 304, 305, 308, 310
Beck, A. T., 268, 269
Beevers, C. G., 269
Bennett, D., 268, 294
Bernstein, A., 270, 289
Blanchard, J. J., 271, 272, 289
Blatt, S. J., 268
Bock, H., 46
Borkenau, P., 271
Borkovec, T. D., 269
Bornstein, B., 277
Bouchard, T. J., 31
Brown, R. A., 269, 289
Brown, S. A., 272
Burdick, M. R., 88
Byrne, B., 298

C

Campbell, M., 271
Cannon, T. D., 272, 294, 303
Carlson, E. B., 270, 279, 300
Carson, R. C., 264
Clark, L. A., 20, 271, 290
Cleland, C., 78, 115, 135, 186, 187, 268, 273, 294, 308, 309, 310

Cleveland, W. S., 90, 156, 207
Cohen, J., 10, 27
Collins, L. M., 271, 289
Cook, C., 276
Cooper, M. C., 48
Cornblatt, B. A., 272, 294
Cororve, M. B., 270
Costa, P. T., 271
Coyne, J. C., 30
Craig, W. M., 272
Cronbach, L. J., 55, 72, 299
Crow, T. J., 289

D

Davis, J. M., 297
Dawes, R. M., 24
De Boeck, P., 50
Dell, P. F., 270
DeSoete, G., 47
Dollaghan, C. A., 277, 295
Dorus, E., 297
Duval, R. D., 320
Dweck, C. S., 291

E

Edens, J. F., 273
Efron, B., 65, 80, 85, 160, 231, 236, 314, 320
Embretson, S. E., 74, 298
Endicott, J., 268
Endler, N. S., 289
Erlenmeyer-Kimling, L., 272, 294
Ernst, D., 33, 306
Everitt, B. S., 45, 47, 48
Eysenck, H. J., 290

F

Fedoroff, I. C., 270, 289
Fincham, F. D., 277
Fiske, A. P., 277
Flett, G. L., 265
Fraley, R. C., 28, 268, 269, 275, 294
Franklin, C. L., 268
Frenkel-Brunswik, E., 291
Frick, P. J., 273, 294

G

Gangestad, S. W., 10, 20, 78, 126, 158, 211, 272, 273, 274, 275, 276, 279, 284
Gelman, S. A., 33
Gibb, B. E., 269, 289
Gibbons, R. D., 297
Gigerenzer, G., 10, 146
Gilbert, P., 268
Giles, J. W., 33
Gleaves, D. H., 270, 274, 278
Goddyn, L., 281
Golden, R. R., 8, 39, 54, 88, 150, 172, 223, 229, 271, 272, 281, 289, 294, 295, 302
Gorsuch, R. L., 298
Grayson, D. A., 39
Green, B. A., 270, 274, 278
Green, B. F., 49, 299
Greene, R. L., 268, 276
Grove, W. M., 8, 25, 48, 54, 70, 78, 87, 90, 92, 93, 139, 218, 223, 246, 268, 305
Guthrie, P., 268, 273

H

Haaga, D. A. F., 265
Hambleton, R. K., 298
Hand, D. J., 45
Hankin, B. L., 268, 269, 294
Harding, J. P., 37
Harris, G. T., 272, 294
Haslam, N., 3, 33, 38, 66, 78, 115, 122, 135, 171, 186, 187, 264, 268, 269, 271, 272, 273, 276, 277, 278, 294, 295, 296, 303, 306, 308, 309, 310
Hay, D. A., 294
Henry, N. W., 49, 299
Heyman, G. D., 33
Hirschfeld, R. M. A., 268
Homann, E., 268
Hoppe, C., 276
Horan, W. P., 271, 272, 289
Hubert, L. J., 47

I, J

Israel, A. C., 300
Jablensky, A., 264

Jackson, D. N., 264, 271
Jahn, T., 272
Jang, K. L., 264, 271
John, S. L., 273
Johnston, T., 276
Joiner, T. E., 173, 195, 196

K

Kagan, J., 270, 273, 275, 294
Kahler, C. W., 269, 289
Kaiser, H. A., 316
Keane, T. M., 70, 77, 80, 85, 106, 115, 128, 135, 152, 170, 173, 174, 219, 256, 269, 305
Keller, F., 269, 272, 289
Keller, M. B., 268
Kessler, R. C., 15, 24, 69, 268, 292, 293
Kim, H., 3, 66, 122, 171, 264, 271, 278
Kinsey, A. C., 276
Klein, C., 272
Klein, D. N., 301
Knowles, E. E., 289
Kocovski, N. L., 289
Korfine, L., 272, 275
Kotov, R., 173, 195, 196, 273, 294
Kraemer, H. C., 15, 24
Krames, L., 265
Kripke, S., 13
Kurtzberg, D., 271, 294
Kutlesic, V., 270
Kwapil, T. R., 272

L

Lahey, B. B., 268, 269, 294
Lazarsfeld, P. F., 49, 299
Lee, W. C., 315
Leen-Feldner, E. W., 270, 289
Lenzenweger, M. F., 229, 244, 270, 272, 275, 294
Leonard, K. E., 277
Levinson, D., 291
Levy, D. L., 297
Levy, F., 294
Levy, S., 291
Lilienfeld, S. O., 300, 301
Livesley, W. J., 264, 271
Lloyd-Richardson, E. E., 269, 289
Loney, B. R., 273, 294
Long, J. S., 298
Lord, F. M., 298
Lorr, M., 47, 48
Lowe, M. R., 270, 278
Lykken, D. T., 31, 88
Lynam, D. R., 294

M

MacCallum, R. C., 27
Malle, B., 291
Maraun, M. D., 281
Marcus, D. K., 273
Marneros, A., 271
Martin, C. E., 276
Martin, N. G., 276
Maxwell, S. E., 10
McCarton, C. M., 271, 294
McCrae, R. R., 271
McGue, M., 31
McHugh, P. R., 264
McLachlan, G. J., 45, 46, 47, 297
McStephen, M., 294
Mednick, S. A., 272, 294, 303
Meehl, P. E., 6, 8, 14, 20, 22, 24, 26, 31, 32, 54, 55, 60, 66, 67, 72, 74, 75, 76, 78, 79, 87, 88, 90, 92, 93, 103, 104, 105, 115, 116, 117, 120, 122, 123, 126, 127, 128, 129, 130, 135, 136, 138, 139, 141, 146, 150, 151, 156, 161, 162, 163, 166, 170, 172, 173, 174, 177, 178, 183, 185, 186, 189, 192, 193, 194, 195, 196, 197, 207, 218, 219, 223, 229, 232, 245, 247, 248, 250, 256, 257, 265, 272, 273, 281, 282, 285, 289, 299, 301, 302, 307, 308, 312, 313, 314, 322
Melancon, J. G., 274
Meron, M., 80, 160, 187, 193, 202, 232, 255, 316, 319
Metalsky, G. I., 268, 269
Meyer, T., 269, 272, 289
Micceri, T., 47, 77, 82, 170, 199, 222, 308
Michener, C. D., 48
Milich, R., 294
Miller, I. W., 269, 289
Miller, M. B., 281, 305
Milligan, G. W., 48
Monahan, J., 24
Mooney, C. Z., 320
Murphy, E. A., 37

Murphy-Eberenz, K. P., 270, 278
Muthén, B. O., 49, 85, 299

N

Netemeyer, R. G., 270
Niaura, R., 269, 289
Noda, A., 15, 24

O

Oakman, J. M., 275
O'Hara, R., 15, 24
Olesen, N., 276
Ostrow, D. G., 297

P

Pandey, G. N., 297
Parnas, J., 272
Peel, D., 45, 46, 47, 297
Perry, R., 271
Pinto, A., 269, 294
Pomeroy, W. B., 276
Porter, M., 296
Pratto, F., 291
Preacher, K. J., 27
Putnam, F. W., 270, 279, 300

Q, R

Quinsey, V. L., 272, 294
Rabian, B., 270, 289
Reich, T., 268
Rice, M. E., 272, 294
Riley, K. C., 270
Riso, L. P., 301
Rodgers, J. L., 315
Rogers, H. J., 298
Rosen, A., 24
Rosen, D. H., 274
Ross, C. A., 270, 300, 302
Rost, J., 275
Rothbart, M., 33
Rothschild, L., 33, 186, 187, 273, 296, 306, 309, 310
Rubin, D. B., 275
Rucker, D. D., 27
Ruscio, A. M., 11, 12, 25, 26, 28, 37, 53, 60, 67, 68, 70, 77, 80, 85, 106, 107, 111, 115, 125, 128, 135, 141, 152, 160, 170, 173, 174, 187, 193, 202, 219, 232, 245, 248, 255, 256, 268, 269, 298, 299, 301, 302, 305, 316, 319
Ruscio, J., 11, 12, 25, 26, 28, 37, 53, 60, 67, 68, 70, 77, 78, 80, 85, 106, 107, 111, 115, 125, 126, 128, 135, 136, 139, 141, 144, 152, 160, 170, 173, 174, 175, 178, 187, 192, 193, 194, 202, 204, 219, 232, 245, 248, 255, 256, 268, 269, 281, 284, 285, 298, 299, 301, 302, 305, 308, 316, 319

S

Sanford, N., 291
Schinka, J. A., 276
Schmidt, N. B., 173, 195, 196
Schoener, G. R., 88
Schroeder, M. L., 264, 271
Schulsinger, F., 272, 294, 303
Schulsinger, H., 272, 294, 303
Sedlmeier, P., 10
Sidanius, J., 291
Skilling, T. A., 272, 294
Slade, T., 268
Slaney, K., 281
Slavney, P. R., 264
Smeets, M. A. M., 270
Snidman, N., 270, 275
Snow, A. C., 270, 278
Snyder, M., 10, 20, 78, 126, 158, 211, 273, 274, 275, 279, 284
Sokal, R. R., 48
Solomon, A., 265
Spieker, S. J., 275, 294
Spitzer, R., 268
Stallworth, L., 291
Stern, H. S., 275
Stickle, T., 270, 289
Stone, C. A., 85
Stroessner, S. J., 291
Strong, D. R., 268, 269, 276, 289
Strube, M. J., 28, 275
Subich, L. M., 270
Swaminathan, H. S., 298
Swets, J. A., 24

T

Taylor, M., 33
Taylor, S., 270, 289
Tellegen, A., 31
Thaw, J., 270
Thompson, R. L., 276
Thurstone, L. L., 116
Tibshirani, R. J., 65, 80, 85, 160, 231, 236, 314, 320
Todd, P. M., 146
Trull, T. J., 268, 273
Tukey, J. W., 156, 207, 211
Tyan, S. H., 88
Tylka, T. L., 270
Tyrka, A., 272, 294, 303

U

Uebersax, J. S., 49
Ullrich, S., 271
Underhill, J. M., 316

V

van Arken, M. A. G., 274
Vasey, M. W., 273, 294
Vaughan, H. G., Jr., 271, 294
von Davier, M., 275
Vredenburg, K., 265

W

Waldman, I. D., 268, 269, 294, 300, 301
Waller, N. G., 8, 28, 54, 60, 87, 116, 117, 120, 122, 136, 138, 139, 141, 150, 156, 162, 166, 170, 172, 183, 185, 186, 197, 207, 218, 223, 229, 270, 275, 279, 281, 285, 300, 302, 316
Ward, J. H., 48
Waters, E., 38, 281, 297, 305, 310
Watson, D., 271, 300, 303
Weems, C., 270, 289
Whisman, M. A., 269, 294
Widiger, T. A., 20, 72, 268, 271, 273, 290
Williams, T. L., 270
Williamson, D. A., 270
Wilson, M., 50
Wolfe, J. M., 296
Womble, L. G., 270
Wood, C., 270, 294
Woodward, S. A., 275, 294
Woody, E. Z., 275
World Health Organization, 22

Y

Yonce, L. J., 8, 54, 55, 66, 75, 78, 87, 103, 104, 105, 115, 120, 122, 123, 127, 129, 130, 135, 139, 146, 151, 172, 173, 177, 178, 192, 193, 194, 195, 218, 219, 223, 232, 245, 248, 250, 256, 257, 285, 308, 312, 313, 314, 322
Young, M., 268

Z

Zachar, P., 32
Zhang, S., 27
Zimmerman, M., 268, 273
Zvolensky, M. J., 270, 289

Subject Index

A

Adaptive calibration, 231–232
Anorexia nervosa (AN), 270
Antisocial personality disorder, 272–273, 289, 294
Anxiety and anxiety disorders, 269–270
  anxiety sensitivity, 270
  posttraumatic stress disorder (PTSD), 269
  social inhibition, 270
  worry, 269
Anxiety sensitivity, 270
Assessment, 25–27, 75, 248
Attachment styles, 275, 294

B

Base rate, *see* Taxon base rate
Base-rate classification, 140, 146 148, 149–150, 176
Bayes' Theorem, 25, 27, 126, 135, 139, 140, 146–148, 150, 176, 183, 322
Bimodality, 37–45, 116–117
Biology, 296–297
Bootstrap, 65–66, 80, 85, 232, 314–320, *see also* Adaptive calibration, Empirical sampling distributions
Borderline personality disorder, 273, 289
Bulimia nervosa (BN), 270

C

Causal explanation, 30–32, 300–304
Checklist of conceptual and methodological issues
  appropriateness of data, 244–252
  articulation of implications, 258–259
  presentation and interpretation of results, 255–258
  scientific justification, 242–244
  variety and implementation of procedures, 253–255
Classification, 19–22, 36–37, 264–265
Cluster analysis, 47–49, 53, 310, *see also* Finite mixture modeling
Cognition, 295–296
Cognitive personality constructs, 290–291
Coherent cut kinetics, 54
Comorbidity, 300–302
Complement, 6
Consistency tests, 161–162, 254–255, 284–286
  analyzing subsamples, 165–166, 285
  base-rate estimates, 171–176, 195–203, 219–222, 284
  case removal, 177–182, 285
  comparison of classified cases, 184–185

Consistency tests *(cont.)*
comparison curve-fit index (CCFI), 188, 195–203, 255
curve-fit indexes, 187–195
distribution of probabilities of taxon membersip, 183–184, 195–197, 285
goodness of fit index (GFI), 185–187, 195–203, 285
inchworm, 152, 166–171, 255
latent parameter estimates, 176–177
Monte Carlo study, 195–203
multiple indicator configurations, 163–164
multiple indicator sets, 165
multiple procedural implementations, 164
multiple taxometric procedures, 162–163, 253
"nose count," 195–203
simulated comparison data, 284–285
Construct validation, 299–300

D

Data appropriateness
a priori examination, 250–251
empirical sampling distributions, 252
post hoc examination, 251
Data sets, illustrative, 39–44, 93–94
Deceptive responding, 276–277
Dementia, 271, 295
Depression, 265–269, 288, 294
Developmental psychopathology, 294–295
Diagnosis, 22–24
Diagnostic algorithm, 22
Diagnostic co-occurrence, *see* Comorbidity
Diatheses, 29
Dichotomization, 27–28
Dimcat, 50–51
Dissociative disorders, 270–271, 300

E

Eating disorders, 270
anorexia nervosa (AN), 270
bulimia nervosa (BN), 270
Emergenesis, 31–32
Empirical sampling distributions, 79–85, 187–195, 231–239, 252, 257–258, 284–285
Envy, 277
Epidemiology, 292–294
Essentialism, 33
Etiology, *see* Causal explanation
Exploration vs. theory-testing, 150

F

Factor analysis, 36–37, 116, 298, 299
False positive rate, 322, *see also* Hitmax
Finite mixture modeling, 45–47, 297

G

Gender identity, 275
General Covariance Mixture Theorem, 123
Graphing
aggregating or averaging curves, 157, 211–218, 256, 282–283
scaling axes, 107–109, 128–130, 142, 156, 255–256
smoothing curves, 106, 156–157, 207–211, 256, 282

H

Hitmax, 92, 125, 139, 176
Hypnotic ability, 275

I

Impulsivity, 275
Indicator, 10, 36
composite input, 93, 99–102, 106, 109–111, 126–127, 132–135, 140–141, 142–144, 211
configurations, 92–93, 94–99, 99–102, 106, 109–111, 118, 126–127, 130, 137–138, 140–144, 149–151, 163–164, 212
content validity, 72–73, 246–247
discriminant validity, 72–73, 246–247
distributions, 47, 77–78, 82, 99–102, 106, 115, 126–127, 135, 152–153, 170, 173, 174, 199, 218–223, 248, 256–257
input, 89, 103, 124
number, 78–79, 162–163, 248–249, 253

output, 89, 103, 124
selection and construction, 73, 75–76, 79, 225, 246–249
subsamples, 127–128, 132, 138–139, 141–142, 151–156, 166–171, 208, 222–223
validity, 39, 74, 76–77, 176–178, 223–224, 225–228, 247, 256
within-group correlations, 11, 75–77, 78–79, 174, 223–228, 247–248, 256
Infant reactivity, 275, 294
Institutional pseudotaxa, 305
Internal replications, 107, 153–154
Interpretational safeguards
empirical sampling distributions, 231–239, 257–258
independent raters, 229–231, 257, 283
presenting as many curves as possible, 228–229, 282–283
Intervals, *see* Indicator, subsamples
Item-response theory, 74–75, 298, 299

J

Jealousy, 277

L

Latent class analysis, 49–50, 57
Latent mode (L-Mode), 116–120
base-rate estimation, 117
empirical illustrations, 118–120
implementation decisions, 117–118
logic, 116–117
systematic study, 120
Latent profile analysis, 49
Lay conceptions, 32–33
Local independence, 49, 57

M

Marital discordance, 277
Maximum covariance (MAXCOV), 123–136
base-rate estimation, 125–126, 136, 321–322
blending with MAXEIG, 148–157
class assignment, 126, 322
contrast with MAXEIG, 137
empirical illustrations, 130–135
implementation decisions, 126–130
logic, 123–125
systematic study, 135–136
Maximum eigenvalue (MAXEIG), 136–146
base-rate estimation, 139, 144–146, 323–325
blending with MAXCOV, 148–157
class assignment, 139–140, 322
contrast with MAXCOV, 137
empirical illustrations, 142–144
implementation decisions, 140–142
logic, 136–139
systematic study, 144–146
Maximum slope (MAXSLOPE), 89–102
base-rate estimation, 92
empirical illustrations, 93–102
implementation decisions, 92–93
logic, 89–92
systematic study, 102
Mean above minus below a cut (MAMBAC), 102–116
base-rate estimation, 104–106, 115–116
empirical illustrations, 109–115
implementation decisions, 106–109
logic, 102–104
systematic study, 115–116
Measurement, *see* Assessment
Mixture models, *see* Finite mixture modeling
Moderator variables, 29–30
Mood disorders, 265–269, 288, 294
Multimodality, 51, *see also* Bimodality

N

Natural kind, 13–14
Neonatal brain dysfunction, 271, 294
Normal personality, 273–276, 290–291
broad traits, 274, 290
cognitive personality constructs, 290–291
specific traits, 274–276
attachment styles, 275, 294
gender identity, 275
hypnotic ability, 275
impulsivity, 275
infant reactivity, 275, 294
self-monitoring, 274–275
sexual orientation, 276
Type A personality, 275

Nuisance correlations, *see* Indicator, within-group correlations
Nuisance covariance, *see* Indicator, within-group correlations
Null hypothesis, 58–59, *see also* Taxometric method, inferential frameworks

O

Observer bias, 38, 297, 305–307
Overlap, *see* Indicator, subsamples

P

Personality, *see* Normal personality
Personality disorders, *see* Psychopathology, personality disorders (PDs)
Posttraumatic stress disorder (PTSD), 269
Pseudodimensions, 307–308
Pseudotaxa, 305–308
Psychopathology, 264–273, 288–289
  anxiety and anxiety disorders, 269–270
    anxiety sensitivity, 270
    posttraumatic stress disorder (PTSD), 269
    social inhibition, 270
    worry, 269
  dissociative disorders, 270–271, 300
  eating disorders, 270
    anorexia nervosa (AN), 270
    bulimia nervosa (BN), 270
  mood disorders, 265–269, 288, 294
  other Axis I conditions, 271
    dementia, 271, 295
    neonatal brain dysfunction, 271, 294
    tardive dyskinesia, 271
  personality disorders (PDs), 271–273, 289–290
    antisocial PD, 272–273, 289, 294
    borderline PD, 273, 289
    psychopathy, 272–273, 289
    schizotypy, 272, 289, 294
  schizophrenia, 288–289, 294
Psychopathy, 272–273, 289

R

Rater bias, *see* Observer bias
Relationship schemas, 277, 295
Response styles, 276–277

S

Sample-specific simulations, *see* Empirical sampling distributions
Sampling
  admixed samples, 69–70, 245–246
  analogue samples, 30
  case removal, 70–72, 246
  populations, 68–69
  sample bias, 38, 84–85
  sample size, 66, 208, 244–245, 282
  split samples, 70, 165–166, 246
Schizophrenia, 288–289, 294
Schizotypy, 272, 289, 294
Self-monitoring, 274–275
Sexual orientation, 276
Simulating comparison data, 232, 284–285, 312–320, *see also* Empirical sampling distributions
Social inhibition, 270
Specific etiology, 31, 302, 304
Specific language impairment, 277, 295
Statistical power, 27–28, 79
Structure
  complex, 11, 12–13, 51–54, 298–299
  dimensional, 6, 312–313
  latent, 6
  manifest, 6
  misconceptions, 9–16
  nontaxonic, 16
  overview of findings, 277–279
  reproducing, 40–41, 312–320
  taxonic, 6, 313

T

Tardive dyskinesia, 271
Taxometric method
  features, 54–58
  future research, 308–309
  implementation in published studies, 279–286
    consistency tests, 284–286
    curve presentation, 282–283
    data sources, 281
    indicator types, 281–282
    latent parameter estimates, 283–284
    sample size, 282

taxometric procedures, 281
inferential framework, 58–61, 81–82
Taxometric procedures
consistency hurdles, 88
frequency in published studies, 281
L-Mode, *see* Latent mode (L-Mode)
MAMBAC, *see* Mean above minus below a cut (MAMBAC)
MAXCOV, *see* Maximum covariance (MAXCOV)
MAXEIG, *see* Maximum eigenvalue (MAXEIG)
MAXSLOPE, *see* Maximum slope (MAXSLOPE)
normal minimum chi-square, 88
Taxon
base rate, 23, 66–68, 244–245
definition, 6–8
environment-mold, 14
small, 66–68, 165–166, 244–245
Threshold model, 31, 302, 304
Type A personality, 275

V

Valid positive rate, 322, *see also* Hitmax
Visual attention, 296

W

Windows, *see* Indicator, subsamples
Worry, 269